The GNU Octave 3.8 Reference Manual - Part 1/2

A catalogue record for this book is available from the Hong Kong Public Libraries.

Published by Samurai Media Limited.

Email: info@samuraimedia.org

ISBN 978-988-13277-3-4

Table of Contents

Preface

Octave was originally intended to be companion software for an undergraduate-level textbook on chemical reactor design being written by James B. Rawlings of the University of Wisconsin-Madison and John G. Ekerdt of the University of Texas.

Clearly, Octave is now much more than just another 'courseware' package with limited utility beyond the classroom. Although our initial goals were somewhat vague, we knew that we wanted to create something that would enable students to solve realistic problems, and that they could use for many things other than chemical reactor design problems. We find that most students pick up the basics of Octave quickly, and are using it confidently in just a few hours.

Although it was originally intended to be used to teach reactor design, it has been used in several other undergraduate and graduate courses in the Chemical Engineering Department at the University of Texas, and the math department at the University of Texas has been using it for teaching differential equations and linear algebra as well. More recently, Octave has been used as the primary computational tool for teaching Stanford's online Machine Learning class (`ml-class.org`) taught by Andrew Ng. Tens of thousands of students participated in the course.

If you find Octave useful, please let us know. We are always interested to find out how Octave is being used.

Virtually everyone thinks that the name Octave has something to do with music, but it is actually the name of one of John W. Eaton's former professors who wrote a famous textbook on chemical reaction engineering, and who was also well known for his ability to do quick 'back of the envelope' calculations. We hope that this software will make it possible for many people to do more ambitious computations just as easily.

Everyone is encouraged to share this software with others under the terms of the GNU General Public License (see Appendix J [Copying], page 857). You are also encouraged to help make Octave more useful by writing and contributing additional functions for it, and by reporting any problems you may have.

Acknowledgements

Many people have contributed to Octave's development. The following people have helped code parts of Octave or aided in various other ways (listed alphabetically).

Ben Abbott	Andy Adler	Adam H. Aitkenhead
Giles Anderson	Joel Andersson	Muthiah Annamalai
Markus Appel	Branden Archer	Marco Atzeri
Shai Ayal	Roger Banks	Ben Barrowes
Alexander Barth	David Bateman	Heinz Bauschke
Julien Bect	Roman Belov	Karl Berry
David Billinghurst	Don Bindner	Jakub Bogusz
Moritz Borgmann	Paul Boven	Richard Bovey
John Bradshaw	Marcus Brinkmann	Max Brister
Remy Bruno	Clemens Buchacher	Ansgar Burchard
Marco Caliari	Daniel Calvelo	John C. Campbell
Juan Pablo Carbajal	Jean-Francois Cardoso	Joao Cardoso

Larrie Carr

David Castelow

Vincent Cautaerts

Clinton Chee

Albert Chin-A-Young

Carsten Clark

Catalin Codreanu

J. D. Cole

Martin Costabel

Michael Creel

Richard Crozier

Jeff Cunningham

Martin Dalecki

Jacob Dawid

Jorge Barros de Abreu

Carlo de Falco

Thomas D. Dean

Philippe Defert

Bill Denney

Fabian Deutsch

Christos Dimitrakakis

Pantxo Diribarne

Vivek Dogra

John Donoghue

David M. Doolin

Carn Draug

Pascal A. Dupuis

John W. Eaton

Dirk Eddelbuettel

Pieter Eendebak

Paul Eggert

Stephen Eglen

Peter Ekberg

Rolf Fabian

Gunnar Farnebck

Stephen Fegan

Ramon Garcia Fernandez

Torsten Finke

Jose Daniel Munoz Frias

Brad Froehle

Castor Fu

Eduardo Gallestey

Walter Gautschi

Klaus Gebhardt

Driss Ghaddab

Nicolo Giorgetti

Arun Giridhar

Michael D. Godfrey

Michael Goffioul

Glenn Golden

Tomislav Goles

Keith Goodman

Brian Gough

Steffen Groot

Etienne Grossmann

David Grundberg

Kyle Guinn

Peter Gustafson

Kai Habel

Patrick Hcker

William P. Y. Hadisoeseno

Jaroslav Hajek

Benjamin Hall

Kim Hansen

Sren Hauberg

Dave Hawthorne

Daniel Heiserer

Martin Helm

Stefan Hepp

Martin Hepperle

Jordi Gutirrez Hermoso

Yozo Hida

Ryan Hinton

Roman Hodek

A. Scottedward Hodel

Richard Allan Holcombe

Tom Holroyd

David Hoover

Kurt Hornik

Christopher Hulbert

Cyril Humbert

John Hunt

Teemu Ikonen

Alan W. Irwin

Geoff Jacobsen

Mats Jansson

Cai Jianming

Steven G. Johnson

Heikki Junes

Matthias Jschke

Atsushi Kajita

Jarkko Kaleva

Mohamed Kamoun

Lute Kamstra

Fotios Kasolis

Thomas Kasper

Joel Keay

Mumit Khan

Paul Kienzle

Aaron A. King

Erik Kjellson

Arno J. Klaassen

Alexander Klein

Geoffrey Knauth

Heine Kolltveit

Ken Kouno

Kacper Kowalik

Daniel Kraft

Nir Krakauer

Aravindh Krishnamoorthy

Oyvind Kristiansen

Artem Krosheninnikov

Piotr Krzyzanowski

Volker Kuhlmann

Tetsuro Kurita

Philipp Kutin

Miroslaw Kwasniak

Rafael Laboissiere

Kai Labusch

Claude Lacoursiere

Walter Landry

Bill Lash

Dirk Laurie

Maurice LeBrun

Friedrich Leisch

Thorsten Liebig

Jyh-miin Lin

Timo Lindfors

Benjamin Lindner

Ross Lippert

David Livings

Sebastien Loisel

Erik de Castro Lopo

Massimo Lorenzin

Emil Lucretiu

Hoxide Ma

Colin Macdonald

James Macnicol

Jens-Uwe Mager

Rob Mahurin

Alexander Mamonov

Ricardo Marranita
Makoto Matsumoto
Laurent Mazet
Jlio Hoffimann Mendes
Petr Mikulik
Antoine Moreau
Victor Munoz
Todd Neal
Felipe G. Nievinski
Kai Noda
Krzesimir Nowak
Thorsten Ohl
Luis F. Ortiz
Scott Pakin
Per Persson
Danilo Piazzalunga
Robert Platt
Orion Poplawski
Francesco Potort
James B. Rawlings
Joshua Redstone
Jens Restemeier
E. Joshua Rigler
Matthew W. Roberts
Fabio Rossi
Kevin Ruland
Olli Saarela
Radek Salac
Alois Schlgl
Nicol N. Schraudolph
Thomas L. Scofield
Vanya Sergeev
Robert T. Short
Julius Smith
Joerg Specht
Richard Stallman
Doug Stewart
Thomas Stuart
Daisuke Takago
Duncan Temple Lang
Georg Thimm
Christophe Tournery
David Turner
Stefan van der Walt
Risto Vanhanen
Ivana Varekova
Thomas Walter

Orestes Mas
Tatsuro Matsuoka
G. D. McBain
Ed Meyer
Mike Miller
Kai P. Mueller
Iain Murray
Philip Nienhuis
Rick Niles
Patrick Noffke
Michael O'Brien
Arno Onken
Carl Osterwisch
Gabriele Pannocchia
Primozz Peterlin
Nicholas Piper
Hans Ekkehard Plesser
Ondrej Popp
Konstantinos Poulios
Eric S. Raymond
Lukas Reichlin
Anthony Richardson
Sander van Rijn
Peter Rosin
Mark van Rossum
Kristian Rumberg
Toni Saarela
Ben Sapp
Michel D. Schmid
Sebastian Schubert
Daniel J. Sebald
Baylis Shanks
Joseph P. Skudlarek
Shan G. Smith
Quentin H. Spencer
Russell Standish
Jonathan Stickel
Ivan Sutoris
Ariel Tankus
Matthew Tenny
Corey Thomasson
Thomas Treichl
Frederick Umminger
Peter Van Wieren
Gregory Vanuxem
Sbastien Villemot
Andreas Weber

Axel Mathi
Christoph Mayer
Ronald van der Meer
Thorsten Meyer
Stefan Monnier
Hannes Mller
Carmen Navarrete
Al Niessner
Takuji Nishimura
Eric Norum
Peter O'Gorman
Valentin Ortega-Clavero
Janne Olavi Paanajrvi
Sylvain Pelissier
Jim Peterson
Elias Pipping
Tom Poage
Jef Poskanzer
Jarno Rajahalme
Balint Reczey
Michael Reifenberger
Jason Riedy
Petter Risholm
Andrew Ross
Joe Rothweiler
Ryan Rusaw
Juhani Saastamoinen
Aleksej Saushev
Julian Schnidder
Ludwig Schwardt
Dmitri A. Sergatskov
Andriy Shinkarchuck
John Smith
Peter L. Sondergaard
Christoph Spiel
Brett Stewart
Judd Storrs
John Swensen
Falk Tannhuser
Kris Thielemans
Olaf Till
Karsten Trulsen
Utkarsh Upadhyay
James R. Van Zandt
Mihas Varantsou
Daniel Wagenaar
Olaf Weber

Thomas Weber	Rik Wehbring	Bob Weigel
Andreas Weingessel	Martin Weiser	Michael Weitzel
David Wells	Fook Fah Yap	Sean Young
Michael Zeising	Federico Zenith	Alex Zvoleff

Special thanks to the following people and organizations for supporting the development of Octave:

- The United States Department of Energy, through grant number DE-FG02-04ER25635.

- Ashok Krishnamurthy, David Hudak, Juan Carlos Chaves, and Stanley C. Ahalt of the Ohio Supercomputer Center.

- The National Science Foundation, through grant numbers CTS-0105360, CTS-9708497, CTS-9311420, CTS-8957123, and CNS-0540147.

- The industrial members of the Texas-Wisconsin Modeling and Control Consortium (TWMCC).

- The Paul A. Elfers Endowed Chair in Chemical Engineering at the University of Wisconsin-Madison.

- Digital Equipment Corporation, for an equipment grant as part of their External Research Program.

- Sun Microsystems, Inc., for an Academic Equipment grant.

- International Business Machines, Inc., for providing equipment as part of a grant to the University of Texas College of Engineering.

- Texaco Chemical Company, for providing funding to continue the development of this software.

- The University of Texas College of Engineering, for providing a Challenge for Excellence Research Supplement, and for providing an Academic Development Funds grant.

- The State of Texas, for providing funding through the Texas Advanced Technology Program under Grant No. 003658-078.

- Noel Bell, Senior Engineer, Texaco Chemical Company, Austin Texas.

- John A. Turner, Group Leader, Continuum Dynamics (CCS-2), Los Alamos National Laboratory, for registering the `octave.org` domain name.

- James B. Rawlings, Professor, University of Wisconsin-Madison, Department of Chemical and Biological Engineering.

- Richard Stallman, for writing GNU.

This project would not have been possible without the GNU software used in and to produce Octave.

Citing Octave in Publications

In view of the many contributions made by numerous developers over many years it is common courtesy to cite Octave in publications when it has been used during the course of research or the preparation of figures. The `citation` function can automatically generate a recommended citation text for Octave or any of its packages. See the help text below on how to use `citation`.

`citation`   [Command]
`citation `*`package`*   [Command]

      Display instructions for citing GNU Octave or its packages in publications.

      When called without an argument, display information on how to cite the core GNU Octave system. When given a package name *package*, display information on citing the specific named package. Note that some packages may not yet have instructions on how to cite them.

      The GNU Octave developers and its active community of package authors have invested a lot of time and effort in creating GNU Octave as it is today. Please give credit where credit is due and cite GNU Octave and its packages when you use them.

How You Can Contribute to Octave

There are a number of ways that you can contribute to help make Octave a better system. Perhaps the most important way to contribute is to write high-quality code for solving new problems, and to make your code freely available for others to use. See Appendix D [Contributing Guidelines], page 819, for detailed information on contributing new code.

If you find Octave useful, consider providing additional funding to continue its development. Even a modest amount of additional funding could make a significant difference in the amount of time that is available for development and support.

Donations supporting Octave development may be made on the web at `https://my.fsf.org/donate/working-together/octave`. These donations also help to support the Free Software Foundation

If you'd prefer to pay by check or money order, you can do so by sending a check to the FSF at the following address:

      Free Software Foundation
      51 Franklin Street, Suite 500
      Boston, MA 02110-1335
      USA

If you pay by check, please be sure to write "GNU Octave" in the memo field of your check.

If you cannot provide funding or contribute code, you can still help make Octave better and more reliable by reporting any bugs you find and by offering suggestions for ways to improve Octave. See Appendix F [Trouble], page 829, for tips on how to write useful bug reports.

Distribution

Octave is *free* software. This means that everyone is free to use it and free to redistribute it on certain conditions. Octave is not, however, in the public domain. It is copyrighted and there are restrictions on its distribution, but the restrictions are designed to ensure that others will have the same freedom to use and redistribute Octave that you have. The precise conditions can be found in the GNU General Public License that comes with Octave and that also appears in Appendix J [Copying], page 857.

To download a copy of Octave, please visit `http://www.octave.org/download.html`.

1 A Brief Introduction to Octave

GNU Octave is a high-level language, primarily intended for numerical computations. It is typically used for such problems as solving linear and nonlinear equations, numerical linear algebra, statistical analysis, and for performing other numerical experiments. It may also be used as a batch-oriented language for automated data processing.

Until recently GNU Octave provided a command-line interface with graphical results displayed in separate windows. The current version (version 3.8, released in late 2013) also provides, by default, a graphical user interface.

GNU Octave is freely redistributable software. You may redistribute it and/or modify it under the terms of the GNU General Public License as published by the Free Software Foundation. The GPL is included in this manual, see Appendix J [Copying], page 857.

This manual provides comprehensive documentation on how to install, run, use, and extend GNU Octave. Additional chapters describe how to report bugs and help contribute code.

This document corresponds to Octave version 3.8.2.

1.1 Running Octave

On most systems, Octave is started with the shell command 'octave'. This, by default, starts the graphical user interface (GUI). The central window in the GUI is the Octave command-line interface. Octave displays an initial message and then a prompt indicating it is ready to accept input. If you have chosen the traditional command-line interface only the command prompt appears. In any case, you can immediately begin typing Octave commands.

If you get into trouble, you can usually interrupt Octave by typing `Control-C` (written `C-c` for short). `C-c` gets its name from the fact that you type it by holding down CTRL and then pressing C. Doing this will normally return you to Octave's prompt.

To exit Octave, type `quit`, or `exit` at the Octave prompt.

On systems that support job control, you can suspend Octave by sending it a SIGTSTP signal, usually by typing `C-z`.

1.2 Simple Examples

The following chapters describe all of Octave's features in detail, but before doing that, it might be helpful to give a sampling of some of its capabilities.

If you are new to Octave, I recommend that you try these examples to begin learning Octave by using it. Lines marked like so, 'octave:13>', are lines you type, ending each with a carriage return. Octave will respond with an answer, or by displaying a graph.

1.2.1 Elementary Calculations

Octave can easily be used for basic numerical calculations. Octave knows about arithmetic operations (+,-,*,/), exponentiation (^), natural logarithms/exponents (log, exp), and the trigonometric functions (sin, cos, ...). Moreover, Octave calculations work on real or imaginary numbers (i,j). In addition, some mathematical constants such as the base of the natural logarithm (e) and the ratio of a circle's circumference to its diameter (pi) are pre-defined.

For example, to verify Euler's Identity,

$$e^{i\pi} = -1$$

type the following which will evaluate to -1 within the tolerance of the calculation.

```
octave:1> exp (i*pi)
```

1.2.2 Creating a Matrix

Vectors and matrices are the basic building blocks for numerical analysis. To create a new matrix and store it in a variable so that you can refer to it later, type the command

```
octave:1> A = [ 1, 1, 2; 3, 5, 8; 13, 21, 34 ]
```

Octave will respond by printing the matrix in neatly aligned columns. Octave uses a comma or space to separate entries in a row, and a semicolon or carriage return to separate one row from the next. Ending a command with a semicolon tells Octave not to print the result of the command. For example,

```
octave:2> B = rand (3, 2);
```

will create a 3 row, 2 column matrix with each element set to a random value between zero and one.

To display the value of a variable, simply type the name of the variable at the prompt. For example, to display the value stored in the matrix B, type the command

```
octave:3> B
```

1.2.3 Matrix Arithmetic

Octave has a convenient operator notation for performing matrix arithmetic. For example, to multiply the matrix A by a scalar value, type the command

```
octave:4> 2 * A
```

To multiply the two matrices A and B, type the command

```
octave:5> A * B
```

and to form the matrix product $A^T A$, type the command

```
octave:6> A' * A
```

1.2.4 Solving Systems of Linear Equations

Systems of linear equations are ubiquitous in numerical analysis. To solve the set of linear equations Ax = b, use the left division operator, '\':

```
x = A \ b
```

This is conceptually equivalent to $A^{-1}b$, but avoids computing the inverse of a matrix directly.

If the coefficient matrix is singular, Octave will print a warning message and compute a minimum norm solution.

A simple example comes from chemistry and the need to obtain balanced chemical equations. Consider the burning of hydrogen and oxygen to produce water.

$$H_2 + O_2 \rightarrow H_2O$$

The equation above is not accurate. The Law of Conservation of Mass requires that the number of molecules of each type balance on the left- and right-hand sides of the equation. Writing the variable overall reaction with individual equations for hydrogen and oxygen one finds:

$$x_1 H_2 + x_2 O_2 \rightarrow H_2O$$

$$H: \quad 2x_1 + 0x_2 \rightarrow 2$$

$$O: \quad 0x_1 + 2x_2 \rightarrow 1$$

The solution in Octave is found in just three steps.

```
octave:1> A = [ 2, 0; 0, 2 ];
octave:2> b = [ 2; 1 ];
octave:3> x = A \ b
```

1.2.5 Integrating Differential Equations

Octave has built-in functions for solving nonlinear differential equations of the form

$$\frac{dx}{dt} = f(x,t), \qquad x(t = t_0) = x_0$$

For Octave to integrate equations of this form, you must first provide a definition of the function $f(x,t)$. This is straightforward, and may be accomplished by entering the function body directly on the command line. For example, the following commands define the right-hand side function for an interesting pair of nonlinear differential equations. Note that while you are entering a function, Octave responds with a different prompt, to indicate that it is waiting for you to complete your input.

```
octave:1> function xdot = f (x, t)
>
>   r = 0.25;
>   k = 1.4;
>   a = 1.5;
>   b = 0.16;
>   c = 0.9;
>   d = 0.8;
>
>   xdot(1) = r*x(1)*(1 - x(1)/k) - a*x(1)*x(2)/(1 + b*x(1));
>   xdot(2) = c*a*x(1)*x(2)/(1 + b*x(1)) - d*x(2);
>
> endfunction
```

Given the initial condition

```
octave:2> x0 = [1; 2];
```

and the set of output times as a column vector (note that the first output time corresponds to the initial condition given above)

```
octave:3> t = linspace (0, 50, 200)';
```

it is easy to integrate the set of differential equations:

```
octave:4> x = lsode ("f", x0, t);
```

The function `lsode` uses the Livermore Solver for Ordinary Differential Equations, described in A. C. Hindmarsh, *ODEPACK, a Systematized Collection of ODE Solvers*, in: Scientific Computing, R. S. Stepleman et al. (Eds.), North-Holland, Amsterdam, 1983, pages 55–64.

1.2.6 Producing Graphical Output

To display the solution of the previous example graphically, use the command

```
octave:1> plot (t, x)
```

If you are using a graphical user interface, Octave will automatically create a separate window to display the plot.

To save a plot once it has been displayed on the screen, use the print command. For example,

```
print -deps foo.eps
```

will create a file called 'foo.eps' that contains a rendering of the current plot in Encapsulated PostScript format. The command

```
help print
```

explains more options for the `print` command and provides a list of additional output file formats.

1.2.7 Editing What You Have Typed

At the Octave prompt, you can recall, edit, and reissue previous commands using Emacs- or vi-style editing commands. The default keybindings use Emacs-style commands. For example, to recall the previous command, press *Control-p* (written *C-p* for short). Doing this will normally bring back the previous line of input. *C-n* will bring up the next line of input, *C-b* will move the cursor backward on the line, *C-f* will move the cursor forward on the line, etc.

A complete description of the command line editing capability is given in this manual, see Section 2.4 [Command Line Editing], page 24.

1.2.8 Help and Documentation

Octave has an extensive help facility. The same documentation that is available in printed form is also available from the Octave prompt, because both forms of the documentation are created from the same input file.

In order to get good help you first need to know the name of the command that you want to use. This name of the function may not always be obvious, but a good place to start is to type `help --list`. This will show you all the operators, keywords, built-in functions, and loadable functions available in the current session of Octave. An alternative is to search the documentation using the `lookfor` function. This function is described in Section 2.3 [Getting Help], page 20.

Once you know the name of the function you wish to use, you can get more help on the function by simply including the name as an argument to help. For example,

```
help plot
```

will display the help text for the plot function.

Octave sends output that is too long to fit on one screen through a pager like less or more. Type a RET to advance one line, a SPC to advance one page, and Q to exit the pager.

The part of Octave's help facility that allows you to read the complete text of the printed manual from within Octave normally uses a separate program called Info. When you invoke Info you will be put into a menu driven program that contains the entire Octave manual. Help for using Info is provided in this manual, see Section 2.3 [Getting Help], page 20.

1.3 Conventions

This section explains the notational conventions that are used in this manual. You may want to skip this section and refer back to it later.

1.3.1 Fonts

Examples of Octave code appear in this font or form: svd (a). Names that represent variables or function arguments appear in this font or form: *first-number*. Commands that you type at the shell prompt appear in this font or form: 'octave --no-init-file'. Commands that you type at the Octave prompt sometimes appear in this font or form: *foo --bar --baz*. Specific keys on your keyboard appear in this font or form: ANY.

1.3.2 Evaluation Notation

In the examples in this manual, results from expressions that you evaluate are indicated with '⇒'. For example:

```
sqrt (2)
     ⇒ 1.4142
```

You can read this as "sqrt (2) evaluates to 1.4142".

In some cases, matrix values that are returned by expressions are displayed like this

```
[1, 2; 3, 4] == [1, 3; 2, 4]
     ⇒ [ 1, 0; 0, 1 ]
```

and in other cases, they are displayed like this

```
eye (3)
     ⇒   1   0   0
          0   1   0
          0   0   1
```

in order to clearly show the structure of the result.

Sometimes to help describe one expression, another expression is shown that produces identical results. The exact equivalence of expressions is indicated with '≡'. For example:

```
rot90 ([1, 2; 3, 4], -1)
≡
rot90 ([1, 2; 3, 4], 3)
≡
rot90 ([1, 2; 3, 4], 7)
```

1.3.3 Printing Notation

Many of the examples in this manual print text when they are evaluated. In this manual the printed text resulting from an example is indicated by '⊣'. The value that is returned by evaluating the expression is displayed with '⇒' (1 in the next example) and follows on a separate line.

```
printf ("foo %s\n", "bar")
     ⊣ foo bar
     ⇒ 1
```

1.3.4 Error Messages

Some examples signal errors. This normally displays an error message on your terminal. Error messages are shown on a line beginning with `error:`.

```
fieldnames ([1, 2; 3, 4])
error: fieldnames: Invalid input argument
```

1.3.5 Format of Descriptions

Functions and commands are described in this manual in a uniform format. The first line of a description contains the name of the item followed by its arguments, if any. The category—function, command, or whatever—is printed next to the right margin. The description follows on succeeding lines, sometimes with examples.

1.3.5.1 A Sample Function Description

In a function description, the name of the function being described appears first. It is followed on the same line by a list of parameters. The names used for the parameters are also used in the body of the description.

Here is a description of an imaginary function `foo`:

`foo (x)`   [Function File]
`foo (x, y)`   [Function File]
`foo (x, y, ...)`   [Function File]
> The function `foo` subtracts x from y, then adds the remaining arguments to the result. If y is not supplied, then the number 19 is used by default.
>
> ```
> foo (1, [3, 5], 3, 9)
>        ⇒ [ 14, 16 ]
> foo (5)
>        ⇒ 14
> ```
> More generally,
> ```
> foo (w, x, y, ...)
> ≡
> x - w + y + ...
> ```

Any parameter whose name contains the name of a type (e.g., *integer* or *matrix*) is expected to be of that type. Parameters named *object* may be of any type. Parameters with other sorts of names (e.g., *new_file*) are discussed specifically in the description of the function. In some sections, features common to parameters of several functions are described at the beginning.

Functions in Octave may be defined in several different ways. The category name for functions may include another name that indicates the way that the function is defined. These additional tags include

Function File

> The function described is defined using Octave commands stored in a text file. See Section 11.9 [Function Files], page 178.

Built-in Function

> The function described is written in a language like C++, C, or Fortran, and is part of the compiled Octave binary.

Loadable Function

> The function described is written in a language like C++, C, or Fortran. On systems that support dynamic linking of user-supplied functions, it may be automatically linked while Octave is running, but only if it is needed. See Appendix A [External Code Interface], page 759.

Mapping Function

> The function described works element-by-element for matrix and vector arguments.

1.3.5.2 A Sample Command Description

Command descriptions have a format similar to function descriptions, except that the word 'Function' is replaced by 'Command'. Commands are functions that may be called without surrounding their arguments in parentheses. For example, here is the description for Octave's `cd` command:

`cd` *dir*   [Command]
`chdir` *dir*   [Command]

> Change the current working directory to *dir*. For example, `cd ~/octave` changes the current working directory to '`~/octave`'. If the directory does not exist, an error message is printed and the working directory is not changed.

2 Getting Started

This chapter explains some of Octave's basic features, including how to start an Octave session, get help at the command prompt, edit the command line, and write Octave programs that can be executed as commands from your shell.

2.1 Invoking Octave from the Command Line

Normally, Octave is used interactively by running the program 'octave' without any arguments. Once started, Octave reads commands from the terminal until you tell it to exit.

You can also specify the name of a file on the command line, and Octave will read and execute the commands from the named file and then exit when it is finished.

You can further control how Octave starts by using the command-line options described in the next section, and Octave itself can remind you of the options available. Type 'octave --help' to display all available options and briefly describe their use ('octave -h' is a shorter equivalent).

2.1.1 Command Line Options

Here is a complete list of the command line options that Octave accepts.

`--built-in-docstrings-file` *filename*

> Specify the name of the file containing documentation strings for the built-in functions of Octave. This value is normally correct and should only need to specified in extraordinary situations.

`--debug`

`-d`
> Enter parser debugging mode. Using this option will cause Octave's parser to print a lot of information about the commands it reads, and is probably only useful if you are actually trying to debug the parser.

`--debug-jit`
> Enable JIT compiler debugging and tracing.

`--doc-cache-file` *filename*
> Specify the name of the doc cache file to use. The value of *filename* specified on the command line will override any value of `OCTAVE_DOC_CACHE_FILE` found in the environment, but not any commands in the system or user startup files that use the `doc_cache_file` function.

`--echo-commands`

`-x`
> Echo commands as they are executed.

`--eval` *code*
> Evaluate *code* and exit when finished unless '--persist' is also specified.

`--exec-path` *path*
> Specify the path to search for programs to run. The value of *path* specified on the command line will override any value of `OCTAVE_EXEC_PATH` found in the environment, but not any commands in the system or user startup files that set the built-in variable `EXEC_PATH`.

`--force-gui`
> Force the graphical user interface (GUI) to start.

`--help`
`-h`
`-?`   Print short help message and exit.

`--image-path` *path*
> Add path to the head of the search path for images. The value of *path* specified on the command line will override any value of `OCTAVE_IMAGE_PATH` found in the environment, but not any commands in the system or user startup files that set the built-in variable `IMAGE_PATH`.

`--info-file` *filename*
> Specify the name of the info file to use. The value of *filename* specified on the command line will override any value of `OCTAVE_INFO_FILE` found in the environment, but not any commands in the system or user startup files that use the `info_file` function.

`--info-program` *program*
> Specify the name of the info program to use. The value of *program* specified on the command line will override any value of `OCTAVE_INFO_PROGRAM` found in the environment, but not any commands in the system or user startup files that use the `info_program` function.

`--interactive`
`-i`   Force interactive behavior. This can be useful for running Octave via a remote shell command or inside an Emacs shell buffer. For another way to run Octave within Emacs, see Appendix H [Emacs Octave Support], page 847.

`--jit-compiler`
> Enable the JIT compiler used for accelerating loops.

`--line-editing`
> Force readline use for command-line editing.

`--no-gui`   Disable the graphical user interface (GUI) and use the command line interface (CLI) instead.

`--no-history`
`-H`   Disable recording of command-line history.

`--no-init-file`
> Don't read the initialization files '~/.octaverc' and '.octaverc'.

`--no-init-path`
> Don't initialize the search path for function files to include default locations.

`--no-line-editing`
> Disable command-line editing.

`--no-site-file`
> Don't read the site-wide 'octaverc' initialization files.

`--no-window-system`

`-W`   Disable use of a windowing system including graphics. This forces a strictly terminal-only environment.

`--norc`

`-f`   Don't read any of the system or user initialization files at startup. This is equivalent to using both of the options '`--no-init-file`' and '`--no-site-file`'.

`--path` *path*

`-p` *path*   Add path to the head of the search path for function files. The value of *path* specified on the command line will override any value of `OCTAVE_PATH` found in the environment, but not any commands in the system or user startup files that set the internal load path through one of the path functions.

`--persist`

              Go to interactive mode after '`--eval`' or reading from a file named on the command line.

`--silent`

`--quiet`

`-q`   Don't print the usual greeting and version message at startup.

`--texi-macros-file` *filename*

              Specify the name of the file containing Texinfo macros for use by makeinfo.

`--traditional`

`--braindead`

              For compatibility with MATLAB, set initial values for user preferences to the following values

```
PS1                               = ">> "
PS2                               = ""
allow_noninteger_range_as_index   = true
beep_on_error                     = true
confirm_recursive_rmdir           = false
crash_dumps_octave_core           = false
save_default_options              = "-mat-binary"
do_braindead_shortcircuit_evaluation = true
fixed_point_format                = true
history_timestamp_format_string   = "%%-- %D %I:%M %p --%%"
page_screen_output                = false
print_empty_dimensions            = false
```

              and disable the following warnings

```
Octave:abbreviated-property-match
Octave:fopen-file-in-path
Octave:function-name-clash
Octave:load-file-in-path
```

              Note that this does not enable the `Octave:matlab-incompatible` warning, which you might want if you want to be told about writing code that works in Octave but not MATLAB (see [warning], page 204, [warning_ids], page 205).

```
--verbose
-V          Turn on verbose output.

--version
-v          Print the program version number and exit.
```

file   Execute commands from *file*. Exit when done unless '`--persist`' is also speci-
            fied.

Octave also includes several functions which return information about the command line,
including the number of arguments and all of the options.

argv ()   [Built-in Function]
        Return the command line arguments passed to Octave. For example, if you invoked
        Octave using the command

            `octave --no-line-editing --silent`

        `argv` would return a cell array of strings with the elements '`--no-line-editing`' and
        '`--silent`'.

        If you write an executable Octave script, `argv` will return the list of arguments passed
        to the script. See Section 2.6 [Executable Octave Programs], page 35, for an example
        of how to create an executable Octave script.

program_name ()   [Built-in Function]
        Return the last component of the value returned by `program_invocation_name`.

        See also: [program_invocation_name], page 18.

program_invocation_name ()   [Built-in Function]
        Return the name that was typed at the shell prompt to run Octave.

        If executing a script from the command line (e.g., `octave foo.m`) or using an ex-
        ecutable Octave script, the program name is set to the name of the script. See
        Section 2.6 [Executable Octave Programs], page 35, for an example of how to create
        an executable Octave script.

        See also: [program_name], page 18.

Here is an example of using these functions to reproduce the command line which invoked
Octave.

```
printf ("%s", program_name ());
arg_list = argv ();
for i = 1:nargin
  printf (" %s", arg_list{i});
endfor
printf ("\n");
```

See Section 6.2.3 [Indexing Cell Arrays], page 114, for an explanation of how to retrieve
objects from cell arrays, and Section 11.2 [Defining Functions], page 167, for information
about the variable `nargin`.

2.1.2 Startup Files

When Octave starts, it looks for commands to execute from the files in the following list. These files may contain any valid Octave commands, including function definitions.

octave-home/share/octave/site/m/startup/octaverc

>   where *octave-home* is the directory in which Octave is installed (the default is '/usr/local'). This file is provided so that changes to the default Octave environment can be made globally for all users at your site for all versions of Octave you have installed. Care should be taken when making changes to this file since all users of Octave at your site will be affected. The default file may be overridden by the environment variable OCTAVE_SITE_INITFILE.

octave-home/share/octave/*version*/m/startup/octaverc

>   where *octave-home* is the directory in which Octave is installed (the default is '/usr/local'), and *version* is the version number of Octave. This file is provided so that changes to the default Octave environment can be made globally for all users of a particular version of Octave. Care should be taken when making changes to this file since all users of Octave at your site will be affected. The default file may be overridden by the environment variable OCTAVE_VERSION_INITFILE.

~/.octaverc

>   This file is used to make personal changes to the default Octave environment.

.octaverc

>   This file can be used to make changes to the default Octave environment for a particular project. Octave searches for this file in the current directory after it reads '~/.octaverc'. Any use of the cd command in the '~/.octaverc' file will affect the directory where Octave searches for '.octaverc'.

>   If you start Octave in your home directory, commands from the file '~/.octaverc' will only be executed once.

A message will be displayed as each of the startup files is read if you invoke Octave with the '--verbose' option but without the '--silent' option.

The dump_prefs function is useful for determining what customizations to Octave are possible and which are in effect.

dump_prefs ()   [Function File]
dump_prefs (*fid*)   [Function File]

>   Dump all of the current user preference variables in a format that can be parsed by Octave later. *fid* is a file descriptor as returned by fopen. If *file* is omitted, the listing is printed to stdout.

2.2 Quitting Octave

exit (*status*)   [Built-in Function]
quit (*status*)   [Built-in Function]

>   Exit the current Octave session. If the optional integer value *status* is supplied, pass that value to the operating system as the Octave's exit status. The default value is zero.

`atexit (`*`fcn`*`)`   [Built-in Function]
`atexit (`*`fcn`*`, `*`flag`*`)`   [Built-in Function]

Register a function to be called when Octave exits. For example,

```
function last_words ()
  disp ("Bye bye");
endfunction
atexit ("last_words");
```

will print the message `"Bye bye"` when Octave exits.

The additional argument *flag* will register or unregister *fcn* from the list of functions to be called when Octave exits. If *flag* is true, the function is registered, and if *flag* is false, it is unregistered. For example, after registering the function `last_words` above,

```
atexit ("last_words", false);
```

will remove the function from the list and Octave will not call `last_words` when it exits.

Note that `atexit` only removes the first occurrence of a function from the list, so if a function was placed in the list multiple times with `atexit`, it must also be removed from the list multiple times.

2.3 Commands for Getting Help

The entire text of this manual is available from the Octave prompt via the command *doc*. In addition, the documentation for individual user-written functions and variables is also available via the *help* command. This section describes the commands used for reading the manual and the documentation strings for user-supplied functions and variables. See Section 11.9 [Function Files], page 178, for more information about how to document the functions you write.

`help `*`name`*   [Command]
`help --list`   [Command]
`help .`   [Command]

Display the help text for *name*. For example, the command *help help* prints a short message describing the `help` command.

Given the single argument `--list`, list all operators, keywords, built-in functions, and loadable functions available in the current session of Octave.

Given the single argument ., list all operators available in the current session of Octave.

If invoked without any arguments, `help` display instructions on how to access help from the command line.

The help command can give you information about operators, but not the comma and semicolons that are used as command separators. To get help for those, you must type *help comma* or *help semicolon*.

See also: [doc], page 21, [lookfor], page 21, [which], page 130.

doc *function_name*   [Command]

>   Display documentation for the function *function_name* directly from an online version of the printed manual, using the GNU Info browser. If invoked without any arguments, the manual is shown from the beginning.
>
>   For example, the command **doc rand** starts the GNU Info browser at the **rand** node in the online version of the manual.
>
>   Once the GNU Info browser is running, help for using it is available using the command *C-h*.
>
>   **See also:** [help], page 20.

lookfor *str*   [Command]
lookfor *-all* *str*   [Command]
[*func, helpstring*] = **lookfor** (*str*)   [Function File]
[*func, helpstring*] = **lookfor** ("*-all*", *str*)   [Function File]

>   Search for the string *str* in all functions found in the current function search path. By default, **lookfor** searches for *str* in the first sentence of the help string of each function found. The entire help text of each function can be searched if the "**-all**" argument is supplied. All searches are case insensitive.
>
>   Called with no output arguments, **lookfor** prints the list of matching functions to the terminal. Otherwise, the output arguments *func* and *helpstring* define the matching functions and the first sentence of each of their help strings.
>
>   The ability of **lookfor** to correctly identify the first sentence of the help text is dependent on the format of the function's help. All Octave core functions are correctly formatted, but the same can not be guaranteed for external packages and user-supplied functions. Therefore, the use of the "**-all**" argument may be necessary to find related functions that are not a part of Octave.
>
>   **See also:** [help], page 20, [doc], page 21, [which], page 130.

To see what is new in the current release of Octave, use the **news** function.

news   [Command]
news *package*   [Command]

>   Display the current NEWS file for Octave or an installed package.
>
>   When called without an argument, display the NEWS file for Octave. When given a package name *package*, display the current NEWS file for that package.

info ()   [Function File]

>   Display contact information for the GNU Octave community.

warranty ()   [Built-in Function]

>   Describe the conditions for copying and distributing Octave.

The following functions can be used to change which programs are used for displaying the documentation, and where the documentation can be found.

val = **info_file** ()   [Built-in Function]
old_val = **info_file** (*new_val*)   [Built-in Function]

`info_file (`*`new_val`*`, "`*`local`*`")`   [Built-in Function]

> Query or set the internal variable that specifies the name of the Octave info file. The default value is '`octave-home/info/octave.info`', in which *octave-home* is the root directory of the Octave installation. The default value may be overridden by the environment variable `OCTAVE_INFO_FILE`, or the command line argument '`--info-file FNAME`'.
>
> When called from inside a function with the `"local"` option, the variable is changed locally for the function and any subroutines it calls. The original variable value is restored when exiting the function.
>
> **See also:** [info_program], page 22, [doc], page 21, [help], page 20, [makeinfo_program], page 22.

`val` `= info_program ()`   [Built-in Function]
`old_val` `= info_program (`*`new_val`*`)`   [Built-in Function]
`info_program (`*`new_val`*`, "`*`local`*`")`   [Built-in Function]

> Query or set the internal variable that specifies the name of the info program to run. The default value is '`octave-home/libexec/octave/version/exec/arch/info`' in which *octave-home* is the root directory of the Octave installation, *version* is the Octave version number, and *arch* is the system type (for example, `i686-pc-linux-gnu`). The default value may be overridden by the environment variable `OCTAVE_INFO_PROGRAM`, or the command line argument '`--info-program NAME`'.
>
> When called from inside a function with the `"local"` option, the variable is changed locally for the function and any subroutines it calls. The original variable value is restored when exiting the function.
>
> **See also:** [info_file], page 21, [doc], page 21, [help], page 20, [makeinfo_program], page 22.

`val` `= makeinfo_program ()`   [Built-in Function]
`old_val` `= makeinfo_program (`*`new_val`*`)`   [Built-in Function]
`makeinfo_program (`*`new_val`*`, "`*`local`*`")`   [Built-in Function]

> Query or set the internal variable that specifies the name of the program that Octave runs to format help text containing Texinfo markup commands. The default value is `makeinfo`.
>
> When called from inside a function with the `"local"` option, the variable is changed locally for the function and any subroutines it calls. The original variable value is restored when exiting the function.
>
> **See also:** [texi_macros_file], page 22, [info_file], page 21, [info_program], page 22, [doc], page 21, [help], page 20.

`val` `= texi_macros_file ()`   [Built-in Function]
`old_val` `= texi_macros_file (`*`new_val`*`)`   [Built-in Function]
`texi_macros_file (`*`new_val`*`, "`*`local`*`")`   [Built-in Function]

> Query or set the internal variable that specifies the name of the file containing Texinfo macros that are prepended to documentation strings

before they are passed to makeinfo. The default value is 'octave-home/share/octave/version/etc/macros.texi', in which *octave-home* is the root directory of the Octave installation, and *version* is the Octave version number. The default value may be overridden by the environment variable `OCTAVE_TEXI_MACROS_FILE`, or the command line argument '`--texi-macros-file FNAME`'.

When called from inside a function with the "`local`" option, the variable is changed locally for the function and any subroutines it calls. The original variable value is restored when exiting the function.

See also: [makeinfo_program], page 22.

`val = doc_cache_file ()`   [Built-in Function]
`old_val = doc_cache_file (new_val)`   [Built-in Function]
`doc_cache_file (new_val, "local")`   [Built-in Function]
    Query or set the internal variable that specifies the name of the Octave documentation cache file. A cache file significantly improves the performance of the `lookfor` command. The default value is 'octave-home/share/octave/version/etc/doc-cache', in which *octave-home* is the root directory of the Octave installation, and *version* is the Octave version number. The default value may be overridden by the environment variable `OCTAVE_DOC_CACHE_FILE`, or the command line argument '`--doc-cache-file FNAME`'.

    When called from inside a function with the "`local`" option, the variable is changed locally for the function and any subroutines it calls. The original variable value is restored when exiting the function.

    See also: [doc_cache_create], page 24, [lookfor], page 21, [info_program], page 22, [doc], page 21, [help], page 20, [makeinfo_program], page 22.

`val = built_in_docstrings_file ()`   [Built-in Function]
`old_val = built_in_docstrings_file (new_val)`   [Built-in Function]
`built_in_docstrings_file (new_val, "local")`   [Built-in Function]
    Query or set the internal variable that specifies the name of the file containing docstrings for built-in Octave functions. The default value is 'octave-home/share/octave/version/etc/built-in-docstrings', in which *octave-home* is the root directory of the Octave installation, and *version* is the Octave version number. The default value may be overridden by the environment variable `OCTAVE_BUILT_IN_DOCSTRINGS_FILE`, or the command line argument '`--built-in-docstrings-file FNAME`'.

    Note: This variable is only used when Octave is initializing itself. Modifying it during a running session of Octave will have no effect.

`val = suppress_verbose_help_message ()`   [Built-in Function]
`old_val = suppress_verbose_help_message (new_val)`   [Built-in Function]
`suppress_verbose_help_message (new_val, "local")`   [Built-in Function]
    Query or set the internal variable that controls whether Octave will add additional help information to the end of the output from the `help` command and usage messages for built-in commands.

When called from inside a function with the `"local"` option, the variable is changed locally for the function and any subroutines it calls. The original variable value is restored when exiting the function.

The following functions are principally used internally by Octave for generating the documentation. They are documented here for completeness and because they may occasionally be useful for users.

`doc_cache_create (out_file, directory)`   [Function File]
> Generate documentation caches for all functions in a given directory.
>
> A documentation cache is generated for all functions in *directory*. The resulting cache is saved in the file *out_file*. The cache is used to speed up `lookfor`.
>
> If no directory is given (or it is the empty matrix), a cache for built-in operators, etc. is generated.
>
> **See also:** [doc_cache_file], page 23, [lookfor], page 21, [path], page 181.

`[text, format] = get_help_text (name)`   [Built-in Function]
> Return the raw help text of function *name*.
>
> The raw help text is returned in *text* and the format in *format* The format is a string which is one of `"texinfo"`, `"html"`, or `"plain text"`.

`[text, format] = get_help_text_from_file (fname)`   [Built-in Function]
> Return the raw help text from the file *fname*.
>
> The raw help text is returned in *text* and the format in *format* The format is a string which is one of `"texinfo"`, `"html"`, or `"plain text"`.

`[text, status] = get_first_help_sentence (name)`   [Function File]
`[text, status] = get_first_help_sentence (name, max_len)`   [Function File]
> Return the first sentence of a function's help text.
>
> The first sentence is defined as the text after the function declaration until either the first period (`"."`) or the first appearance of two consecutive newlines (`"\n\n"`). The text is truncated to a maximum length of *max_len*, which defaults to 80.
>
> The optional output argument *status* returns the status reported by `makeinfo`. If only one output argument is requested, and *status* is non-zero, a warning is displayed.
>
> As an example, the first sentence of this help text is
>
> ```
> get_first_help_sentence ("get_first_help_sentence")
> ⊣ ans = Return the first sentence of a function's help text.
> ```

2.4 Command Line Editing

Octave uses the GNU Readline library to provide an extensive set of command-line editing and history features. Only the most common features are described in this manual. In addition, all of the editing functions can be bound to different key strokes at the user's discretion. This manual assumes no changes from the default Emacs bindings. See the GNU Readline Library manual for more information on customizing Readline and for a complete feature list.

To insert printing characters (letters, digits, symbols, etc.), simply type the character. Octave will insert the character at the cursor and advance the cursor forward.

Many of the command-line editing functions operate using control characters. For example, the character `Control-a` moves the cursor to the beginning of the line. To type `C-a`, hold down CTRL and then press A. In the following sections, control characters such as `Control-a` are written as `C-a`.

Another set of command-line editing functions use Meta characters. To type `M-u`, hold down the META key and press U. Depending on the keyboard, the META key may be labeled ALT or even WINDOWS. If your terminal does not have a META key, you can still type Meta characters using two-character sequences starting with `ESC`. Thus, to enter `M-u`, you would type ESC U. The `ESC` character sequences are also allowed on terminals with real Meta keys. In the following sections, Meta characters such as `Meta-u` are written as `M-u`.

2.4.1 Cursor Motion

The following commands allow you to position the cursor.

`C-b`   Move back one character.

`C-f`   Move forward one character.

BACKSPACE
        Delete the character to the left of the cursor.

DEL   Delete the character underneath the cursor.

`C-d`   Delete the character underneath the cursor.

`M-f`   Move forward a word.

`M-b`   Move backward a word.

`C-a`   Move to the start of the line.

`C-e`   Move to the end of the line.

`C-l`   Clear the screen, reprinting the current line at the top.

`C-_`
`C-/`   Undo the last action. You can undo all the way back to an empty line.

`M-r`   Undo all changes made to this line. This is like typing the 'undo' command enough times to get back to the beginning.

The above table describes the most basic possible keystrokes that you need in order to do editing of the input line. On most terminals, you can also use the left and right arrow keys in place of `C-f` and `C-b` to move forward and backward.

Notice how `C-f` moves forward a character, while `M-f` moves forward a word. It is a loose convention that control keystrokes operate on characters while meta keystrokes operate on words.

The function `clc` will allow you to clear the screen from within Octave programs.

`clc ()`   [Built-in Function]
`home ()`   [Built-in Function]
        Clear the terminal screen and move the cursor to the upper left corner.

2.4.2 Killing and Yanking

Killing text means to delete the text from the line, but to save it away for later use, usually by *yanking* it back into the line. If the description for a command says that it 'kills' text, then you can be sure that you can get the text back in a different (or the same) place later.

Here is the list of commands for killing text.

C-k   Kill the text from the current cursor position to the end of the line.

M-d   Kill from the cursor to the end of the current word, or if between words, to the end of the next word.

M-DEL   Kill from the cursor to the start of the previous word, or if between words, to the start of the previous word.

C-w   Kill from the cursor to the previous whitespace. This is different than M-DEL because the word boundaries differ.

And, here is how to *yank* the text back into the line. Yanking means to copy the most-recently-killed text from the kill buffer.

C-y   Yank the most recently killed text back into the buffer at the cursor.

M-y   Rotate the kill-ring, and yank the new top. You can only do this if the prior command is C-y or M-y.

When you use a kill command, the text is saved in a *kill-ring*. Any number of consecutive kills save all of the killed text together, so that when you yank it back, you get it in one clean sweep. The kill ring is not line specific; the text that you killed on a previously typed line is available to be yanked back later, when you are typing another line.

2.4.3 Commands For Changing Text

The following commands can be used for entering characters that would otherwise have a special meaning (e.g., TAB, C-q, etc.), or for quickly correcting typing mistakes.

C-q
C-v   Add the next character that you type to the line verbatim. This is how to insert things like C-q for example.

M-TAB   Insert a tab character.

C-t   Drag the character before the cursor forward over the character at the cursor, also moving the cursor forward. If the cursor is at the end of the line, then transpose the two characters before it.

M-t   Drag the word behind the cursor past the word in front of the cursor moving the cursor over that word as well.

M-u   Uppercase the characters following the cursor to the end of the current (or following) word, moving the cursor to the end of the word.

M-l   Lowercase the characters following the cursor to the end of the current (or following) word, moving the cursor to the end of the word.

M-c   Uppercase the character following the cursor (or the beginning of the next word if the cursor is between words), moving the cursor to the end of the word.

2.4.4 Letting Readline Type For You

The following commands allow Octave to complete command and variable names for you.

TAB　　　　Attempt to do completion on the text before the cursor. Octave can complete the names of commands and variables.

M-?　　　　List the possible completions of the text before the cursor.

`val = completion_append_char ()`   [Built-in Function]
`old_val = completion_append_char (`*new_val*`)`   [Built-in Function]
`completion_append_char (`*new_val*`, "`*local*`")`   [Built-in Function]
　　　Query or set the internal character variable that is appended to successful command-line completion attempts. The default value is `" "` (a single space).

　　　When called from inside a function with the `"local"` option, the variable is changed locally for the function and any subroutines it calls. The original variable value is restored when exiting the function.

`completion_matches (`*hint*`)`   [Built-in Function]
　　　Generate possible completions given *hint*.

　　　This function is provided for the benefit of programs like Emacs which might be controlling Octave and handling user input. The current command number is not incremented when this function is called. This is a feature, not a bug.

2.4.5 Commands For Manipulating The History

Octave normally keeps track of the commands you type so that you can recall previous commands to edit or execute them again. When you exit Octave, the most recent commands you have typed, up to the number specified by the variable `history_size`, are saved in a file. When Octave starts, it loads an initial list of commands from the file named by the variable `history_file`.

　　　Here are the commands for simple browsing and searching the history list.

LFD
RET　　　　Accept the current line regardless of where the cursor is. If the line is non-empty, add it to the history list. If the line was a history line, then restore the history line to its original state.

C-p　　　　Move 'up' through the history list.

C-n　　　　Move 'down' through the history list.

M-<　　　　Move to the first line in the history.

M->　　　　Move to the end of the input history, i.e., the line you are entering!

C-r　　　　Search backward starting at the current line and moving 'up' through the history as necessary. This is an incremental search.

C-s　　　　Search forward starting at the current line and moving 'down' through the history as necessary.

　　　On most terminals, you can also use the up and down arrow keys in place of *C-p* and *C-n* to move through the history list.

In addition to the keyboard commands for moving through the history list, Octave provides three functions for viewing, editing, and re-running chunks of commands from the history list.

`history`	[Command]
`history opt1 ...`	[Command]
`h = history ()`	[Built-in Function]
`h = history (opt1, ...)`	[Built-in Function]

If invoked with no arguments, `history` displays a list of commands that you have executed. Valid options are:

> `n`
> `-n`   Display only the most recent n lines of history.
>
> `-c`   Clear the history list.
>
> `-q`   Don't number the displayed lines of history. This is useful for cutting and pasting commands using the X Window System.
>
> `-r file`   Read the file *file*, appending its contents to the current history list. If the name is omitted, use the default history file (normally '`~/.octave_hist`').
>
> `-w file`   Write the current history to the file *file*. If the name is omitted, use the default history file (normally '`~/.octave_hist`').

For example, to display the five most recent commands that you have typed without displaying line numbers, use the command *history -q 5*.

If invoked with a single output argument, the history will be saved to that argument as a cell string and will not be output to screen.

`edit_history`	[Command]
`edit_history cmd_number`	[Command]
`edit_history first last`	[Command]

Edit the history list using the editor named by the variable `EDITOR`.

The commands to be edited are first copied to a temporary file. When you exit the editor, Octave executes the commands that remain in the file. It is often more convenient to use `edit_history` to define functions rather than attempting to enter them directly on the command line. The block of commands is executed as soon as you exit the editor. To avoid executing any commands, simply delete all the lines from the buffer before leaving the editor.

When invoked with no arguments, edit the previously executed command; With one argument, edit the specified command *cmd_number*; With two arguments, edit the list of commands between *first* and *last*. Command number specifiers may also be negative where -1 refers to the most recently executed command. The following are equivalent and edit the most recently executed command.

> `edit_history`
> `edit_history -1`

When using ranges, specifying a larger number for the first command than the last command reverses the list of commands before they are placed in the buffer to be edited.

See also: [run_history], page 29.

run_history   [Command]
run_history *cmd_number*   [Command]
run_history *first last*   [Command]
    Run commands from the history list.

    When invoked with no arguments, run the previously executed command; With one argument, run the specified command *cmd_number*; With two arguments, run the list of commands between *first* and *last*. Command number specifiers may also be negative where -1 refers to the most recently executed command. For example, the command

```
run_history
    OR
run_history -1
```

executes the most recent command again. The command

```
run_history 13 169
```

executes commands 13 through 169.

    Specifying a larger number for the first command than the last command reverses the list of commands before executing them. For example:

```
disp (1)
disp (2)
run_history -1 -2
⇒
2
1
```

    See also: [edit_history], page 28.

Octave also allows you customize the details of when, where, and how history is saved.

val = history_save ()   [Built-in Function]
old_val = history_save (*new_val*)   [Built-in Function]
history_save (*new_val*, "*local*")   [Built-in Function]
    Query or set the internal variable that controls whether commands entered on the command line are saved in the history file.

    When called from inside a function with the "local" option, the variable is changed locally for the function and any subroutines it calls. The original variable value is restored when exiting the function.

    See also: [history_control], page 29, [history_file], page 30, [history_size], page 30, [history_timestamp_format_string], page 30.

val = history_control ()   [Built-in Function]
old_val = history_control (*new_val*)   [Built-in Function]
    Query or set the internal variable that specifies how commands are saved to the history list. The default value is an empty character string, but may be overridden by the environment variable OCTAVE_HISTCONTROL.

    The value of history_control is a colon-separated list of values controlling how commands are saved on the history list. If the list of values includes ignorespace,

lines which begin with a space character are not saved in the history list. A value of
ignoredups causes lines matching the previous history entry to not be saved. A value
of **ignoreboth** is shorthand for **ignorespace** and **ignoredups**. A value of **erasedups**
causes all previous lines matching the current line to be removed from the history list
before that line is saved. Any value not in the above list is ignored. If **history_**
control is the empty string, all commands are saved on the history list, subject to
the value of **history_save**.

> **See also:** [history_file], page 30, [history_size], page 30, [history_timestamp_format_string],
> page 30, [history_save], page 29.

`val = history_file ()`   [Built-in Function]
`old_val = history_file (new_val)`   [Built-in Function]
> Query or set the internal variable that specifies the name of the file used to store
> command history. The default value is '`~/.octave_hist`', but may be overridden by
> the environment variable `OCTAVE_HISTFILE`.

> **See also:** [history_size], page 30, [history_save], page 29, [history_timestamp_format_string],
> page 30.

`val = history_size ()`   [Built-in Function]
`old_val = history_size (new_val)`   [Built-in Function]
> Query or set the internal variable that specifies how many entries to store in the
> history file. The default value is `1000`, but may be overridden by the environment
> variable `OCTAVE_HISTSIZE`.

> **See also:** [history_file], page 30, [history_timestamp_format_string], page 30,
> [history_save], page 29.

`val = history_timestamp_format_string ()`   [Built-in Function]
`old_val = history_timestamp_format_string (new_val)`   [Built-in Function]
`history_timestamp_format_string (new_val, "local")`   [Built-in Function]
> Query or set the internal variable that specifies the format string for the comment
> line that is written to the history file when Octave exits. The format string is passed
> to `strftime`. The default value is

> ```
>     "# Octave VERSION, %a %b %d %H:%M:%S %Y %Z <USER@HOST>"
> ```

> When called from inside a function with the `"local"` option, the variable is changed
> locally for the function and any subroutines it calls. The original variable value is
> restored when exiting the function.

> **See also:** [strftime], page 699, [history_file], page 30, [history_size], page 30,
> [history_save], page 29.

`val = EDITOR ()`   [Built-in Function]
`old_val = EDITOR (new_val)`   [Built-in Function]
`EDITOR (new_val, "local")`   [Built-in Function]
> Query or set the internal variable that specifies the default text editor.

> The default value is taken from the environment variable `EDITOR` when Octave starts.
> If the environment variable is not initialized, `EDITOR` will be set to `"emacs"`.

When called from inside a function with the `"local"` option, the variable is changed locally for the function and any subroutines it calls. The original variable value is restored when exiting the function.

See also: [edit], page 179, [edit_history], page 28.

2.4.6 Customizing `readline`

Octave uses the GNU Readline library for command-line editing and history features. Readline is very flexible and can be modified through a configuration file of commands (See the GNU Readline library for the exact command syntax). The default configuration file is normally '`~/.inputrc`'.

Octave provides two commands for initializing Readline and thereby changing the command line behavior.

`readline_read_init_file (file)`   [Built-in Function]
> Read the readline library initialization file *file*. If *file* is omitted, read the default initialization file (normally '`~/.inputrc`').
>
> See Section "Readline Init File" in *GNU Readline Library*, for details.
>
> **See also:** [readline_re_read_init_file], page 31.

`readline_re_read_init_file ()`   [Built-in Function]
> Re-read the last readline library initialization file that was read. See Section "Readline Init File" in *GNU Readline Library*, for details.
>
> **See also:** [readline_read_init_file], page 31.

2.4.7 Customizing the Prompt

The following variables are available for customizing the appearance of the command-line prompts. Octave allows the prompt to be customized by inserting a number of backslash-escaped special characters that are decoded as follows:

'`\t`'   The time.

'`\d`'   The date.

'`\n`'   Begins a new line by printing the equivalent of a carriage return followed by a line feed.

'`\s`'   The name of the program (usually just '`octave`').

'`\w`'   The current working directory.

'`\W`'   The basename of the current working directory.

'`\u`'   The username of the current user.

'`\h`'   The hostname, up to the first '.'.

'`\H`'   The hostname.

'`\#`'   The command number of this command, counting from when Octave starts.

'`\!`'   The history number of this command. This differs from '`\#`' by the number of commands in the history list when Octave starts.

'\$'   If the effective UID is 0, a '#', otherwise a '$'.

'\nnn'   The character whose character code in octal is *nnn*.

'\\'   A backslash.

val = PS1 ()   [Built-in Function]
old_val = PS1 (*new_val*)   [Built-in Function]
PS1 (*new_val*, "*local*")   [Built-in Function]
> Query or set the primary prompt string. When executing interactively, Octave dis-
> plays the primary prompt when it is ready to read a command.
>
> The default value of the primary prompt string is `"octave:\#> "`. To change it, use
> a command like
>
>   PS1 ("\\u@\\H> ")
>
> which will result in the prompt 'boris@kremvax> ' for the user 'boris' logged in
> on the host 'kremvax.kgb.su'. Note that two backslashes are required to enter a
> backslash into a double-quoted character string. See Chapter 5 [Strings], page 65.
>
> You can also use ANSI escape sequences if your terminal supports them. This can be
> useful for coloring the prompt. For example,
>
>   PS1 ("\\[\\033[01;31m\\]\\s:\\#> \\[\\033[0m\\]")
>
> will give the default Octave prompt a red coloring.
>
> When called from inside a function with the `"local"` option, the variable is changed
> locally for the function and any subroutines it calls. The original variable value is
> restored when exiting the function.
>
> **See also:** [PS2], page 32, [PS4], page 32.

val = PS2 ()   [Built-in Function]
old_val = PS2 (*new_val*)   [Built-in Function]
PS2 (*new_val*, "*local*")   [Built-in Function]
> Query or set the secondary prompt string. The secondary prompt is printed when
> Octave is expecting additional input to complete a command. For example, if you are
> typing a `for` loop that spans several lines, Octave will print the secondary prompt at
> the beginning of each line after the first. The default value of the secondary prompt
> string is `"> "`.
>
> When called from inside a function with the `"local"` option, the variable is changed
> locally for the function and any subroutines it calls. The original variable value is
> restored when exiting the function.
>
> **See also:** [PS1], page 32, [PS4], page 32.

val = PS4 ()   [Built-in Function]
old_val = PS4 (*new_val*)   [Built-in Function]
PS4 (*new_val*, "*local*")   [Built-in Function]
> Query or set the character string used to prefix output produced when echoing com-
> mands is enabled. The default value is `"+ "`. See Section 2.4.8 [Diary and Echo
> Commands], page 33, for a description of echoing commands.

When called from inside a function with the `"local"` option, the variable is changed locally for the function and any subroutines it calls. The original variable value is restored when exiting the function.

See also: [echo], page 33, [echo_executing_commands], page 33, [PS1], page 32, [PS2], page 32.

2.4.8 Diary and Echo Commands

Octave's diary feature allows you to keep a log of all or part of an interactive session by recording the input you type and the output that Octave produces in a separate file.

`diary`	[Command]
`diary` *on*	[Command]
`diary` *off*	[Command]
`diary` *filename*	[Command]

Record a list of all commands *and* the output they produce, mixed together just as they appear on the terminal.

Valid options are:

on   Start recording a session in a file called 'diary' in the current working directory.

off   Stop recording the session in the diary file.

filename   Record the session in the file named *filename*.

With no arguments, `diary` toggles the current diary state.

See also: [history], page 28.

Sometimes it is useful to see the commands in a function or script as they are being evaluated. This can be especially helpful for debugging some kinds of problems.

`echo` *options*   [Command]

Control whether commands are displayed as they are executed. Valid options are:

`on`   Enable echoing of commands as they are executed in script files.

`off`   Disable echoing of commands as they are executed in script files.

`on all`   Enable echoing of commands as they are executed in script files and functions.

`off all`   Disable echoing of commands as they are executed in script files and functions.

With no arguments, `echo` toggles the current echo state.

`val = echo_executing_commands ()`	[Built-in Function]
`old_val = echo_executing_commands (`*new_val*`)`	[Built-in Function]
`echo_executing_commands (`*new_val*`, "`*local*`")`	[Built-in Function]

Query or set the internal variable that controls the echo state. It may be the sum of the following values:

1   Echo commands read from script files.

2   Echo commands from functions.

4   Echo commands read from command line.

More than one state can be active at once. For example, a value of 3 is equivalent to the command *echo on all*.

The value of `echo_executing_commands` may be set by the *echo* command or the command line option '`--echo-commands`'.

When called from inside a function with the `"local"` option, the variable is changed locally for the function and any subroutines it calls. The original variable value is restored when exiting the function.

2.5 How Octave Reports Errors

Octave reports two kinds of errors for invalid programs.

A *parse error* occurs if Octave cannot understand something you have typed. For example, if you misspell a keyword,

```
octave:13> function y = f (x) y = x***2; endfunction
```

Octave will respond immediately with a message like this:

```
parse error:

  syntax error

>>> function y = f (x) y = x***2; endfunction
                            ^
```

For most parse errors, Octave uses a caret ('^') to mark the point on the line where it was unable to make sense of your input. In this case, Octave generated an error message because the keyword for exponentiation (**) was misspelled. It marked the error at the third '*' because the code leading up to this was correct but the final '*' was not understood.

Another class of error message occurs at evaluation time. These errors are called *run-time errors*, or sometimes *evaluation errors*, because they occur when your program is being *run*, or *evaluated*. For example, if after correcting the mistake in the previous function definition, you type

```
octave:13> f ()
```

Octave will respond with

```
error: 'x' undefined near line 1 column 24
error: called from:
error:   f at line 1, column 22
```

This error message has several parts, and gives quite a bit of information to help you locate the source of the error. The messages are generated from the point of the innermost error, and provide a traceback of enclosing expressions and function calls.

In the example above, the first line indicates that a variable named 'x' was found to be undefined near line 1 and column 24 of some function or expression. For errors occurring within functions, lines are counted from the beginning of the file containing the function definition. For errors occurring outside of an enclosing function, the line number indicates the input line number, which is usually displayed in the primary prompt string.

The second and third lines of the error message indicate that the error occurred within the function `f`. If the function `f` had been called from within another function, for example, `g`, the list of errors would have ended with one more line:

```
error:   g at line 1, column 17
```

These lists of function calls make it fairly easy to trace the path your program took before the error occurred, and to correct the error before trying again.

2.6 Executable Octave Programs

Once you have learned Octave, you may want to write self-contained Octave scripts, using the '#!' script mechanism. You can do this on GNU systems and on many Unix systems[1].

Self-contained Octave scripts are useful when you want to write a program which users can invoke without knowing that the program is written in the Octave language. Octave scripts are also used for batch processing of data files. Once an algorithm has been developed and tested in the interactive portion of Octave, it can be committed to an executable script and used again and again on new data files.

As a trivial example of an executable Octave script, you might create a text file named 'hello', containing the following lines:

```
#! octave-interpreter-name -qf
# a sample Octave program
printf ("Hello, world!\n");
```

(where *octave-interpreter-name* should be replaced with the full path and name of your Octave binary). Note that this will only work if '#!' appears at the very beginning of the file. After making the file executable (with the **chmod** command on Unix systems), you can simply type:

```
hello
```

at the shell, and the system will arrange to run Octave as if you had typed:

```
octave hello
```

The line beginning with '#!' lists the full path and filename of an interpreter to be run, and an optional initial command line argument to pass to that interpreter. The operating system then runs the interpreter with the given argument and the full argument list of the executed program. The first argument in the list is the full file name of the Octave executable. The rest of the argument list will either be options to Octave, or data files, or both. The '-qf' options are usually specified in stand-alone Octave programs to prevent them from printing the normal startup message, and to keep them from behaving differently depending on the contents of a particular user's '~/.octaverc' file. See Section 2.1 [Invoking Octave from the Command Line], page 15.

Note that some operating systems may place a limit on the number of characters that are recognized after '#!'. Also, the arguments appearing in a '#!' line are parsed differently by various shells/systems. The majority of them group all the arguments together in one string and pass it to the interpreter as a single argument. In this case, the following script:

---

[1] The '#!' mechanism works on Unix systems derived from Berkeley Unix, System V Release 4, and some System V Release 3 systems.

```
#! octave-interpreter-name -q -f # comment
```

is equivalent to typing at the command line:

```
octave "-q -f # comment"
```

which will produce an error message. Unfortunately, it is not possible for Octave to determine whether it has been called from the command line or from a '#!' script, so some care is needed when using the '#!' mechanism.

Note that when Octave is started from an executable script, the built-in function **argv** returns a cell array containing the command line arguments passed to the executable Octave script, not the arguments passed to the Octave interpreter on the '#!' line of the script. For example, the following program will reproduce the command line that was used to execute the script, not '-qf'.

```
#! /bin/octave -qf
printf ("%s", program_name ());
arg_list = argv ();
for i = 1:nargin
  printf (" %s", arg_list{i});
endfor
printf ("\n");
```

2.7 Comments in Octave Programs

A *comment* is some text that is included in a program for the sake of human readers, and which is NOT an executable part of the program. Comments can explain what the program does, and how it works. Nearly all programming languages have provisions for comments, because programs are typically hard to understand without them.

2.7.1 Single Line Comments

In the Octave language, a comment starts with either the sharp sign character, '#', or the percent symbol '%' and continues to the end of the line. Any text following the sharp sign or percent symbol is ignored by the Octave interpreter and not executed. The following example shows whole line and partial line comments.

```
function countdown
  # Count down for main rocket engines
  disp (3);
  disp (2);
  disp (1);
  disp ("Blast Off!");  # Rocket leaves pad
endfunction
```

2.7.2 Block Comments

Entire blocks of code can be commented by enclosing the code between matching '#{' and '#}' or '%{' and '%}' markers. For example,

```
function quick_countdown
  # Count down for main rocket engines
  disp (3);
 #{
  disp (2);
  disp (1);
 #}
  disp ("Blast Off!");  # Rocket leaves pad
endfunction
```

will produce a very quick countdown from '3' to "Blast Off" as the lines "disp (2);" and "disp (1);" won't be executed.

The block comment markers must appear alone as the only characters on a line (excepting whitespace) in order to be parsed correctly.

2.7.3 Comments and the Help System

The help command (see Section 2.3 [Getting Help], page 20) is able to find the first block of comments in a function and return those as a documentation string. This means that the same commands used to get help on built-in functions are available for properly formatted user-defined functions. For example, after defining the function f below,

```
function xdot = f (x, t)

# usage: f (x, t)
#
# This function defines the right-hand
# side functions for a set of nonlinear
# differential equations.

  r = 0.25;
  ...
endfunction
```

the command *help f* produces the output

```
usage: f (x, t)

This function defines the right-hand
side functions for a set of nonlinear
differential equations.
```

Although it is possible to put comment lines into keyboard-composed, throw-away Octave programs, it usually isn't very useful because the purpose of a comment is to help you or another person understand the program at a later time.

The help parser currently only recognizes single line comments (see Section 2.7.1 [Single Line Comments], page 36) and not block comments for the initial help text.

3 Data Types

All versions of Octave include a number of built-in data types, including real and complex scalars and matrices, character strings, a data structure type, and an array that can contain all data types.

It is also possible to define new specialized data types by writing a small amount of C++ code. On some systems, new data types can be loaded dynamically while Octave is running, so it is not necessary to recompile all of Octave just to add a new type. See Appendix A [External Code Interface], page 759, for more information about Octave's dynamic linking capabilities. Section 3.2 [User-defined Data Types], page 43 describes what you must do to define a new data type for Octave.

typeinfo ()   [Built-in Function]
typeinfo (*expr*)   [Built-in Function]
> Return the type of the expression *expr*, as a string. If *expr* is omitted, return a cell array of strings containing all the currently installed data types.

> **See also:** [class], page 39, [isa], page 39.

3.1 Built-in Data Types

The standard built-in data types are real and complex scalars and matrices, ranges, character strings, a data structure type, and cell arrays. Additional built-in data types may be added in future versions. If you need a specialized data type that is not currently provided as a built-in type, you are encouraged to write your own user-defined data type and contribute it for distribution in a future release of Octave.

The data type of a variable can be determined and changed through the use of the following functions.

classname = class (*obj*)   [Function File]
class (*s*, *id*)   [Function File]
class (*s*, *id*, *p*, . . .)   [Function File]
> Return the class of the object *obj* or create a class with fields from structure *s* and name (string) *id*. Additional arguments name a list of parent classes from which the new class is derived.

> **See also:** [typeinfo], page 39, [isa], page 39.

isa (*obj*, *classname*)   [Function File]
> Return true if *obj* is an object from the class *classname*.

> *classname* may also be one of the following class categories:

> "float"   Floating point value comprising classes "double" and "single".

> "integer"
>> Integer value comprising classes (u)int8, (u)int16, (u)int32, (u)int64.

> "numeric"
>> Numeric value comprising either a floating point or integer value.

> **See also:** [class], page 39, [typeinfo], page 39.

cast (*val*, *type*)   [Function File]
    Convert *val* to data type *type*.

    See also: [int8], page 54, [uint8], page 54, [int16], page 54, [uint16], page 54, [int32], page 54, [uint32], page 55, [int64], page 55, [uint64], page 55, [double], page 47.

typecast (*x*, *class*)   [Built-in Function]
    Return a new array *y* resulting from interpreting the data of *x* in memory as data of the numeric class *class*. Both the class of *x* and *class* must be one of the built-in numeric classes:

```
"logical"
"char"
"int8"
"int16"
"int32"
"int64"
"uint8"
"uint16"
"uint32"
"uint64"
"double"
"single"
"double complex"
"single complex"
```

    the last two are reserved for *class*; they indicate that a complex-valued result is requested. Complex arrays are stored in memory as consecutive pairs of real numbers. The sizes of integer types are given by their bit counts. Both logical and char are typically one byte wide; however, this is not guaranteed by C++. If your system is IEEE conformant, single and double should be 4 bytes and 8 bytes wide, respectively. "logical" is not allowed for *class*. If the input is a row vector, the return value is a row vector, otherwise it is a column vector. If the bit length of *x* is not divisible by that of *class*, an error occurs.

    An example of the use of typecast on a little-endian machine is

```
x = uint16 ([1, 65535]);
typecast (x, "uint8")
⇒ [   1,   0, 255, 255]
```

    See also: [cast], page 40, [bitunpack], page 41, [bitpack], page 41, [swapbytes], page 40.

swapbytes (*x*)   [Function File]
    Swap the byte order on values, converting from little endian to big endian and vice versa. For example:

```
swapbytes (uint16 (1:4))
⇒ [   256   512   768  1024]
```

    See also: [typecast], page 40, [cast], page 40.

y = bitpack (*x*, *class*)   [Built-in Function]

> Return a new array *y* resulting from interpreting an array *x* as raw bit patterns for data of the numeric class *class*. *class* must be one of the built-in numeric classes:
>
>   "char"
>   "int8"
>   "int16"
>   "int32"
>   "int64"
>   "uint8"
>   "uint16"
>   "uint32"
>   "uint64"
>   "double"
>   "single"
>
> The number of elements of *x* should be divisible by the bit length of *class*. If it is not, excess bits are discarded. Bits come in increasing order of significance, i.e., `x(1)` is bit 0, `x(2)` is bit 1, etc. The result is a row vector if *x* is a row vector, otherwise it is a column vector.
>
> **See also:** [bitunpack], page 41, [typecast], page 40.

y = bitunpack (*x*)   [Built-in Function]

> Return an array *y* corresponding to the raw bit patterns of *x*. *x* must belong to one of the built-in numeric classes:
>
>   "char"
>   "int8"
>   "int16"
>   "int32"
>   "int64"
>   "uint8"
>   "uint16"
>   "uint32"
>   "uint64"
>   "double"
>   "single"
>
> The result is a row vector if *x* is a row vector; otherwise, it is a column vector.
>
> **See also:** [bitpack], page 41, [typecast], page 40.

3.1.1 Numeric Objects

Octave's built-in numeric objects include real, complex, and integer scalars and matrices. All built-in floating point numeric data is currently stored as double precision numbers. On systems that use the IEEE floating point format, values in the range of approximately 2.2251×10^{-308} to 1.7977×10^{308} can be stored, and the relative precision is approximately 2.2204×10^{-16}. The exact values are given by the variables `realmin`, `realmax`, and `eps`, respectively.

Matrix objects can be of any size, and can be dynamically reshaped and resized. It is easy to extract individual rows, columns, or submatrices using a variety of powerful indexing features. See Section 8.1 [Index Expressions], page 131.

See Chapter 4 [Numeric Data Types], page 47, for more information.

3.1.2 Missing Data

It is possible to represent missing data explicitly in Octave using NA (short for "Not Available"). Missing data can only be represented when data is represented as floating point numbers. In this case missing data is represented as a special case of the representation of NaN.

NA	[Built-in Function]
NA (*n*)	[Built-in Function]
NA (*n*, *m*)	[Built-in Function]
NA (*n*, *m*, *k*, ...)	[Built-in Function]
NA (..., *class*)	[Built-in Function]

Return a scalar, matrix, or N-dimensional array whose elements are all equal to the special constant used to designate missing values.

Note that NA always compares not equal to NA (NA != NA). To find NA values, use the `isna` function.

When called with no arguments, return a scalar with the value 'NA'. When called with a single argument, return a square matrix with the dimension specified. When called with more than one scalar argument the first two arguments are taken as the number of rows and columns and any further arguments specify additional matrix dimensions. The optional argument *class* specifies the return type and may be either `"double"` or `"single"`.

See also: [isna], page 42.

isna (*x*)   [Mapping Function]

Return a logical array which is true where the elements of *x* are NA (missing) values and false where they are not. For example:

```
isna ([13, Inf, NA, NaN])
    ⇒ [ 0, 0, 1, 0 ]
```

See also: [isnan], page 380, [isinf], page 380, [isfinite], page 380.

3.1.3 String Objects

A character string in Octave consists of a sequence of characters enclosed in either double-quote or single-quote marks. Internally, Octave currently stores strings as matrices of characters. All the indexing operations that work for matrix objects also work for strings.

See Chapter 5 [Strings], page 65, for more information.

3.1.4 Data Structure Objects

Octave's data structure type can help you to organize related objects of different types. The current implementation uses an associative array with indices limited to strings, but the syntax is more like C-style structures.

See Section 6.1 [Structures], page 97, for more information.

3.1.5 Cell Array Objects

A Cell Array in Octave is general array that can hold any number of different data types.

See Section 6.2 [Cell Arrays], page 110, for more information.

3.2 User-defined Data Types

Someday I hope to expand this to include a complete description of Octave's mechanism for managing user-defined data types. Until this feature is documented here, you will have to make do by reading the code in the 'ov.h', 'ops.h', and related files from Octave's 'src' directory.

3.3 Object Sizes

The following functions allow you to determine the size of a variable or expression. These functions are defined for all objects. They return −1 when the operation doesn't make sense. For example, Octave's data structure type doesn't have rows or columns, so the rows and columns functions return −1 for structure arguments.

ndims (a)   [Built-in Function]
> Return the number of dimensions of a. For any array, the result will always be larger than or equal to 2. Trailing singleton dimensions are not counted.
>
>   ndims (ones (4, 1, 2, 1))
>   ⇒ 3

See also: [size], page 44.

columns (a)   [Built-in Function]
> Return the number of columns of a.
>
> **See also:** [rows], page 43, [size], page 44, [length], page 44, [numel], page 43, [isscalar], page 62, [isvector], page 62, [ismatrix], page 62.

rows (a)   [Built-in Function]
> Return the number of rows of a.
>
> **See also:** [columns], page 43, [size], page 44, [length], page 44, [numel], page 43, [isscalar], page 62, [isvector], page 62, [ismatrix], page 62.

numel (a)   [Built-in Function]
numel (a, *idx1*, *idx2*, ...)   [Built-in Function]
> Return the number of elements in the object a. Optionally, if indices *idx1*, *idx2*, ... are supplied, return the number of elements that would result from the indexing
>
>   a(*idx1*, *idx2*, ...)
>
> Note that the indices do not have to be numerical. For example,
>
>   a = 1;
>   b = ones (2, 3);
>   numel (a, b)
>
> will return 6, as this is the number of ways to index with *b*.

This method is also called when an object appears as lvalue with cs-list indexing, i.e., `object{...}` or `object(...).field`.

See also: [size], page 44.

length (`a`)   [Built-in Function]
    Return the length of the object *a*.

    The length is 0 for empty objects, 1 for scalars, and the number of elements for vectors. For matrix objects, the length is the number of rows or columns, whichever is greater (this odd definition is used for compatibility with MATLAB).

    See also: [numel], page 43, [size], page 44.

size (`a`)   [Built-in Function]
size (`a`, *dim*)   [Built-in Function]
    Return the number of rows and columns of *a*.

    With one input argument and one output argument, the result is returned in a row vector. If there are multiple output arguments, the number of rows is assigned to the first, and the number of columns to the second, etc. For example:

```
size ([1, 2; 3, 4; 5, 6])
    ⇒ [ 3, 2 ]

[nr, nc] = size ([1, 2; 3, 4; 5, 6])
    ⇒ nr = 3
    ⇒ nc = 2
```

    If given a second argument, `size` will return the size of the corresponding dimension. For example,

```
size ([1, 2; 3, 4; 5, 6], 2)
    ⇒ 2
```

    returns the number of columns in the given matrix.

    See also: [numel], page 43, [ndims], page 43, [length], page 44, [rows], page 43, [columns], page 43.

isempty (`a`)   [Built-in Function]
    Return true if *a* is an empty matrix (any one of its dimensions is zero). Otherwise, return false.

    See also: [isnull], page 44, [isa], page 39.

isnull (`x`)   [Built-in Function]
    Return true if *x* is a special null matrix, string, or single quoted string. Indexed assignment with such a value on the right-hand side should delete array elements. This function should be used when overloading indexed assignment for user-defined classes instead of `isempty`, to distinguish the cases:

`A(I) = []`   This should delete elements if `I` is nonempty.

`X = []; A(I) = X`
        This should give an error if `I` is nonempty.

    See also: [isempty], page 44, [isindex], page 134.

`sizeof (val)`   [Built-in Function]
> Return the size of *val* in bytes.
>
> **See also:** [whos], page 126.

`size_equal (a, b, ...)`   [Built-in Function]
> Return true if the dimensions of all arguments agree. Trailing singleton dimensions
> are ignored. Called with a single or no argument, size_equal returns true.
>
> **See also:** [size], page 44, [numel], page 43, [ndims], page 43.

`squeeze (x)`   [Built-in Function]
> Remove singleton dimensions from *x* and return the result. Note that for compatibility
> with MATLAB, all objects have a minimum of two dimensions and row vectors are left
> unchanged.
>
> **See also:** [reshape], page 385.

4 Numeric Data Types

A *numeric constant* may be a scalar, a vector, or a matrix, and it may contain complex values.

The simplest form of a numeric constant, a scalar, is a single number that can be an integer, a decimal fraction, a number in scientific (exponential) notation, or a complex number. Note that by default numeric constants are represented within Octave in double-precision floating point format (complex constants are stored as pairs of double-precision floating point values). It is, however, possible to represent real integers as described in Section 4.4 [Integer Data Types], page 54. Here are some examples of real-valued numeric constants, which all have the same value:

```
105
1.05e+2
1050e-1
```

To specify complex constants, you can write an expression of the form

```
3 + 4i
3.0 + 4.0i
0.3e1 + 40e-1i
```

all of which are equivalent. The letter 'i' in the previous example stands for the pure imaginary constant, defined as $\sqrt{-1}$.

For Octave to recognize a value as the imaginary part of a complex constant, a space must not appear between the number and the 'i'. If it does, Octave will print an error message, like this:

```
octave:13> 3 + 4 i

parse error:

  syntax error

>>> 3 + 4 i
          ^
```

You may also use 'j', 'I', or 'J' in place of the 'i' above. All four forms are equivalent.

double (*x*)   [Built-in Function]
    Convert *x* to double precision type.

    See also: [single], page 53.

complex (*x*)   [Built-in Function]
complex (*re*, *im*)   [Built-in Function]
    Return a complex result from real arguments. With 1 real argument *x*, return the complex result `x + 0i`. With 2 real arguments, return the complex result *re* + *im*. `complex` can often be more convenient than expressions such as `a + i*b`. For example:

```
complex ([1, 2], [3, 4])
   ⇒ [ 1 + 3i   2 + 4i ]
```

    See also: [real], page 409, [imag], page 409, [iscomplex], page 62, [abs], page 409, [arg], page 409.

4.1 Matrices

It is easy to define a matrix of values in Octave. The size of the matrix is determined automatically, so it is not necessary to explicitly state the dimensions. The expression

    a = [1, 2; 3, 4]

results in the matrix

$$a = \begin{bmatrix} 1 & 2 \\ 3 & 4 \end{bmatrix}$$

Elements of a matrix may be arbitrary expressions, provided that the dimensions all make sense when combining the various pieces. For example, given the above matrix, the expression

    [a, a]

produces the matrix

    ans =

        1   2   1   2
        3   4   3   4

but the expression

    [a, 1]

produces the error

    error: number of rows must match (1 != 2) near line 13, column 6

(assuming that this expression was entered as the first thing on line 13, of course).

Inside the square brackets that delimit a matrix expression, Octave looks at the surrounding context to determine whether spaces and newline characters should be converted into element and row separators, or simply ignored, so an expression like

    a = [1 2
          3 4]

will work. However, some possible sources of confusion remain. For example, in the expression

    [1 - 1]

the '-' is treated as a binary operator and the result is the scalar 0, but in the expression

    [1 -1]

the '-' is treated as a unary operator and the result is the vector [1, -1]. Similarly, the expression

    [sin (pi)]

will be parsed as

    [sin, (pi)]

and will result in an error since the **sin** function will be called with no arguments. To get around this, you must omit the space between **sin** and the opening parenthesis, or enclose the expression in a set of parentheses:

    [(sin (pi))]

Whitespace surrounding the single quote character ('`'`', used as a transpose operator and for delimiting character strings) can also cause confusion. Given **a = 1**, the expression

```
[ 1 a' ]
```

results in the single quote character being treated as a transpose operator and the result is the vector [1, 1], but the expression

```
[ 1 a ' ]
```

produces the error message

```
parse error:

  syntax error

>>> [ 1 a ' ]
                 ^
```

because not doing so would cause trouble when parsing the valid expression

```
[ a 'foo' ]
```

For clarity, it is probably best to always use commas and semicolons to separate matrix elements and rows.

The maximum number of elements in a matrix is fixed when Octave is compiled. The allowable number can be queried with the function `sizemax`. Note that other factors, such as the amount of memory available on your machine, may limit the maximum size of matrices to something smaller.

`sizemax ()`   [Built-in Function]
  Return the largest value allowed for the size of an array. If Octave is compiled with 64-bit indexing, the result is of class int64, otherwise it is of class int32. The maximum array size is slightly smaller than the maximum value allowable for the relevant class as reported by `intmax`.

  See also: [intmax], page 55.

When you type a matrix or the name of a variable whose value is a matrix, Octave responds by printing the matrix in with neatly aligned rows and columns. If the rows of the matrix are too large to fit on the screen, Octave splits the matrix and displays a header before each section to indicate which columns are being displayed. You can use the following variables to control the format of the output.

`val = output_max_field_width ()`   [Built-in Function]
`old_val = output_max_field_width (new_val)`   [Built-in Function]
`output_max_field_width (new_val, "local")`   [Built-in Function]
  Query or set the internal variable that specifies the maximum width of a numeric output field.

  When called from inside a function with the `"local"` option, the variable is changed locally for the function and any subroutines it calls. The original variable value is restored when exiting the function.

  See also: [format], page 224, [fixed_point_format], page 50, [output_precision], page 50.

`val = output_precision ()`   [Built-in Function]
`old_val = output_precision (new_val)`   [Built-in Function]
`output_precision (new_val, "local")`   [Built-in Function]
> Query or set the internal variable that specifies the minimum number of significant figures to display for numeric output.
>
> When called from inside a function with the `"local"` option, the variable is changed locally for the function and any subroutines it calls. The original variable value is restored when exiting the function.
>
> **See also:** [format], page 224, [fixed_point_format], page 50, [output_max_field_width], page 49.

It is possible to achieve a wide range of output styles by using different values of `output_precision` and `output_max_field_width`. Reasonable combinations can be set using the `format` function. See Section 14.1 [Basic Input and Output], page 223.

`val = split_long_rows ()`   [Built-in Function]
`old_val = split_long_rows (new_val)`   [Built-in Function]
`split_long_rows (new_val, "local")`   [Built-in Function]
> Query or set the internal variable that controls whether rows of a matrix may be split when displayed to a terminal window. If the rows are split, Octave will display the matrix in a series of smaller pieces, each of which can fit within the limits of your terminal width and each set of rows is labeled so that you can easily see which columns are currently being displayed. For example:

```
octave:13> rand (2,10)
ans =

Columns 1 through 6:

   0.75883   0.93290   0.40064   0.43818   0.94958   0.16467
   0.75697   0.51942   0.40031   0.61784   0.92309   0.40201

Columns 7 through 10:

   0.90174   0.11854   0.72313   0.73326
   0.44672   0.94303   0.56564   0.82150
```

> When called from inside a function with the `"local"` option, the variable is changed locally for the function and any subroutines it calls. The original variable value is restored when exiting the function.
>
> **See also:** [format], page 224.

Octave automatically switches to scientific notation when values become very large or very small. This guarantees that you will see several significant figures for every value in a matrix. If you would prefer to see all values in a matrix printed in a fixed point format, you can set the built-in variable `fixed_point_format` to a nonzero value. But doing so is not recommended, because it can produce output that can easily be misinterpreted.

val = fixed_point_format ()	[Built-in Function]
old_val = fixed_point_format (*new_val*)	[Built-in Function]
fixed_point_format (*new_val*, "*local*")	[Built-in Function]

Query or set the internal variable that controls whether Octave will use a scaled format to print matrix values.

The scaled format prints a scaling factor on the first line of output chosen such that the largest matrix element can be written with a single leading digit. For example:

```
logspace (1, 7, 5)'
ans =

   1.0e+07  *

   0.00000
   0.00003
   0.00100
   0.03162
   1.00000
```

Notice that the first value appears to be 0 when it is actually 1. Because of the possibilty for confusion you should be careful about enabling `fixed_point_format`.

When called from inside a function with the "`local`" option, the variable is changed locally for the function and any subroutines it calls. The original variable value is restored when exiting the function.

See also: [format], page 224, [output_max_field_width], page 49, [output_precision], page 50.

4.1.1 Empty Matrices

A matrix may have one or both dimensions zero, and operations on empty matrices are handled as described by Carl de Boor in *An Empty Exercise*, SIGNUM, Volume 25, pages 2-6, 1990 and C. N. Nett and W. M. Haddad, in *A System-Theoretic Appropriate Realization of the Empty Matrix Concept*, IEEE Transactions on Automatic Control, Volume 38, Number 5, May 1993. Briefly, given a scalar s, an $m \times n$ matrix $M_{m \times n}$, and an $m \times n$ empty matrix $[]_{m \times n}$ (with either one or both dimensions equal to zero), the following are true:

$$s \cdot []_{m \times n} = []_{m \times n} \cdot s = []_{m \times n}$$
$$[]_{m \times n} + []_{m \times n} = []_{m \times n}$$
$$[]_{0 \times m} \cdot M_{m \times n} = []_{0 \times n}$$
$$M_{m \times n} \cdot []_{n \times 0} = []_{m \times 0}$$
$$[]_{m \times 0} \cdot []_{0 \times n} = 0_{m \times n}$$

By default, dimensions of the empty matrix are printed along with the empty matrix symbol, '[]'. The built-in variable `print_empty_dimensions` controls this behavior.

val = print_empty_dimensions ()	[Built-in Function]
old_val = print_empty_dimensions (*new_val*)	[Built-in Function]

`print_empty_dimensions` (*new_val*, "*local*")   [Built-in Function]

> Query or set the internal variable that controls whether the dimensions of empty matrices are printed along with the empty matrix symbol, '[]'. For example, the expression
>
>   zeros (3, 0)
>
> will print
>
>   ans = [](3x0)
>
> When called from inside a function with the "`local`" option, the variable is changed locally for the function and any subroutines it calls. The original variable value is restored when exiting the function.
>
> **See also:** [format], page 224.

Empty matrices may also be used in assignment statements as a convenient way to delete rows or columns of matrices. See Section 8.6 [Assignment Expressions], page 145.

When Octave parses a matrix expression, it examines the elements of the list to determine whether they are all constants. If they are, it replaces the list with a single matrix constant.

4.2 Ranges

A *range* is a convenient way to write a row vector with evenly spaced elements. A range expression is defined by the value of the first element in the range, an optional value for the increment between elements, and a maximum value which the elements of the range will not exceed. The base, increment, and limit are separated by colons (the ':' character) and may contain any arithmetic expressions and function calls. If the increment is omitted, it is assumed to be 1. For example, the range

    1 : 5

defines the set of values '[1, 2, 3, 4, 5]', and the range

    1 : 3 : 5

defines the set of values '[1, 4]'.

Although a range constant specifies a row vector, Octave does *not* convert range constants to vectors unless it is necessary to do so. This allows you to write a constant like '1 : 10000' without using 80,000 bytes of storage on a typical 32-bit workstation.

A common example of when it does become necessary to convert ranges into vectors occurs when they appear within a vector (i.e., inside square brackets). For instance, whereas

    x = 0 : 0.1 : 1;

defines x to be a variable of type **range** and occupies 24 bytes of memory, the expression

    y = [0 : 0.1 : 1];

defines y to be of type **matrix** and occupies 88 bytes of memory.

Note that the upper (or lower, if the increment is negative) bound on the range is not always included in the set of values, and that ranges defined by floating point values can produce surprising results because Octave uses floating point arithmetic to compute the values in the range. If it is important to include the endpoints of a range and the number of elements is known, you should use the **linspace** function instead (see Section 16.3 [Special Utility Matrices], page 391).

When adding a scalar to a range, subtracting a scalar from it (or subtracting a range from a scalar) and multiplying by scalar, Octave will attempt to avoid unpacking the range and keep the result as a range, too, if it can determine that it is safe to do so. For instance, doing

```
a = 2*(1:1e7) - 1;
```

will produce the same result as '`1:2:2e7-1`', but without ever forming a vector with ten million elements.

Using zero as an increment in the colon notation, as '`1:0:1`' is not allowed, because a division by zero would occur in determining the number of range elements. However, ranges with zero increment (i.e., all elements equal) are useful, especially in indexing, and Octave allows them to be constructed using the built-in function *ones*. Note that because a range must be a row vector, '`ones (1, 10)`' produces a range, while '`ones (10, 1)`' does not.

When Octave parses a range expression, it examines the elements of the expression to determine whether they are all constants. If they are, it replaces the range expression with a single range constant.

4.3 Single Precision Data Types

Octave includes support for single precision data types, and most of the functions in Octave accept single precision values and return single precision answers. A single precision variable is created with the **single** function.

single (*x*)   [Built-in Function]
        Convert *x* to single precision type.

        See also: [double], page 47.

for example:

```
sngl = single (rand (2, 2))
     ⇒ sngl =
        0.37569    0.92982
        0.11962    0.50876
class (sngl)
     ⇒ single
```

Many functions can also return single precision values directly. For example

```
ones (2, 2, "single")
zeros (2, 2, "single")
eye (2, 2, "single")
rand (2, 2, "single")
NaN (2, 2, "single")
NA (2, 2, "single")
Inf (2, 2, "single")
```

will all return single precision matrices.

4.4 Integer Data Types

Octave supports integer matrices as an alternative to using double precision. It is possible to use both signed and unsigned integers represented by 8, 16, 32, or 64 bits. It should be noted that most computations require floating point data, meaning that integers will often change type when involved in numeric computations. For this reason integers are most often used to store data, and not for calculations.

In general most integer matrices are created by casting existing matrices to integers. The following example shows how to cast a matrix into 32 bit integers.

```
float = rand (2, 2)
    ⇒ float = 0.37569   0.92982
              0.11962   0.50876
integer = int32 (float)
    ⇒ integer = 0   1
                0   1
```

As can be seen, floating point values are rounded to the nearest integer when converted.

isinteger (*x*)   [Built-in Function]
Return true if *x* is an integer object (int8, uint8, int16, etc.). Note that isinteger (14) is false because numeric constants in Octave are double precision floating point values.

See also: [isfloat], page 61, [ischar], page 66, [islogical], page 63, [isnumeric], page 61, [isa], page 39.

int8 (*x*)   [Built-in Function]
Convert *x* to 8-bit integer type.

See also: [uint8], page 54, [int16], page 54, [uint16], page 54, [int32], page 54, [uint32], page 55, [int64], page 55, [uint64], page 55.

uint8 (*x*)   [Built-in Function]
Convert *x* to unsigned 8-bit integer type.

See also: [int8], page 54, [int16], page 54, [uint16], page 54, [int32], page 54, [uint32], page 55, [int64], page 55, [uint64], page 55.

int16 (*x*)   [Built-in Function]
Convert *x* to 16-bit integer type.

See also: [int8], page 54, [uint8], page 54, [uint16], page 54, [int32], page 54, [uint32], page 55, [int64], page 55, [uint64], page 55.

uint16 (*x*)   [Built-in Function]
Convert *x* to unsigned 16-bit integer type.

See also: [int8], page 54, [uint8], page 54, [int16], page 54, [int32], page 54, [uint32], page 55, [int64], page 55, [uint64], page 55.

int32 (*x*)   [Built-in Function]
Convert *x* to 32-bit integer type.

See also: [int8], page 54, [uint8], page 54, [int16], page 54, [uint16], page 54, [uint32], page 55, [int64], page 55, [uint64], page 55.

uint32 (*x*)   [Built-in Function]

> Convert *x* to unsigned 32-bit integer type.
>
> **See also:** [int8], page 54, [uint8], page 54, [int16], page 54, [uint16], page 54, [int32], page 54, [int64], page 55, [uint64], page 55.

int64 (*x*)   [Built-in Function]

> Convert *x* to 64-bit integer type.
>
> **See also:** [int8], page 54, [uint8], page 54, [int16], page 54, [uint16], page 54, [int32], page 54, [uint32], page 55, [uint64], page 55.

uint64 (*x*)   [Built-in Function]

> Convert *x* to unsigned 64-bit integer type.
>
> **See also:** [int8], page 54, [uint8], page 54, [int16], page 54, [uint16], page 54, [int32], page 54, [uint32], page 55, [int64], page 55.

intmax (*type*)   [Built-in Function]

> Return the largest integer that can be represented in an integer type. The variable *type* can be
>
> | int8 | signed 8-bit integer. |
> |--------|--------------------------|
> | int16 | signed 16-bit integer. |
> | int32 | signed 32-bit integer. |
> | int64 | signed 64-bit integer. |
> | uint8 | unsigned 8-bit integer. |
> | uint16 | unsigned 16-bit integer. |
> | uint32 | unsigned 32-bit integer. |
> | uint64 | unsigned 64-bit integer. |
>
> The default for *type* is int32.
>
> **See also:** [intmin], page 55, [flintmax], page 56, [bitmax], page 57.

intmin (*type*)   [Built-in Function]

> Return the smallest integer that can be represented in an integer type. The variable *type* can be
>
> | int8 | signed 8-bit integer. |
> |--------|--------------------------|
> | int16 | signed 16-bit integer. |
> | int32 | signed 32-bit integer. |
> | int64 | signed 64-bit integer. |
> | uint8 | unsigned 8-bit integer. |
> | uint16 | unsigned 16-bit integer. |
> | uint32 | unsigned 32-bit integer. |

`uint64`   unsigned 64-bit integer.

The default for *type* is `int32`.

See also: [intmax], page 55, [flintmax], page 56, [bitmax], page 57.

`flintmax ()`   [Built-in Function]
`flintmax ("`*double*`")`   [Built-in Function]
`flintmax ("`*single*`")`   [Built-in Function]

Return the largest integer that can be represented consecutively in a floating point value. The default class is `"double"`, but `"single"` is a valid option. On IEEE-754 compatible systems, `flintmax` is 2^53 for `"double"` and 2^24 for `"single"`.

See also: [bitmax], page 57, [intmax], page 55, [realmax], page 432, [realmin], page 432.

4.4.1 Integer Arithmetic

While many numerical computations can't be carried out in integers, Octave does support basic operations like addition and multiplication on integers. The operators `+`, `-`, `.*`, and `./` work on integers of the same type. So, it is possible to add two 32 bit integers, but not to add a 32 bit integer and a 16 bit integer.

When doing integer arithmetic one should consider the possibility of underflow and overflow. This happens when the result of the computation can't be represented using the chosen integer type. As an example it is not possible to represent the result of $10 - 20$ when using unsigned integers. Octave makes sure that the result of integer computations is the integer that is closest to the true result. So, the result of $10 - 20$ when using unsigned integers is zero.

When doing integer division Octave will round the result to the nearest integer. This is different from most programming languages, where the result is often floored to the nearest integer. So, the result of `int32 (5) ./ int32 (8)` is 1.

`idivide (`*x*`, `*y*`, `*op*`)`   [Function File]

Integer division with different rounding rules.

The standard behavior of integer division such as `a ./ b` is to round the result to the nearest integer. This is not always the desired behavior and `idivide` permits integer element-by-element division to be performed with different treatment for the fractional part of the division as determined by the *op* flag. *op* is a string with one of the values:

`"fix"`   Calculate `a ./ b` with the fractional part rounded towards zero.

`"round"`   Calculate `a ./ b` with the fractional part rounded towards the nearest integer.

`"floor"`   Calculate `a ./ b` with the fractional part rounded towards negative infinity.

`"ceil"`   Calculate `a ./ b` with the fractional part rounded towards positive infinity.

If *op* is not given it defaults to `"fix"`. An example demonstrating these rounding rules is

```
idivide (int8 ([-3, 3]), int8 (4), "fix")
    ⇒ int8 ([0, 0])
idivide (int8 ([-3, 3]), int8 (4), "round")
    ⇒ int8 ([-1, 1])
idivide (int8 ([-3, 3]), int8 (4), "floor")
    ⇒ int8 ([-1, 0])
idivide (int8 ([-3, 3]), int8 (4), "ceil")
    ⇒ int8 ([0, 1])
```

See also: [ldivide], page 139, [rdivide], page 140.

4.5 Bit Manipulations

Octave provides a number of functions for the manipulation of numeric values on a bit by bit basis. The basic functions to set and obtain the values of individual bits are `bitset` and `bitget`.

C = bitset (A, n)   [Function File]
C = bitset (A, n, val)   [Function File]
    Set or reset bit(s) n of unsigned integers in A. $val = 0$ resets and $val = 1$ sets the bits. The lowest significant bit is: $n = 1$. All variables must be the same size or scalars.

```
dec2bin (bitset (10, 1))
    ⇒ 1011
```

    See also: [bitand], page 58, [bitor], page 58, [bitxor], page 58, [bitget], page 57, [bitcmp], page 58, [bitshift], page 58, [bitmax], page 57.

c = bitget (A, n)   [Function File]
    Return the status of bit(s) n of unsigned integers in A the lowest significant bit is $n = 1$.

```
bitget (100, 8:-1:1)
    ⇒ 0  1  1  0  0  1  0  0
```

    See also: [bitand], page 58, [bitor], page 58, [bitxor], page 58, [bitset], page 57, [bitcmp], page 58, [bitshift], page 58, [bitmax], page 57.

The arguments to all of Octave's bitwise operations can be scalar or arrays, except for `bitcmp`, whose k argument must a scalar. In the case where more than one argument is an array, then all arguments must have the same shape, and the bitwise operator is applied to each of the elements of the argument individually. If at least one argument is a scalar and one an array, then the scalar argument is duplicated. Therefore

```
bitget (100, 8:-1:1)
```

is the same as

```
bitget (100 * ones (1, 8), 8:-1:1)
```

It should be noted that all values passed to the bit manipulation functions of Octave are treated as integers. Therefore, even though the example for `bitset` above passes the floating point value 10, it is treated as the bits [1, 0, 1, 0] rather than the bits of the native floating point format representation of 10.

As the maximum value that can be represented by a number is important for bit manipulation, particularly when forming masks, Octave supplies the function `bitmax`.

`bitmax ()`   [Built-in Function]
`bitmax ("double")`   [Built-in Function]
`bitmax ("single")`   [Built-in Function]
> Return the largest integer that can be represented within a floating point value. The default class is `"double"`, but `"single"` is a valid option. On IEEE-754 compatible systems, `bitmax` is $2^{53} - 1$ for `"double"` and $2^{24} - 1$ for `"single"`.
>
> **See also:** [flintmax], page 56, [intmax], page 55, [realmax], page 432, [realmin], page 432.

This is the double precision version of the function `intmax`, previously discussed.

Octave also includes the basic bitwise 'and', 'or', and 'exclusive or' operators.

`bitand (x, y)`   [Built-in Function]
> Return the bitwise AND of non-negative integers. x, y must be in the range [0,bitmax]
>
> **See also:** [bitor], page 58, [bitxor], page 58, [bitset], page 57, [bitget], page 57, [bitcmp], page 58, [bitshift], page 58, [bitmax], page 57.

`bitor (x, y)`   [Built-in Function]
> Return the bitwise OR of non-negative integers. x, y must be in the range [0,bitmax]
>
> **See also:** [bitor], page 58, [bitxor], page 58, [bitset], page 57, [bitget], page 57, [bitcmp], page 58, [bitshift], page 58, [bitmax], page 57.

`bitxor (x, y)`   [Built-in Function]
> Return the bitwise XOR of non-negative integers. x, y must be in the range [0,bitmax]
>
> **See also:** [bitand], page 58, [bitor], page 58, [bitset], page 57, [bitget], page 57, [bitcmp], page 58, [bitshift], page 58, [bitmax], page 57.

The bitwise 'not' operator is a unary operator that performs a logical negation of each of the bits of the value. For this to make sense, the mask against which the value is negated must be defined. Octave's bitwise 'not' operator is `bitcmp`.

`bitcmp (A, k)`   [Function File]
> Return the k-bit complement of integers in A. If k is omitted k = log2 (bitmax) + 1 is assumed.
>
> ```
>       bitcmp (7,4)
>         ⇒ 8
>       dec2bin (11)
>         ⇒ 1011
>       dec2bin (bitcmp (11, 6))
>         ⇒ 110100
> ```
>
> **See also:** [bitand], page 58, [bitor], page 58, [bitxor], page 58, [bitset], page 57, [bitget], page 57, [bitcmp], page 58, [bitshift], page 58, [bitmax], page 57.

Octave also includes the ability to left-shift and right-shift values bitwise.

bitshift (*a*, *k*)   [Built-in Function]
bitshift (*a*, *k*, *n*)   [Built-in Function]

> Return a *k* bit shift of *n*-digit unsigned integers in *a*. A positive *k* leads to a left shift; A negative value to a right shift. If *n* is omitted it defaults to log2(bitmax)+1. *n* must be in the range [1,log2(bitmax)+1] usually [1,33].
>
> ```
> bitshift (eye (3), 1)
> ⇒
> 2 0 0
> 0 2 0
> 0 0 2
>
> bitshift (10, [-2, -1, 0, 1, 2])
>    ⇒ 2   5   10   20   40
> ```
>
> **See also:** [bitand], page 58, [bitor], page 58, [bitxor], page 58, [bitset], page 57, [bitget], page 57, [bitcmp], page 58, [bitmax], page 57.

Bits that are shifted out of either end of the value are lost. Octave also uses arithmetic shifts, where the sign bit of the value is kept during a right shift. For example:

```
bitshift (-10, -1)
⇒ -5
bitshift (int8 (-1), -1)
⇒ -1
```

Note that `bitshift (int8 (-1), -1)` is -1 since the bit representation of -1 in the int8 data type is [1, 1, 1, 1, 1, 1, 1, 1].

4.6 Logical Values

Octave has built-in support for logical values, i.e., variables that are either `true` or `false`. When comparing two variables, the result will be a logical value whose value depends on whether or not the comparison is true.

The basic logical operations are &, |, and !, which correspond to "Logical And", "Logical Or", and "Logical Negation". These operations all follow the usual rules of logic.

It is also possible to use logical values as part of standard numerical calculations. In this case `true` is converted to 1, and `false` to 0, both represented using double precision floating point numbers. So, the result of `true*22 - false/6` is 22.

Logical values can also be used to index matrices and cell arrays. When indexing with a logical array the result will be a vector containing the values corresponding to `true` parts of the logical array. The following example illustrates this.

```
data = [ 1, 2; 3, 4 ];
idx = (data <= 2);
data(idx)
      ⇒ ans = [ 1; 2 ]
```

Instead of creating the `idx` array it is possible to replace `data(idx)` with `data( data <= 2 )` in the above code.

Logical values can also be constructed by casting numeric objects to logical values, or by using the `true` or `false` functions.

`logical` (*x*)   [Built-in Function]
> Convert the numeric object *x* to logical type.
>
> Any non-zero values will be converted to true (1) while zero values will be converted to false (0). The non-numeric value NaN cannot be converted and will produce an error.
>
> Compatibility Note: Octave accepts complex values as input, whereas MATLAB issues an error.
>
> **See also:** [double], page 47, [single], page 53, [char], page 69.

`true` (*x*)   [Built-in Function]
`true` (*n*, *m*)   [Built-in Function]
`true` (*n*, *m*, *k*, ...)   [Built-in Function]
> Return a matrix or N-dimensional array whose elements are all logical 1. If invoked with a single scalar integer argument, return a square matrix of the specified size. If invoked with two or more scalar integer arguments, or a vector of integer values, return an array with given dimensions.
>
> **See also:** [false], page 60.

`false` (*x*)   [Built-in Function]
`false` (*n*, *m*)   [Built-in Function]
`false` (*n*, *m*, *k*, ...)   [Built-in Function]
> Return a matrix or N-dimensional array whose elements are all logical 0. If invoked with a single scalar integer argument, return a square matrix of the specified size. If invoked with two or more scalar integer arguments, or a vector of integer values, return an array with given dimensions.
>
> **See also:** [true], page 60.

4.7 Promotion and Demotion of Data Types

Many operators and functions can work with mixed data types. For example,

```
uint8 (1) + 1
    ⇒ 2
```

where the above operator works with an 8-bit integer and a double precision value and returns an 8-bit integer value. Note that the type is demoted to an 8-bit integer, rather than promoted to a double precision value as might be expected. The reason is that if Octave promoted values in expressions like the above with all numerical constants would need to be explicitly cast to the appropriate data type like

```
uint8 (1) + uint8 (1)
    ⇒ 2
```

which becomes difficult for the user to apply uniformly and might allow hard to find bugs to be introduced. The same applies to single precision values where a mixed operation such as

```
single (1) + 1
    ⇒ 2
```

returns a single precision value. The mixed operations that are valid and their returned data types are

Mixed Operation	Result
double OP single	single
double OP integer	integer
double OP char	double
double OP logical	double
single OP integer	integer
single OP char	single
single OP logical	single

The same logic applies to functions with mixed arguments such as

```
min (single (1), 0)
  ⇒ 0
```

where the returned value is single precision.

In the case of mixed type indexed assignments, the type is not changed. For example,

```
x = ones (2, 2);
x(1, 1) = single (2)
  ⇒ x = 2   1
        1   1
```

where x remains of the double precision type.

4.8 Predicates for Numeric Objects

Since the type of a variable may change during the execution of a program, it can be necessary to do type checking at run-time. Doing this also allows you to change the behavior of a function depending on the type of the input. As an example, this naive implementation of abs returns the absolute value of the input if it is a real number, and the length of the input if it is a complex number.

```
function a = abs (x)
  if (isreal (x))
    a = sign (x) .* x;
  elseif (iscomplex (x))
    a = sqrt (real(x).^2 + imag(x).^2);
  endif
endfunction
```

The following functions are available for determining the type of a variable.

isnumeric (x)   [Built-in Function]
    Return true if x is a numeric object, i.e., an integer, real, or complex array. Logical and character arrays are not considered to be numeric.

    See also: [isinteger], page 54, [isfloat], page 61, [isreal], page 62, [iscomplex], page 62, [islogical], page 63, [ischar], page 66, [iscell], page 111, [isstruct], page 105, [isa], page 39.

isfloat (x)   [Built-in Function]
    Return true if x is a floating-point numeric object. Objects of class double or single are floating-point objects.

See also: [isinteger], page 54, [ischar], page 66, [islogical], page 63, [isnumeric], page 61, [isa], page 39.

isreal (*x*)   [Built-in Function]
> Return true if *x* is a non-complex matrix or scalar. For compatibility with MATLAB, this includes logical and character matrices.
>
> See also: [iscomplex], page 62, [isnumeric], page 61, [isa], page 39.

iscomplex (*x*)   [Built-in Function]
> Return true if *x* is a complex-valued numeric object.
>
> See also: [isreal], page 62, [isnumeric], page 61, [islogical], page 63, [ischar], page 66, [isfloat], page 61, [isa], page 39.

ismatrix (*a*)   [Built-in Function]
> Return true if *a* is a numeric, logical, or character matrix. Scalars (1x1 matrices) and vectors (1xN or Nx1 matrices) are subsets of the more general N-dimensional matrix and `ismatrix` will return true for these objects as well.
>
> See also: [isscalar], page 62, [isvector], page 62, [iscell], page 111, [isstruct], page 105, [issparse], page 493, [isa], page 39.

isvector (*x*)   [Function File]
> Return true if *x* is a vector. A vector is a 2-D array where one of the dimensions is equal to 1. As a consequence a 1x1 array, or scalar, is also a vector.
>
> See also: [isscalar], page 62, [ismatrix], page 62, [size], page 44, [rows], page 43, [columns], page 43, [length], page 44.

isrow (*x*)   [Function File]
> Return true if *x* is a row vector.
>
> See also: [iscolumn], page 62, [isscalar], page 62, [isvector], page 62, [ismatrix], page 62.

iscolumn (*x*)   [Function File]
> Return true if *x* is a column vector.
>
> See also: [isrow], page 62, [isscalar], page 62, [isvector], page 62, [ismatrix], page 62.

isscalar (*x*)   [Function File]
> Return true if *x* is a scalar.
>
> See also: [isvector], page 62, [ismatrix], page 62.

issquare (*x*)   [Function File]
> Return true if *x* is a square matrix.
>
> See also: [isscalar], page 62, [isvector], page 62, [ismatrix], page 62, [size], page 44.

issymmetric (*x*)   [Function File]
issymmetric (*x, tol*)   [Function File]
> Return true if *x* is a symmetric matrix within the tolerance specified by *tol*. The default tolerance is zero (uses faster code). Matrix *x* is considered symmetric if `norm (x - x.', Inf) / norm (x, Inf) < tol`.
>
> See also: [ishermitian], page 63, [isdefinite], page 63.

`ishermitian (x)` [Function File]
`ishermitian (x, tol)` [Function File]

> Return true if x is Hermitian within the tolerance specified by *tol*. The default tolerance is zero (uses faster code). Matrix x is considered symmetric if `norm (x - x', Inf) / norm (x, Inf) < tol`.
>
> **See also:** [issymmetric], page 62, [isdefinite], page 63.

`isdefinite (x)` [Function File]
`isdefinite (x, tol)` [Function File]

> Return 1 if x is symmetric positive definite within the tolerance specified by *tol* or 0 if x is symmetric positive semidefinite. Otherwise, return -1. If *tol* is omitted, use a tolerance of `100 * eps * norm (x, "fro")`
>
> **See also:** [issymmetric], page 62, [ishermitian], page 63.

`islogical (x)` [Built-in Function]
`isbool (x)` [Built-in Function]

> Return true if x is a logical object.
>
> **See also:** [isfloat], page 61, [isinteger], page 54, [ischar], page 66, [isnumeric], page 61, [isa], page 39.

`isprime (x)` [Function File]

> Return a logical array which is true where the elements of x are prime numbers and false where they are not.
>
> If the maximum value in x is very large, then you should be using special purpose factorization code.
>
> ```
> isprime (1:6)
>    ⇒ [0, 1, 1, 0, 1, 0]
> ```
>
> **See also:** [primes], page 421, [factor], page 420, [gcd], page 420, [lcm], page 420.

If instead of knowing properties of variables, you wish to know which variables are defined and to gather other information about the workspace itself, see Section 7.3 [Status of Variables], page 126.

5 Strings

A *string constant* consists of a sequence of characters enclosed in either double-quote or single-quote marks. For example, both of the following expressions

```
"parrot"
'parrot'
```

represent the string whose contents are 'parrot'. Strings in Octave can be of any length.

Since the single-quote mark is also used for the transpose operator (see Section 8.3 [Arithmetic Ops], page 137) but double-quote marks have no other purpose in Octave, it is best to use double-quote marks to denote strings.

Strings can be concatenated using the notation for defining matrices. For example, the expression

```
[ "foo" , "bar" , "baz" ]
```

produces the string whose contents are 'foobarbaz'. See Chapter 4 [Numeric Data Types], page 47, for more information about creating matrices.

5.1 Escape Sequences in String Constants

In double-quoted strings, the backslash character is used to introduce *escape sequences* that represent other characters. For example, '\n' embeds a newline character in a double-quoted string and '\"' embeds a double quote character. In single-quoted strings, backslash is not a special character. Here is an example showing the difference:

```
toascii ("\n")
    ⇒ 10
toascii ('\n')
    ⇒ [ 92 110 ]
```

Here is a table of all the escape sequences used in Octave (within double quoted strings). They are the same as those used in the C programming language.

\\   Represents a literal backslash, '\'.

\"   Represents a literal double-quote character, '"'.

\'   Represents a literal single-quote character, '''.

\0   Represents the null character, control-@, ASCII code 0.

\a   Represents the "alert" character, control-g, ASCII code 7.

\b   Represents a backspace, control-h, ASCII code 8.

\f   Represents a formfeed, control-l, ASCII code 12.

\n   Represents a newline, control-j, ASCII code 10.

\r   Represents a carriage return, control-m, ASCII code 13.

\t   Represents a horizontal tab, control-i, ASCII code 9.

\v   Represents a vertical tab, control-k, ASCII code 11.

nnn   Represents the octal value *nnn*, where *nnn* are one to three digits between 0 and 7. For example, the code for the ASCII ESC (escape) character is '\033'.

`\xhh...`   Represents the hexadecimal value *hh*, where *hh* are hexadecimal digits ('0' through '9' and either 'A' through 'F' or 'a' through 'f'). Like the same construct in ANSI C, the escape sequence continues until the first non-hexadecimal digit is seen. However, using more than two hexadecimal digits produces undefined results.

In a single-quoted string there is only one escape sequence: you may insert a single quote character using two single quote characters in succession. For example,

```
'I can''t escape'
    ⇒ I can't escape
```

In scripts the two different string types can be distinguished if necessary by using `is_dq_string` and `is_sq_string`.

`is_dq_string (x)`   [Built-in Function]
     Return true if *x* is a double-quoted character string.

     See also: [is_sq_string], page 66, [ischar], page 66.

`is_sq_string (x)`   [Built-in Function]
     Return true if *x* is a single-quoted character string.

     See also: [is_dq_string], page 66, [ischar], page 66.

5.2 Character Arrays

The string representation used by Octave is an array of characters, so internally the string `"dddddddddd"` is actually a row vector of length 10 containing the value 100 in all places (100 is the ASCII code of `"d"`). This lends itself to the obvious generalization to character matrices. Using a matrix of characters, it is possible to represent a collection of same-length strings in one variable. The convention used in Octave is that each row in a character matrix is a separate string, but letting each column represent a string is equally possible.

The easiest way to create a character matrix is to put several strings together into a matrix.

```
collection = [ "String #1"; "String #2" ];
```

This creates a 2-by-9 character matrix.

The function `ischar` can be used to test if an object is a character matrix.

`ischar (x)`   [Built-in Function]
     Return true if *x* is a character array.

     See also: [isfloat], page 61, [isinteger], page 54, [islogical], page 63, [isnumeric], page 61, [iscellstr], page 117, [isa], page 39.

To test if an object is a string (i.e., a character vector and not a character matrix) you can use the `ischar` function in combination with the `isvector` function as in the following example:

```
ischar (collection)
   ⇒ 1

ischar (collection) && isvector (collection)
   ⇒ 0

ischar ("my string") && isvector ("my string")
   ⇒ 1
```

One relevant question is, what happens when a character matrix is created from strings of different length. The answer is that Octave puts blank characters at the end of strings shorter than the longest string. It is possible to use a different character than the blank character using the `string_fill_char` function.

val = string_fill_char ()	[Built-in Function]
old_val = string_fill_char (*new_val*)	[Built-in Function]
string_fill_char (*new_val*, "*local*")	[Built-in Function]

Query or set the internal variable used to pad all rows of a character matrix to the same length; It must be a single character. The default value is " " (a single space). For example:

```
string_fill_char ("X");
[ "these"; "are"; "strings" ]
    ⇒   "theseXX"
        "areXXXX"
        "strings"
```

When called from inside a function with the `"local"` option, the variable is changed locally for the function and any subroutines it calls. The original variable value is restored when exiting the function.

This shows a problem with character matrices. It simply isn't possible to represent strings of different lengths. The solution is to use a cell array of strings, which is described in Section 6.2.4 [Cell Arrays of Strings], page 116.

5.3 Creating Strings

The easiest way to create a string is, as illustrated in the introduction, to enclose a text in double-quotes or single-quotes. It is however possible to create a string without actually writing a text. The function `blanks` creates a string of a given length consisting only of blank characters (ASCII code 32).

blanks (*n*)	[Function File]

Return a string of *n* blanks, for example:

```
blanks (10);
whos ans
    ⇒
        Attr Name        Size                    Bytes  Class
        ==== ====        ====                    =====  =====
             ans         1x10                       10  char
```

See also: [repmat], page 392.

5.3.1 Concatenating Strings

Strings can be concatenated using matrix notation (see Chapter 5 [Strings], page 65, Section 5.2 [Character Arrays], page 66) which is often the most natural method. For example:

```
fullname = [fname ".txt"];
email = ["<" user "@" domain ">"];
```

In each case it is easy to see what the final string will look like. This method is also the most efficient. When using matrix concatenation the parser immediately begins joining the strings without having to process the overhead of a function call and the input validation of the associated function.

Nevertheless, there are several other functions for concatenating string objects which can be useful in specific circumstances: `char`, `strvcat`, `strcat`, and `cstrcat`. Finally, the general purpose concatenation functions can be used: see [cat], page 384, [horzcat], page 384, and [vertcat], page 385.

- All string concatenation functions except `cstrcat` convert numerical input into character data by taking the corresponding ASCII character for each element, as in the following example:

```
char ([98, 97, 110, 97, 110, 97])
    ⇒ banana
```

- `char` and `strvcat` concatenate vertically, while `strcat` and `cstrcat` concatenate horizontally. For example:

```
char ("an apple", "two pears")
    ⇒ an apple
      two pears

strcat ("oc", "tave", " is", " good", " for you")
    ⇒ octave is good for you
```

- `char` generates an empty row in the output for each empty string in the input. `strvcat`, on the other hand, eliminates empty strings.

```
char ("orange", "green", "", "red")
    ⇒ orange
      green

      red

strvcat ("orange", "green", "", "red")
    ⇒ orange
      green
      red
```

- All string concatenation functions except `cstrcat` also accept cell array data (see Section 6.2 [Cell Arrays], page 110). `char` and `strvcat` convert cell arrays into character arrays, while `strcat` concatenates within the cells of the cell arrays:

```
char ({"red", "green", "", "blue"})
    ⇒ red
      green

      blue

strcat ({"abc"; "ghi"}, {"def"; "jkl"})
    ⇒
      {
        [1,1] = abcdef
        [2,1] = ghijkl
      }
```

- strcat removes trailing white space in the arguments (except within cell arrays), while cstrcat leaves white space untouched. Both kinds of behavior can be useful as can be seen in the examples:

```
strcat (["dir1";"directory2"], ["/";"/"], ["file1";"file2"])
    ⇒ dir1/file1
      directory2/file2

cstrcat (["thirteen apples"; "a banana"], [" 5$";" 1$"])
    ⇒ thirteen apples 5$
      a banana        1$
```

Note that in the above example for cstrcat, the white space originates from the internal representation of the strings in a string array (see Section 5.2 [Character Arrays], page 66).

char (*x*)   [Built-in Function]
char (*x, ...*)   [Built-in Function]
char (*s1, s2, ...*)   [Built-in Function]
char (*cell_array*)   [Built-in Function]

Create a string array from one or more numeric matrices, character matrices, or cell arrays. Arguments are concatenated vertically. The returned values are padded with blanks as needed to make each row of the string array have the same length. Empty input strings are significant and will concatenated in the output.

For numerical input, each element is converted to the corresponding ASCII character. A range error results if an input is outside the ASCII range (0-255).

For cell arrays, each element is concatenated separately. Cell arrays converted through char can mostly be converted back with cellstr. For example:

```
char ([97, 98, 99], "", {"98", "99", 100}, "str1", ["ha", "lf"])
    ⇒ ["abc    "
       "       "
       "98     "
       "99     "
       "d      "
       "str1   "
       "half   "]
```

See also: [strvcat], page 70, [cellstr], page 117.

strvcat (*x*)   [Built-in Function]
strvcat (*x, . . .*)   [Built-in Function]
strvcat (*s1, s2, . . .*)   [Built-in Function]
strvcat (*cell_array*)   [Built-in Function]
> Create a character array from one or more numeric matrices, character matrices, or cell arrays. Arguments are concatenated vertically. The returned values are padded with blanks as needed to make each row of the string array have the same length. Unlike `char`, empty strings are removed and will not appear in the output.
>
> For numerical input, each element is converted to the corresponding ASCII character. A range error results if an input is outside the ASCII range (0-255).
>
> For cell arrays, each element is concatenated separately. Cell arrays converted through `strvcat` can mostly be converted back with `cellstr`. For example:

```
strvcat ([97, 98, 99], "", {"98", "99", 100}, "str1", ["ha", "lf"])
    ⇒ ["abc    "
       "98     "
       "99     "
       "d      "
       "str1   "
       "half   "]
```

> **See also:** [char], page 69, [strcat], page 70, [cstrcat], page 71.

strcat (*s1, s2, . . .*)   [Function File]
> Return a string containing all the arguments concatenated horizontally. If the arguments are cell strings, `strcat` returns a cell string with the individual cells concatenated. For numerical input, each element is converted to the corresponding ASCII character. Trailing white space for any character string input is eliminated before the strings are concatenated. Note that cell string values do **not** have whitespace trimmed.
>
> For example:

```
strcat ("|", " leading space is preserved", "|")
    ⇒ | leading space is preserved|

strcat ("|", "trailing space is eliminated ", "|")
    ⇒ |trailing space is eliminated|

strcat ("homogeneous space |", "  ", "| is also eliminated")
    ⇒ homogeneous space || is also eliminated

s = [ "ab"; "cde" ];
strcat (s, s, s)
    ⇒
          "ababab   "
          "cdecdecde"
```

```
s = { "ab"; "cd " };
strcat (s, s, s)
      ⇒
          {
            [1,1] = ababab
            [2,1] = cd cd cd
          }
```

See also: [cstrcat], page 71, [char], page 69, [strvcat], page 70.

cstrcat (*s1*, *s2*, ...)   [Function File]
    Return a string containing all the arguments concatenated horizontally. Trailing white
    space is preserved. For example:

```
cstrcat ("ab   ", "cd")
      ⇒ "ab   cd"

s = [ "ab"; "cde" ];
cstrcat (s, s, s)
      ⇒ "ab ab ab "
        "cdecdecde"
```

See also: [strcat], page 70, [char], page 69, [strvcat], page 70.

5.3.2 Converting Numerical Data to Strings

Apart from the string concatenation functions (see Section 5.3.1 [Concatenating Strings],
page 68) which cast numerical data to the corresponding ASCII characters, there are several
functions that format numerical data as strings. `mat2str` and `num2str` convert real or
complex matrices, while `int2str` converts integer matrices. `int2str` takes the real part
of complex values and round fractional values to integer. A more flexible way to format
numerical data as strings is the `sprintf` function (see Section 14.2.4 [Formatted Output],
page 243, [sprintf], page 244).

s = mat2str (*x*, *n*)   [Function File]
s = mat2str (*x*, *n*, "*class*")   [Function File]
    Format real, complex, and logical matrices as strings. The returned string may be
    used to reconstruct the original matrix by using the `eval` function.

    The precision of the values is given by *n*. If *n* is a scalar then both real and imaginary
    parts of the matrix are printed to the same precision. Otherwise *n*(1) defines the
    precision of the real part and *n*(2) defines the precision of the imaginary part. The
    default for *n* is 15.

    If the argument "class" is given then the class of *x* is included in the string in such
    a way that `eval` will result in the construction of a matrix of the same class.

```
mat2str ([ -1/3 + i/7; 1/3 - i/7 ], [4 2])
    ⇒ "[-0.3333+0.14i;0.3333-0.14i]"

mat2str ([ -1/3 +i/7; 1/3 -i/7 ], [4 2])
    ⇒ "[-0.3333+0i 0+0.14i;0.3333+0i -0-0.14i]"

mat2str (int16 ([1 -1]), "class")
    ⇒ "int16([1 -1])"

mat2str (logical (eye (2)))
    ⇒ "[true false;false true]"

isequal (x, eval (mat2str (x)))
    ⇒ 1
```

See also: [sprintf], page 244, [num2str], page 72, [int2str], page 73.

num2str (*x*)   [Function File]
num2str (*x*, *precision*)   [Function File]
num2str (*x*, *format*)   [Function File]

Convert a number (or array) to a string (or a character array). The optional second argument may either give the number of significant digits (*precision*) to be used in the output or a format template string (*format*) as in **sprintf** (see Section 14.2.4 [Formatted Output], page 243). **num2str** can also handle complex numbers.

Examples:

```
num2str (123.456)
    ⇒ "123.46"

num2str (123.456, 4)
    ⇒ "123.5"

s = num2str ([1, 1.34; 3, 3.56], "%5.1f")
    ⇒ s =
       1.0  1.3
       3.0  3.6
whos s
    ⇒
       Attr Name        Size                      Bytes  Class
       ==== ====        ====                      =====  =====
            s           2x8                          16  char

num2str (1.234 + 27.3i)
    ⇒ "1.234+27.3i"
```

Notes:

For MATLAB compatibility, leading spaces are stripped before returning the string.

The **num2str** function is not very flexible. For better control over the results, use **sprintf** (see Section 14.2.4 [Formatted Output], page 243).

For complex x, the format string may only contain one output conversion specification and nothing else. Otherwise, results will be unpredictable.

See also: [sprintf], page 244, [int2str], page 73, [mat2str], page 71.

int2str (*n*)   [Function File]
    Convert an integer (or array of integers) to a string (or a character array).

```
int2str (123)
    ⇒ "123"

s = int2str ([1, 2, 3; 4, 5, 6])
    ⇒ s =
        1  2  3
        4  5  6

whos s
    ⇒
    Attr Name         Size                    Bytes  Class
    ==== ====         ====                    =====  =====
         s            2x7                        14  char
```

This function is not very flexible. For better control over the results, use **sprintf** (see Section 14.2.4 [Formatted Output], page 243).

See also: [sprintf], page 244, [num2str], page 72, [mat2str], page 71.

5.4 Comparing Strings

Since a string is a character array, comparisons between strings work element by element as the following example shows:

```
GNU = "GNU's Not UNIX";
spaces = (GNU == " ")
    ⇒ spaces =
        0  0  0  0  0  1  0  0  0  1  0  0  0  0
```

To determine if two strings are identical it is necessary to use the **strcmp** function. It compares complete strings and is case sensitive. **strncmp** compares only the first N characters (with N given as a parameter). **strcmpi** and **strncmpi** are the corresponding functions for case-insensitive comparison.

strcmp (*s1*, *s2*)   [Built-in Function]
    Return 1 if the character strings *s1* and *s2* are the same, and 0 otherwise.

    If either *s1* or *s2* is a cell array of strings, then an array of the same size is returned, containing the values described above for every member of the cell array. The other argument may also be a cell array of strings (of the same size or with only one element), char matrix or character string.

    Caution: For compatibility with MATLAB, Octave's strcmp function returns 1 if the character strings are equal, and 0 otherwise. This is just the opposite of the corresponding C library function.

See also: [strcmpi], page 74, [strncmp], page 74, [strncmpi], page 74.

strncmp (*s1*, *s2*, *n*)   [Built-in Function]
    Return 1 if the first *n* characters of strings *s1* and *s2* are the same, and 0 otherwise.

```
strncmp ("abce", "abcd", 3)
     ⇒ 1
```

    If either *s1* or *s2* is a cell array of strings, then an array of the same size is returned, containing the values described above for every member of the cell array. The other argument may also be a cell array of strings (of the same size or with only one element), char matrix or character string.

```
strncmp ("abce", {"abcd", "bca", "abc"}, 3)
     ⇒ [1, 0, 1]
```

    Caution: For compatibility with MATLAB, Octave's strncmp function returns 1 if the character strings are equal, and 0 otherwise. This is just the opposite of the corresponding C library function.

    See also: [strncmpi], page 74, [strcmp], page 73, [strcmpi], page 74.

strcmpi (*s1*, *s2*)   [Built-in Function]
    Return 1 if the character strings *s1* and *s2* are the same, disregarding case of alphabetic characters, and 0 otherwise.

    If either *s1* or *s2* is a cell array of strings, then an array of the same size is returned, containing the values described above for every member of the cell array. The other argument may also be a cell array of strings (of the same size or with only one element), char matrix or character string.

    Caution: For compatibility with MATLAB, Octave's strcmp function returns 1 if the character strings are equal, and 0 otherwise. This is just the opposite of the corresponding C library function.

    Caution: National alphabets are not supported.

    See also: [strcmp], page 73, [strncmp], page 74, [strncmpi], page 74.

strncmpi (*s1*, *s2*, *n*)   [Built-in Function]
    Return 1 if the first *n* character of *s1* and *s2* are the same, disregarding case of alphabetic characters, and 0 otherwise.

    If either *s1* or *s2* is a cell array of strings, then an array of the same size is returned, containing the values described above for every member of the cell array. The other argument may also be a cell array of strings (of the same size or with only one element), char matrix or character string.

    Caution: For compatibility with MATLAB, Octave's strncmpi function returns 1 if the character strings are equal, and 0 otherwise. This is just the opposite of the corresponding C library function.

    Caution: National alphabets are not supported.

    See also: [strncmp], page 74, [strcmp], page 73, [strcmpi], page 74.

validstr = **validatestring** (*str*, *strarray*)   [Function File]
validstr = **validatestring** (*str*, *strarray*, *funcname*)   [Function File]
validstr = **validatestring** (*str*, *strarray*, *funcname*,   [Function File]
    varname)

validstr = validatestring (..., *position*)   [Function File]
> Verify that *str* is an element, or substring of an element, in *strarray*.
>
> When *str* is a character string to be tested, and *strarray* is a cellstr of valid values, then *validstr* will be the validated form of *str* where validation is defined as *str* being a member or substring of *validstr*. This is useful for both verifying and expanding short options, such as `"r"`, to their longer forms, such as `"red"`. If *str* is a substring of *validstr*, and there are multiple matches, the shortest match will be returned if all matches are substrings of each other. Otherwise, an error will be raised because the expansion of *str* is ambiguous. All comparisons are case insensitive.
>
> The additional inputs *funcname*, *varname*, and *position* are optional and will make any generated validation error message more specific.
>
> Examples:
>
> ```
> validatestring ("r", {"red", "green", "blue"})
> ⇒ "red"
> ```
>
> ```
> validatestring ("b", {"red", "green", "blue", "black"})
> ⇒ error: validatestring: multiple unique matches were found for 'b':
>    blue, black
> ```
>
> **See also:** [strcmp], page 73, [strcmpi], page 74.

5.5 Manipulating Strings

Octave supports a wide range of functions for manipulating strings. Since a string is just a matrix, simple manipulations can be accomplished using standard operators. The following example shows how to replace all blank characters with underscores.

```
quote = ...
  "First things first, but not necessarily in that order";
quote( quote == " " ) = "_"
⇒ quote =
    First_things_first,_but_not_necessarily_in_that_order
```

For more complex manipulations, such as searching, replacing, and general regular expressions, the following functions come with Octave.

deblank (*s*)   [Function File]
> Remove trailing whitespace and nulls from *s*. If *s* is a matrix, *deblank* trims each row to the length of longest string. If *s* is a cell array of strings, operate recursively on each string element.
>
> Examples:
>
> ```
> deblank ("    abc   ")
>       ⇒ "    abc"
> ```
>
> ```
> deblank (["  abc   "; "    def   "])
>       ⇒ ["  abc  " ; "    def"]
> ```
>
> **See also:** [strtrim], page 76.

`strtrim` (*s*)   [Function File]
> Remove leading and trailing whitespace from *s*. If *s* is a matrix, *strtrim* trims each row to the length of longest string. If *s* is a cell array of strings, operate recursively on each string element. For example:

>> ```
>> strtrim ("    abc   ")
>>     ⇒  "abc"
>> ```

>> ```
>> strtrim ([" abc   "; "   def    "])
>>     ⇒  ["abc "  ; "  def"]
>> ```

> **See also:** [deblank], page 75.

`strtrunc` (*s, n*)   [Function File]
> Truncate the character string *s* to length *n*. If *s* is a character matrix, then the number of columns is adjusted. If *s* is a cell array of strings, then the operation is performed on each cell element and the new cell array is returned.

`findstr` (*s, t*)   [Function File]
`findstr` (*s, t, overlap*)   [Function File]
> Return the vector of all positions in the longer of the two strings *s* and *t* where an occurrence of the shorter of the two starts. If the optional argument *overlap* is true, the returned vector can include overlapping positions (this is the default). For example:

>> ```
>> findstr ("ababab", "a")
>>     ⇒ [1, 3, 5];
>> findstr ("abababa", "aba", 0)
>>     ⇒ [1, 5]
>> ```

> **Caution:** `findstr` is scheduled for deprecation. Use `strfind` in all new code.

> **See also:** [strfind], page 77, [strmatch], page 78, [strcmp], page 73, [strncmp], page 74, [strcmpi], page 74, [strncmpi], page 74, [find], page 381.

idx = `strchr` (*str, chars*)   [Function File]
idx = `strchr` (*str, chars, n*)   [Function File]
idx = `strchr` (*str, chars, n, direction*)   [Function File]
[*i, j*] = `strchr` (...)   [Function File]
> Search for the string *str* for occurrences of characters from the set *chars*. The return value(s), as well as the *n* and *direction* arguments behave identically as in `find`.

> This will be faster than using regexp in most cases.

> **See also:** [find], page 381.

`index` (*s, t*)   [Function File]
`index` (*s, t, direction*)   [Function File]
> Return the position of the first occurrence of the string *t* in the string *s*, or 0 if no occurrence is found. *s* may also be a string array or cell array of strings.

> For example:

>> ```
>> index ("Teststring", "t")
>>     ⇒ 4
>> ```

If *direction* is `"first"`, return the first element found. If *direction* is `"last"`, return the last element found.

See also: [find], page 381, [rindex], page 77.

rindex (*s*, *t*)   [Function File]
  Return the position of the last occurrence of the character string *t* in the character string *s*, or 0 if no occurrence is found. *s* may also be a string array or cell array of strings.

  For example:

```
rindex ("Teststring", "t")
    ⇒ 6
```

  The `rindex` function is equivalent to `index` with *direction* set to `"last"`.

See also: [find], page 381, [index], page 76.

idx = strfind (*str, pattern*)   [Built-in Function]
idx = strfind (*cellstr, pattern*)   [Built-in Function]
idx = strfind (..., "*overlaps*", *val*)   [Built-in Function]
  Search for *pattern* in the string *str* and return the starting index of every such occurrence in the vector *idx*.

  If there is no such occurrence, or if *pattern* is longer than *str*, then *idx* is the empty array `[]`. The optional argument `"overlaps"` determines whether the pattern can match at every position in *str* (true), or only for unique occurrences of the complete pattern (false). The default is true.

  If a cell array of strings *cellstr* is specified then *idx* is a cell array of vectors, as specified above.

  Examples:

```
strfind ("ababa017", "aba")
    ⇒ [1, 3, 5]

strfind ("ababa017", "aba", "overlaps", false)
    ⇒ [1, 5]

strfind ({"ababa017", "bebebe", "ab"}, "aba")
    ⇒
      {
        [1,1] =

          1   3   5

        [1,2] = [](1x0)
        [1,3] = [](1x0)
      }
```

See also: [findstr], page 76, [strmatch], page 78, [regexp], page 84, [regexpi], page 87, [find], page 381.

str = strjoin (*cstr*)   [Function File]
str = strjoin (*cstr*, *delimiter*)   [Function File]
> Join the elements of the cell string array, *cstr*, into a single string.
>
> If no *delimiter* is specified, the elements of *cstr* separated by a space.
>
> If *delimiter* is specified as a string, the cell string array is joined using the string.
> Escape sequences are supported.
>
> If *delimiter* is a cell string array whose length is one less than *cstr*, then the elements of
> *cstr* are joined by interleaving the cell string elements of *delimiter*. Escape sequences
> are not supported.
>
> > strjoin ({'Octave','Scilab','Lush','Yorick'}, '*')
> > ⇒ 'Octave*Scilab*Lush*Yorick'

See also: strsplit.

strmatch (*s*, *A*)   [Function File]
strmatch (*s*, *A*, "*exact*")   [Function File]
> Return indices of entries of *A* which begin with the string *s*. The second argument *A*
> must be a string, character matrix, or a cell array of strings. If the third argument
> "exact" is not given, then *s* only needs to match *A* up to the length of *s*. Trailing
> spaces and nulls in *s* and *A* are ignored when matching.
>
> For example:
>
> > strmatch ("apple", "apple juice")
> > ⇒ 1
>
> > strmatch ("apple", ["apple   "; "apple juice"; "an apple"])
> > ⇒ [1; 2]
>
> > strmatch ("apple", ["apple   "; "apple juice"; "an apple"], "exact")
> > ⇒ [1]

Caution: strmatch is scheduled for deprecation. Use strncmp (normal case), or
strcmp ("exact" case), or regexp in all new code.

See also: [strfind], page 77, [findstr], page 76, [strcmp], page 73, [strncmp], page 74,
[strcmpi], page 74, [strncmpi], page 74, [find], page 381.

[*tok*, *rem*] = strtok (*str*)   [Function File]
[*tok*, *rem*] = strtok (*str*, *delim*)   [Function File]
> Find all characters in the string *str* up to, but not including, the first character
> which is in the string *delim*. If *rem* is requested, it contains the remainder of the
> string, starting at the first delimiter. Leading delimiters are ignored. If *delim* is not
> specified, whitespace is assumed. *str* may also be a cell array of strings in which case
> the function executes on every individual string and returns a cell array of tokens and
> remainders.
>
> Examples:

```
strtok ("this is the life")
    ⇒ "this"

[tok, rem] = strtok ("14*27+31", "+-*/")
        ⇒
            tok = 14
            rem = *27+31
```

See also: [index], page 76, [strsplit], page 79, [strchr], page 76, [isspace], page 94.

`[cstr] = strsplit (s)`	[Function File]
`[cstr] = strsplit (s, del)`	[Function File]
`[cstr] = strsplit (..., name, value)`	[Function File]
`[cstr, matches] = strsplit (...)`	[Function File]

Split the string *s* using the delimiters specified by *del* and return a cell string array of substrings. If a delimiter is not specified the string, *s*, is split at whitespace. The delimiter, *del* may be a string, a scalar cell string, or cell string array. By default, consecutive delimiters in the input string *s* are collapsed into one.

The second output, *matches*, returns the delimiters which were matched in the original string.

Example:

```
strsplit ("a b c")
    ⇒
        {
            [1,1] = a
            [1,2] = b
            [1,3] = c
        }

strsplit ("a,b,c", ",")
    ⇒
        {
            [1,1] = a
            [1,2] = b
            [1,3] = c
        }

strsplit ("a foo b,bar c", {"\s", "foo", "bar"})
    ⇒
        {
            [1,1] = a
            [1,2] = b
            [1,3] = c
        }

strsplit ("a,,b, c", {",", " "}, false)
    ⇒
```

```
          {
            [1,1] = a
            [1,2] =
            [1,3] = b
            [1,4] =
            [1,5] = c
          }
```

Supported *name/value* pair arguments are;

- *collapsedelimiters* may take the value of *true* or *false* with the default being *false*.
- *delimitertype* may take the value of `simple` or `regularexpression`. The default is *delimitertype* is `simple`.

Example:

```
    strsplit ("a foo b,bar c", ",|\\s|foo|bar", "delimitertype", "regularexpress
        ⇒
          {
            [1,1] = a
            [1,2] = b
            [1,3] = c
          }

    strsplit ("a,,b, c", "[, ]", false, "delimitertype", "regularexpression")█
        ⇒
          {
            [1,1] = a
            [1,2] =
            [1,3] = b
            [1,4] =
            [1,5] = c
          }

    strsplit ("a,\t,b, c", {',', '\s'}, "delimitertype", "regularexpression")█
        ⇒
          {
            [1,1] = a
            [1,2] = b
            [1,3] = c
          }

    strsplit ("a,\t,b, c", {',', ' ', '\t'}, "collapsedelimiters", false)█
        ⇒
          {
            [1,1] = a
            [1,2] =
            [1,3] =
            [1,4] = b
```

```
          [1,5] =
          [1,6] = c
     }
```

See also: [ostrsplit], page 81, [strjoin], page 78, [strtok], page 78, [regexp], page 84.

`[cstr] = ostrsplit (s, sep)`   [Function File]
`[cstr] = ostrsplit (s, sep, strip_empty)`   [Function File]

Split the string *s* using one or more separators *sep* and return a cell array of strings. Consecutive separators and separators at boundaries result in empty strings, unless *strip_empty* is true. The default value of *strip_empty* is false.

2-D character arrays are split at separators and at the original column boundaries.

Example:

```
ostrsplit ("a,b,c", ",")
        ⇒
           {
             [1,1] = a
             [1,2] = b
             [1,3] = c
           }

ostrsplit (["a,b" ; "cde"], ",")
        ⇒
           {
             [1,1] = a
             [1,2] = b
             [1,3] = cde
           }
```

See also: [strsplit], page 79, [strtok], page 78.

`[a, ...] = strread (str)`   [Function File]
`[a, ...] = strread (str, format)`   [Function File]
`[a, ...] = strread (str, format, format_repeat)`   [Function File]
`[a, ...] = strread (str, format, prop1, value1, ...)`   [Function File]
`[a, ...] = strread (str, format, format_repeat, prop1,`   [Function File]
`      value1, ...)`

Read data from a string.

The string *str* is split into words that are repeatedly matched to the specifiers in *format*. The first word is matched to the first specifier, the second to the second specifier and so forth. If there are more words than specifiers, the process is repeated until all words have been processed.

The string *format* describes how the words in *str* should be parsed. It may contain any combination of the following specifiers:

`%s`   The word is parsed as a string.

`%f`
`%n`   The word is parsed as a number and converted to double.

`%d`

`%u`   The word is parsed as a number and converted to int32.

`%*`, `'%*f'`, `'%*s`

                   The word is skipped.

                   For %s and %d, %f, %n, %u and the associated %*s ... specifiers an
                   optional width can be specified as %Ns, etc. where N is an integer > 1.
                   For %f, format specifiers like %N.Mf are allowed.

`literals`   In addition the format may contain literal character strings; these will be
              skipped during reading.

Parsed word corresponding to the first specifier are returned in the first output argument and likewise for the rest of the specifiers.

By default, *format* is `"%f"`, meaning that numbers are read from *str*. This will do if
str contains only numeric fields.

For example, the string

```
str = "\
Bunny Bugs    5.5\n\
Duck Daffy   -7.5e-5\n\
Penguin Tux    6"
```

can be read using

```
[a, b, c] = strread (str, "%s %s %f");
```

Optional numeric argument *format_repeat* can be used for limiting the number of
items read:

-1   (default) read all of the string until the end.

N   Read N times *nargout* items. 0 (zero) is an acceptable value for *for-
                   mat_repeat*.

The behavior of `strread` can be changed via property-value pairs. The following
properties are recognized:

`"commentstyle"`

                   Parts of *str* are considered comments and will be skipped. *value* is the
                   comment style and can be any of the following.

                   • `"shell"` Everything from # characters to the nearest end-of-line is
                     skipped.

                   • `"c"` Everything between /* and */ is skipped.

                   • `"c++"` Everything from // characters to the nearest end-of-line is
                     skipped.

                   • `"matlab"` Everything from % characters to the nearest end-of-line is
                     skipped.

                   • user-supplied. Two options: (1) One string, or 1x1 cell string: Skip
                     everything to the right of it; (2) 2x1 cell string array: Everything
                     between the left and right strings is skipped.

"delimiter"
> Any character in *value* will be used to split *str* into words (default value = any whitespace).

"emptyvalue":
> Value to return for empty numeric values in non-whitespace delimited data. The default is NaN. When the data type does not support NaN (int32 for example), then default is zero.

"multipledelimsasone"
> Treat a series of consecutive delimiters, without whitespace in between, as a single delimiter. Consecutive delimiter series need not be vertically "aligned".

"treatasempty"
> Treat single occurrences (surrounded by delimiters or whitespace) of the string(s) in *value* as missing values.

"returnonerror"
> If *value* true (1, default), ignore read errors and return normally. If false (0), return an error.

"whitespace"
> Any character in *value* will be interpreted as whitespace and trimmed; the string defining whitespace must be enclosed in double quotes for proper processing of special characters like \t. The default value for whitespace = " \b\r\n\t" (note the space). Unless whitespace is set to " (empty) AND at least one "%s" format conversion specifier is supplied, a space is always part of whitespace.

When the number of words in *str* doesn't match an exact multiple of the number of format conversion specifiers, strread's behavior depends on the last character of *str*:

last character = "\n"
> Data columns are padded with empty fields or Nan so that all columns have equal length

last character is not "\n"
> Data columns are not padded; strread returns columns of unequal length

See also: [textscan], page 236, [textread], page 236, [load], page 232, [dlmread], page 235, [fscanf], page 248.

---

newstr = strrep (*str*, *ptn*, *rep*)   [Built-in Function]
newstr = strrep (*cellstr*, *ptn*, *rep*)   [Built-in Function]
newstr = strrep (..., "*overlaps*", *val*)   [Built-in Function]
> Replace all occurrences of the pattern *ptn* in the string *str* with the string *rep* and return the result.

The optional argument "overlaps" determines whether the pattern can match at every position in *str* (true), or only for unique occurrences of the complete pattern (false). The default is true.

s may also be a cell array of strings, in which case the replacement is done for each element and a cell array is returned.

Example:

```
strrep ("This is a test string", "is", "&%$")
    ⇒  "Th&%$ &%$ a test string"
```

See also: [regexprep], page 87, [strfind], page 77, [findstr], page 76.

substr (*s*, *offset*) [Function File]
substr (*s*, *offset*, *len*) [Function File]

Return the substring of *s* which starts at character number *offset* and is *len* characters long.

Position numbering for offsets begins with 1. If *offset* is negative, extraction starts that far from the end of the string.

If *len* is omitted, the substring extends to the end of *S*. A negative value for *len* extracts to within *len* characters of the end of the string

Examples:

```
substr ("This is a test string", 6, 9)
    ⇒ "is a test"
substr ("This is a test string", -11)
    ⇒ "test string"
substr ("This is a test string", -11, -7)
    ⇒ "test"
```

This function is patterned after the equivalent function in Perl.

[*s*, *e*, *te*, *m*, *t*, *nm*, *sp*] = regexp (*str*, *pat*) [Built-in Function]
[...] = regexp (*str*, *pat*, "*opt1*", ...) [Built-in Function]

Regular expression string matching. Search for *pat* in *str* and return the positions and substrings of any matches, or empty values if there are none.

The matched pattern *pat* can include any of the standard regex operators, including:

. Match any character

* + ? {} Repetition operators, representing

  * Match zero or more times

  + Match one or more times

  ? Match zero or one times

  {*n*} Match exactly *n* times

  {*n*,} Match *n* or more times

  {*m*,*n*} Match between *m* and *n* times

[...] [^...]

List operators. The pattern will match any character listed between "[" and "]". If the first character is "^" then the pattern is inverted and any character except those listed between brackets will match.

Escape sequences defined below can also be used inside list operators. For example, a template for a floating point number might be [-+.\d]+.

() (?:)   Grouping operator. The first form, parentheses only, also creates a token.

| Alternation operator. Match one of a choice of regular expressions. The alternatives must be delimited by the grouping operator () above.

^ $ Anchoring operators. Requires pattern to occur at the start (^) or end ($) of the string.

In addition, the following escaped characters have special meaning.

\d   Match any digit

\D   Match any non-digit

\s   Match any whitespace character

\S   Match any non-whitespace character

\w   Match any word character

\W   Match any non-word character

\<   Match the beginning of a word

\>   Match the end of a word

\B   Match within a word

Implementation Note: For compatibility with MATLAB, ordinary escape sequences (e.g., "\n" => newline) are processed in *pat* regardless of whether *pat* has been defined within single quotes. Use a second backslash to stop interpolation of the escape sequence (e.g., "\\n") or use the **regexptranslate** function.

The outputs of **regexp** default to the order given below

s   The start indices of each matching substring

e   The end indices of each matching substring

te   The extents of each matched token surrounded by (...) in *pat*

m   A cell array of the text of each match

t   A cell array of the text of each token matched

nm   A structure containing the text of each matched named token, with the name being used as the fieldname. A named token is denoted by (?<name>...).

sp   A cell array of the text not returned by match, i.e., what remains if you split the string based on *pat*.

Particular output arguments, or the order of the output arguments, can be selected by additional *opt* arguments. These are strings and the correspondence between the output arguments and the optional argument are

'start'	*s*
'end'	*e*
'tokenExtents'	*te*
'match'	*m*

'tokens'	t
'names'	nm
'split'	sp

Additional arguments are summarized below.

'once'   Return only the first occurrence of the pattern.

'matchcase'

Make the matching case sensitive. (default)

Alternatively, use (?-i) in the pattern.

'ignorecase'

Ignore case when matching the pattern to the string.

Alternatively, use (?i) in the pattern.

'stringanchors'

Match the anchor characters at the beginning and end of the string. (default)

Alternatively, use (?-m) in the pattern.

'lineanchors'

Match the anchor characters at the beginning and end of the line.

Alternatively, use (?m) in the pattern.

'dotall'   The pattern . matches all characters including the newline character. (default)

Alternatively, use (?s) in the pattern.

'dotexceptnewline'

The pattern . matches all characters except the newline character.

Alternatively, use (?-s) in the pattern.

'literalspacing'

All characters in the pattern, including whitespace, are significant and are used in pattern matching. (default)

Alternatively, use (?-x) in the pattern.

'freespacing'

The pattern may include arbitrary whitespace and also comments beginning with the character '#'.

Alternatively, use (?x) in the pattern.

'noemptymatch'

Zero-length matches are not returned. (default)

'emptymatch'

Return zero-length matches.

regexp ('a', 'b*', 'emptymatch') returns [1 2] because there are zero or more 'b' characters at positions 1 and end-of-string.

See also: [regexpi], page 87, [strfind], page 77, [regexprep], page 87.

[*s*, *e*, *te*, *m*, *t*, *nm*, *sp*] = regexpi (*str*, *pat*)   [Built-in Function]
[...] = regexpi (*str*, *pat*, "*opt1*", ...)   [Built-in Function]
> Case insensitive regular expression string matching. Search for *pat* in *str* and return the positions and substrings of any matches, or empty values if there are none. See [regexp], page 84, for details on the syntax of the search pattern.

> **See also:** [regexp], page 84.

outstr = regexprep (*string*, *pat*, *repstr*)   [Built-in Function]
outstr = regexprep (*string*, *pat*, *repstr*, "*opt1*", ...)   [Built-in Function]
> Replace occurrences of pattern *pat* in *string* with *repstr*.

> The pattern is a regular expression as documented for **regexp**. See [regexp], page 84.

> The replacement string may contain $i, which substitutes for the ith set of parentheses in the match string. For example,

>> regexprep ("Bill Dunn", '(\w+) (\w+)', '$2, $1')

> returns "Dunn, Bill"

> Options in addition to those of **regexp** are

> 'once'   Replace only the first occurrence of *pat* in the result.

> 'warnings'
>> This option is present for compatibility but is ignored.

> Implementation Note: For compatibility with MATLAB, ordinary escape sequences (e.g., "\n" => newline) are processed in both *pat* and *repstr* regardless of whether they were defined within single quotes. Use a second backslash to stop interpolation of the escape sequence (e.g., "\\n") or use the **regexptranslate** function.

> **See also:** [regexp], page 84, [regexpi], page 87, [strrep], page 83.

regexptranslate (*op*, *s*)   [Function File]
> Translate a string for use in a regular expression. This may include either wildcard replacement or special character escaping. The behavior is controlled by *op* which can take the following values

> "wildcard"
>> The wildcard characters ., *, and ? are replaced with wildcards that are appropriate for a regular expression. For example:

>>> regexptranslate ("wildcard", "*.m")
>>>>   ⇒ ".*\.m"

> "escape"   The characters $.?[], that have special meaning for regular expressions are escaped so that they are treated literally. For example:

>>> regexptranslate ("escape", "12.5")
>>>>   ⇒ "12\.5"

> **See also:** [regexp], page 84, [regexpi], page 87, [regexprep], page 87.

untabify (*t*)   [Function File]
untabify (*t*, *tw*)   [Function File]

untabify (*t*, *tw*, *deblank*)   [Function File]

> Replace TAB characters in *t*, with spaces. The tab width is specified by *tw*, or defaults to eight. The input, *t*, may be either a 2-D character array, or a cell array of character strings. The output is the same class as the input.
>
> If the optional argument *deblank* is true, then the spaces will be removed from the end of the character data.
>
> The following example reads a file and writes an untabified version of the same file with trailing spaces stripped.
>
> ```
> fid = fopen ("tabbed_script.m");
> text = char (fread (fid, "uchar")');
> fclose (fid);
> fid = fopen ("untabified_script.m", "w");
> text = untabify (strsplit (text, "\n"), 8, true);
> fprintf (fid, "%s\n", text{:});
> fclose (fid);
> ```
>
> **See also:** [strjust], page 91, [strsplit], page 79, [deblank], page 75.

5.6 String Conversions

Octave supports various kinds of conversions between strings and numbers. As an example, it is possible to convert a string containing a hexadecimal number to a floating point number.

```
hex2dec ("FF")
     ⇒ 255
```

bin2dec (*s*)   [Function File]

> Return the decimal number corresponding to the binary number represented by the string *s*. For example:
>
> ```
> bin2dec ("1110")
>      ⇒ 14
> ```
>
> Spaces are ignored during conversion and may be used to make the binary number more readable.
>
> ```
> bin2dec ("1000 0001")
>      ⇒ 129
> ```
>
> If *s* is a string matrix, return a column vector with one converted number per row of *s*; Invalid rows evaluate to NaN.
>
> If *s* is a cell array of strings, return a column vector with one converted number per cell element in *s*.
>
> **See also:** [dec2bin], page 88, [base2dec], page 90, [hex2dec], page 89.

dec2bin (*d*, *len*)   [Function File]

> Return a binary number corresponding to the non-negative integer *d*, as a string of ones and zeros. For example:
>
> ```
> dec2bin (14)
>      ⇒ "1110"
> ```

If *d* is a matrix or cell array, return a string matrix with one row per element in *d*, padded with leading zeros to the width of the largest value.

The optional second argument, *len*, specifies the minimum number of digits in the result.

See also: [bin2dec], page 88, [dec2base], page 89, [dec2hex], page 89.

dec2hex (*d*, *len*)   [Function File]
Return the hexadecimal string corresponding to the non-negative integer *d*. For example:

        dec2hex (2748)
            ⇒ "ABC"

If *d* is a matrix or cell array, return a string matrix with one row per element in *d*, padded with leading zeros to the width of the largest value.

The optional second argument, *len*, specifies the minimum number of digits in the result.

See also: [hex2dec], page 89, [dec2base], page 89, [dec2bin], page 88.

hex2dec (*s*)   [Function File]
Return the integer corresponding to the hexadecimal number represented by the string *s*. For example:

        hex2dec ("12B")
            ⇒ 299
        hex2dec ("12b")
            ⇒ 299

If *s* is a string matrix, return a column vector with one converted number per row of *s*; Invalid rows evaluate to NaN.

If *s* is a cell array of strings, return a column vector with one converted number per cell element in *s*.

See also: [dec2hex], page 89, [base2dec], page 90, [bin2dec], page 88.

dec2base (*d*, *base*)   [Function File]
dec2base (*d*, *base*, *len*)   [Function File]
Return a string of symbols in base *base* corresponding to the non-negative integer *d*.

        dec2base (123, 3)
            ⇒ "11120"

If *d* is a matrix or cell array, return a string matrix with one row per element in *d*, padded with leading zeros to the width of the largest value.

If *base* is a string then the characters of *base* are used as the symbols for the digits of *d*. Space (' ') may not be used as a symbol.

        dec2base (123, "aei")
            ⇒ "eeeia"

The optional third argument, *len*, specifies the minimum number of digits in the result.

See also: [base2dec], page 90, [dec2bin], page 88, [dec2hex], page 89.

base2dec (*s*, *base*)   [Function File]
> Convert *s* from a string of digits in base *base* to a decimal integer (base 10).
>
>> base2dec ("11120", 3)
>>   ⇒ 123
>
> If *s* is a string matrix, return a column vector with one value per row of *s*. If a row contains invalid symbols then the corresponding value will be NaN.
>
> If *s* is a cell array of strings, return a column vector with one value per cell element in *s*.
>
> If *base* is a string, the characters of *base* are used as the symbols for the digits of *s*. Space (' ') may not be used as a symbol.
>
>> base2dec ("yyyzx", "xyz")
>>   ⇒ 123
>
> **See also:** [dec2base], page 89, [bin2dec], page 88, [hex2dec], page 89.

s = num2hex (*n*)   [Built-in Function]
> Typecast a double or single precision number or vector to a 8 or 16 character hexadecimal string of the IEEE 754 representation of the number. For example:
>
>> num2hex ([-1, 1, e, Inf])
>>   ⇒ "bff0000000000000
>>   3ff0000000000000
>>   4005bf0a8b145769
>>   7ff0000000000000"
>
> If the argument *n* is a single precision number or vector, the returned string has a length of 8. For example:
>
>> num2hex (single ([-1, 1, e, Inf]))
>>   ⇒ "bf800000
>>   3f800000
>>   402df854
>>   7f800000"
>
> **See also:** [hex2num], page 90, [hex2dec], page 89, [dec2hex], page 89.

n = hex2num (*s*)   [Built-in Function]
n = hex2num (*s*, *class*)   [Built-in Function]
> Typecast the 16 character hexadecimal character string to an IEEE 754 double precision number. If fewer than 16 characters are given the strings are right padded with '0' characters.
>
> Given a string matrix, hex2num treats each row as a separate number.
>
>> hex2num (["4005bf0a8b145769"; "4024000000000000"])
>>   ⇒ [2.7183; 10.000]
>
> The optional argument *class* can be passed as the string "single" to specify that the given string should be interpreted as a single precision number. In this case, *s* should be an 8 character hexadecimal string. For example:
>
>> hex2num (["402df854"; "41200000"], "single")
>>   ⇒ [2.7183; 10.000]
>
> **See also:** [num2hex], page 90, [hex2dec], page 89, [dec2hex], page 89.

str2double (*s*)   [Built-in Function]

Convert a string to a real or complex number.

The string must be in one of the following formats where a and b are real numbers and the complex unit is 'i' or 'j':

- a + bi
- a + b*i
- a + i*b
- bi + a
- b*i + a
- i*b + a

If present, a and/or b are of the form [+-]d[,.]d[[eE][+-]d] where the brackets indicate optional arguments and 'd' indicates zero or more digits. The special input values `Inf`, `NaN`, and `NA` are also accepted.

s may be a character string, character matrix, or cell array. For character arrays the conversion is repeated for every row, and a double or complex array is returned. Empty rows in *s* are deleted and not returned in the numeric array. For cell arrays each character string element is processed and a double or complex array of the same dimensions as *s* is returned.

For unconvertible scalar or character string input `str2double` returns a NaN. Similarly, for character array input `str2double` returns a NaN for any row of *s* that could not be converted. For a cell array, `str2double` returns a NaN for any element of *s* for which conversion fails. Note that numeric elements in a mixed string/numeric cell array are not strings and the conversion will fail for these elements and return NaN.

`str2double` can replace `str2num`, and it avoids the security risk of using `eval` on unknown data.

See also: [str2num], page 92.

strjust (*s*)   [Function File]
strjust (*s*, *pos*)   [Function File]

Return the text, *s*, justified according to *pos*, which may be `"left"`, `"center"`, or `"right"`. If *pos* is omitted it defaults to `"right"`.

Null characters are replaced by spaces. All other character data are treated as non-white space.

Example:

```
strjust (["a"; "ab"; "abc"; "abcd"])
    ⇒
        "   a"
        "  ab"
        " abc"
        "abcd"
```

See also: [deblank], page 75, [strrep], page 83, [strtrim], page 76, [untabify], page 87.

x = str2num (*s*)   [Function File]
[*x*, *state*] = str2num (*s*)   [Function File]
>   Convert the string (or character array) *s* to a number (or an array). Examples:

```
str2num ("3.141596")
    ⇒ 3.141596


str2num (["1, 2, 3"; "4, 5, 6"])
    ⇒ 1   2   3
       4   5   6
```

>   The optional second output, *state*, is logically true when the conversion is successful. If the conversion fails the numeric output, *x*, is empty and *state* is false.

>   **Caution:** As str2num uses the eval function to do the conversion, str2num will execute any code contained in the string *s*. Use str2double for a safer and faster conversion.

>   For cell array of strings use str2double.

>   **See also:** [str2double], page 91, [eval], page 151.

toascii (*s*)   [Mapping Function]
>   Return ASCII representation of *s* in a matrix. For example:

```
toascii ("ASCII")
    ⇒ [ 65, 83, 67, 73, 73 ]
```

>   **See also:** [char], page 69.

tolower (*s*)   [Mapping Function]
lower (*s*)   [Mapping Function]
>   Return a copy of the string or cell string *s*, with each uppercase character replaced by the corresponding lowercase one; non-alphabetic characters are left unchanged. For example:

```
tolower ("MiXeD cAsE 123")
    ⇒ "mixed case 123"
```

>   **See also:** [toupper], page 92.

toupper (*s*)   [Mapping Function]
upper (*s*)   [Mapping Function]
>   Return a copy of the string or cell string *s*, with each lowercase character replaced by the corresponding uppercase one; non-alphabetic characters are left unchanged. For example:

```
toupper ("MiXeD cAsE 123")
    ⇒ "MIXED CASE 123"
```

>   **See also:** [tolower], page 92.

do_string_escapes (*string*)   [Built-in Function]
>   Convert special characters in *string* to their escaped forms.

undo_string_escapes (*s*)   [Built-in Function]
> Convert special characters in strings back to their escaped forms. For example, the expression
>
>   bell = "\a";
>
> assigns the value of the alert character (control-g, ASCII code 7) to the string variable `bell`. If this string is printed, the system will ring the terminal bell (if it is possible). This is normally the desired outcome. However, sometimes it is useful to be able to print the original representation of the string, with the special characters replaced by their escape sequences. For example,
>
>   octave:13> undo_string_escapes (bell)
>   ans = \a
>
> replaces the unprintable alert character with its printable representation.

5.7 Character Class Functions

Octave also provides the following character class test functions patterned after the functions in the standard C library. They all operate on string arrays and return matrices of zeros and ones. Elements that are nonzero indicate that the condition was true for the corresponding character in the string array. For example:

    isalpha ("!Q@WERT^Y&")
        ⇒ [0, 1, 0, 1, 1, 1, 1, 0, 1, 0]

isalnum (*s*)   [Mapping Function]
> Return a logical array which is true where the elements of *s* are letters or digits and false where they are not. This is equivalent to (`isalpha (s) | isdigit (s)`).
>
> **See also:** [isalpha], page 93, [isdigit], page 94, [ispunct], page 94, [isspace], page 94, [iscntrl], page 94.

isalpha (*s*)   [Mapping Function]
> Return a logical array which is true where the elements of *s* are letters and false where they are not. This is equivalent to (`islower (s) | isupper (s)`).
>
> **See also:** [isdigit], page 94, [ispunct], page 94, [isspace], page 94, [iscntrl], page 94, [isalnum], page 93, [islower], page 93, [isupper], page 94.

isletter (*s*)   [Function File]
> Return a logical array which is true where the elements of *s* are letters and false where they are not. This is an alias for the `isalpha` function.
>
> **See also:** [isalpha], page 93, [isdigit], page 94, [ispunct], page 94, [isspace], page 94, [iscntrl], page 94, [isalnum], page 93.

islower (*s*)   [Mapping Function]
> Return a logical array which is true where the elements of *s* are lowercase letters and false where they are not.
>
> **See also:** [isupper], page 94, [isalpha], page 93, [isletter], page 93, [isalnum], page 93.

isupper (s)   [Mapping Function]

>   Return a logical array which is true where the elements of s are uppercase letters and
>   false where they are not.
>
>   **See also:** [islower], page 93, [isalpha], page 93, [isletter], page 93, [isalnum], page 93.

isdigit (s)   [Mapping Function]

>   Return a logical array which is true where the elements of s are decimal digits (0-9)
>   and false where they are not.
>
>   **See also:** [isxdigit], page 94, [isalpha], page 93, [isletter], page 93, [ispunct], page 94,
>   [isspace], page 94, [iscntrl], page 94.

isxdigit (s)   [Mapping Function]

>   Return a logical array which is true where the elements of s are hexadecimal digits
>   (0-9 and a-fA-F).
>
>   **See also:** [isdigit], page 94.

ispunct (s)   [Mapping Function]

>   Return a logical array which is true where the elements of s are punctuation characters
>   and false where they are not.
>
>   **See also:** [isalpha], page 93, [isdigit], page 94, [isspace], page 94, [iscntrl], page 94.

isspace (s)   [Mapping Function]

>   Return a logical array which is true where the elements of s are whitespace characters
>   (space, formfeed, newline, carriage return, tab, and vertical tab) and false where they
>   are not.
>
>   **See also:** [iscntrl], page 94, [ispunct], page 94, [isalpha], page 93, [isdigit], page 94.

iscntrl (s)   [Mapping Function]

>   Return a logical array which is true where the elements of s are control characters
>   and false where they are not.
>
>   **See also:** [ispunct], page 94, [isspace], page 94, [isalpha], page 93, [isdigit], page 94.

isgraph (s)   [Mapping Function]

>   Return a logical array which is true where the elements of s are printable characters
>   (but not the space character) and false where they are not.
>
>   **See also:** [isprint], page 94.

isprint (s)   [Mapping Function]

>   Return a logical array which is true where the elements of s are printable characters
>   (including the space character) and false where they are not.
>
>   **See also:** [isgraph], page 94.

isascii (s)   [Mapping Function]

>   Return a logical array which is true where the elements of s are ASCII characters (in
>   the range 0 to 127 decimal) and false where they are not.

isstrprop (*str*, *prop*)   [Function File]
  Test character string properties. For example:

        isstrprop ("abc123", "alpha")
        ⇒ [1, 1, 1, 0, 0, 0]

  If *str* is a cell array, `isstrpop` is applied recursively to each element of the cell array.

  Numeric arrays are converted to character strings.

  The second argument *prop* must be one of

  "alpha"   True for characters that are alphabetic (letters).

  "alnum"
  "alphanum"
              True for characters that are alphabetic or digits.

  "lower"   True for lowercase letters.

  "upper"   True for uppercase letters.

  "digit"   True for decimal digits (0-9).

  "xdigit"   True for hexadecimal digits (a-fA-F0-9).

  "space"
  "wspace"   True for whitespace characters (space, formfeed, newline, carriage return,
              tab, vertical tab).

  "punct"   True for punctuation characters (printing characters except space or letter
              or digit).

  "cntrl"   True for control characters.

  "graph"
  "graphic"
              True for printing characters except space.

  "print"   True for printing characters including space.

  "ascii"   True for characters that are in the range of ASCII encoding.

  See also: [isalpha], page 93, [isalnum], page 93, [islower], page 93, [isupper], page 94,
  [isdigit], page 94, [isxdigit], page 94, [isspace], page 94, [ispunct], page 94, [iscntrl],
  page 94, [isgraph], page 94, [isprint], page 94, [isascii], page 94.

6 Data Containers

Octave includes support for two different mechanisms to contain arbitrary data types in the same variable. Structures, which are C-like, and are indexed with named fields, and cell arrays, where each element of the array can have a different data type and or shape. Multiple input arguments and return values of functions are organized as another data container, the comma separated list.

6.1 Structures

Octave includes support for organizing data in structures. The current implementation uses an associative array with indices limited to strings, but the syntax is more like C-style structures.

6.1.1 Basic Usage and Examples

Here are some examples of using data structures in Octave.

Elements of structures can be of any value type. For example, the three expressions

```
x.a = 1;
x.b = [1, 2; 3, 4];
x.c = "string";
```

create a structure with three elements. The '.' character separates the structure name from the field name and indicates to Octave that this variable is a structure. To print the value of the structure you can type its name, just as for any other variable:

```
x
    ⇒ x =
      {
          a = 1
          b =

            1  2
            3  4

          c = string
      }
```

Note that Octave may print the elements in any order.

Structures may be copied just like any other variable:

```
y = x
        ⇒ y =
            {
                a = 1
                b =

                    1   2
                    3   4

                c = string
            }
```

Since structures are themselves values, structure elements may reference other structures. The following statements change the value of the element **b** of the structure **x** to be a data structure containing the single element **d**, which has a value of 3.

```
x.b.d = 3;
x.b
        ⇒ ans =
            {
                d = 3
            }

x
        ⇒ x =
            {
                a = 1
                b =
                {
                    d = 3
                }

                c = string
            }
```

Note that when Octave prints the value of a structure that contains other structures, only a few levels are displayed. For example:

```
a.b.c.d.e = 1;
a
    ⇒  a =
        {
          b =
          {
            c =
            {
                1x1 struct array containing the fields:

                d: 1x1 struct
            }
          }
        }
```

This prevents long and confusing output from large deeply nested structures. The number of levels to print for nested structures may be set with the function `struct_levels_to_print`, and the function `print_struct_array_contents` may be used to enable printing of the contents of structure arrays.

val = struct_levels_to_print ()   [Built-in Function]
old_val = struct_levels_to_print (*new_val*)   [Built-in Function]
struct_levels_to_print (*new_val*, "*local*")   [Built-in Function]
  Query or set the internal variable that specifies the number of structure levels to display.

  When called from inside a function with the "local" option, the variable is changed locally for the function and any subroutines it calls. The original variable value is restored when exiting the function.

  See also: [print_struct_array_contents], page 99.

val = print_struct_array_contents ()   [Built-in Function]
old_val = print_struct_array_contents (*new_val*)   [Built-in Function]
print_struct_array_contents (*new_val*, "*local*")   [Built-in Function]
  Query or set the internal variable that specifies whether to print struct array contents.

  If true, values of struct array elements are printed. This variable does not affect scalar structures whose elements are always printed. In both cases, however, printing will be limited to the number of levels specified by *struct_levels_to_print*.

  When called from inside a function with the "local" option, the variable is changed locally for the function and any subroutines it calls. The original variable value is restored when exiting the function.

  See also: [struct_levels_to_print], page 99.

Functions can return structures. For example, the following function separates the real and complex parts of a matrix and stores them in two elements of the same structure variable.

```
function y = f (x)
  y.re = real (x);
  y.im = imag (x);
endfunction
```

When called with a complex-valued argument, f returns the data structure containing the real and imaginary parts of the original function argument.

```
f (rand (2) + rand (2) * I)
    ⇒ ans =
       {
          im =

             0.26475   0.14828
             0.18436   0.83669


          re =

             0.040239   0.242160
             0.238081   0.402523


       }
```

Function return lists can include structure elements, and they may be indexed like any other variable. For example:

```
[ x.u, x.s(2:3,2:3), x.v ] = svd ([1, 2; 3, 4]);
x
    ⇒ x =
       {
          u =

             -0.40455   -0.91451
             -0.91451    0.40455

          s =

             0.00000   0.00000   0.00000
             0.00000   5.46499   0.00000
             0.00000   0.00000   0.36597

          v =

             -0.57605   0.81742
             -0.81742  -0.57605


       }
```

It is also possible to cycle through all the elements of a structure in a loop, using a special form of the for statement (see Section 10.5.1 [Looping Over Structure Elements], page 161).

6.1.2 Structure Arrays

A structure array is a particular instance of a structure, where each of the fields of the structure is represented by a cell array. Each of these cell arrays has the same dimensions. Conceptually, a structure array can also be seen as an array of structures with identical fields. An example of the creation of a structure array is

```
x(1).a = "string1";
x(2).a = "string2";
x(1).b = 1;
x(2).b = 2;
```

which creates a 2-by-1 structure array with two fields. Another way to create a structure array is with the **struct** function (see Section 6.1.3 [Creating Structures], page 102). As previously, to print the value of the structure array, you can type its name:

```
x
    ⇒ x =
      {
          1x2 struct array containing the fields:

            a

            b
      }
```

Individual elements of the structure array can be returned by indexing the variable like *x*(1), which returns a structure with two fields:

```
x(1)
    ⇒ ans =
      {
          a = string1
          b = 1
      }
```

Furthermore, the structure array can return a comma separated list of field values (see Section 6.3 [Comma Separated Lists], page 118), if indexed by one of its own field names. For example:

```
x.a
    ⇒
        ans = string1
        ans = string2
```

Here is another example, using this comma separated list on the left-hand side of an assignment:

```
[x.a] = deal ("new string1", "new string2");
  x(1).a
      ⇒ ans = new string1
  x(2).a
      ⇒ ans = new string2
```

Just as for numerical arrays, it is possible to use vectors as indices (see Section 8.1 [Index Expressions], page 131):

```
x(3:4) = x(1:2);
[x([1,3]).a] = deal ("other string1", "other string2");
x.a
    ⇒
        ans = other string1
        ans = new string2
        ans = other string2
        ans = new string2
```

The function `size` will return the size of the structure. For the example above

```
size (x)
    ⇒ ans =

        1   4
```

Elements can be deleted from a structure array in a similar manner to a numerical array, by assigning the elements to an empty matrix. For example

```
in = struct ("call1", {x, Inf, "last"},
             "call2", {x, Inf, "first"})
    ⇒ in =
        {
            1x3 struct array containing the fields:

                call1
                call2
        }

in(1) = [];
in.call1
    ⇒
        ans = Inf
        ans = last
```

6.1.3 Creating Structures

Besides the index operator ".", Octave can use dynamic naming "(var)" or the `struct` function to create structures. Dynamic naming uses the string value of a variable as the field name. For example:

```
a = "field2";
x.a = 1;
x.(a) = 2;
x
    ⇒ x =
        {
            a =  1
            field2 =  2
        }
```

Dynamic indexing also allows you to use arbitrary strings, not merely valid Octave identifiers (note that this does not work on MATLAB):

```
a = "long field with spaces (and funny char$)";
x.a = 1;
x.(a) = 2;
x
    ⇒ x =
      {
        a =  1
        long field with spaces (and funny char$) =  2
      }
```

The warning id `Octave:matlab-incompatible` can be enabled to warn about this usage. See [warning_ids], page 205.

More realistically, all of the functions that operate on strings can be used to build the correct field name before it is entered into the data structure.

```
names = ["Bill"; "Mary"; "John"];
ages  = [37; 26; 31];
for i = 1:rows (names)
   database.(names(i,:)) = ages(i);
endfor
database
    ⇒ database =
      {
        Bill =   37
        Mary =   26
        John =   31
      }
```

The third way to create structures is the **struct** command. **struct** takes pairs of arguments, where the first argument in the pair is the fieldname to include in the structure and the second is a scalar or cell array, representing the values to include in the structure or structure array. For example:

```
struct ("field1", 1, "field2", 2)
⇒ ans =
      {
        field1 =  1
        field2 =  2
      }
```

If the values passed to **struct** are a mix of scalar and cell arrays, then the scalar arguments are expanded to create a structure array with a consistent dimension. For example:

```
s = struct ("field1", {1, "one"}, "field2", {2, "two"},
        "field3", 3);
s.field1
    ⇒
        ans =  1
        ans = one

s.field2
    ⇒
        ans =  2
        ans = two

s.field3
    ⇒
        ans =  3
        ans =  3
```

If you want to create a struct which contains a cell array as an individual field, you must wrap it in another cell array as shown in the following example:

```
struct ("field1", {{1, "one"}}, "field2", 2)
    ⇒ ans =
        {
          field1 =

          {
            [1,1] =  1
            [1,2] = one
          }

          field2 =  2
        }
```

s = **struct** ()   [Built-in Function]
s = **struct** (*field1, value1, field2, value2, ...*)   [Built-in Function]
s = **struct** (*obj*)   [Built-in Function]

Create a scalar or array structure and initialize its values. The *field1*, *field2*, ... variables are strings specifying the names of the fields and the *value1*, *value2*, ... variables can be of any type.

If the values are cell arrays, create a structure array and initialize its values. The dimensions of each cell array of values must match. Singleton cells and non-cell values are repeated so that they fill the entire array. If the cells are empty, create an empty structure array with the specified field names.

If the argument is an object, return the underlying struct.

Observe that the syntax is optimized for struct **arrays**. Consider the following examples:

```
struct ("foo", 1)
  ⇒ scalar structure containing the fields:
    foo =  1

struct ("foo", {})
  ⇒ 0x0 struct array containing the fields:
    foo

struct ("foo", { {} })
  ⇒ scalar structure containing the fields:
    foo = {}(0x0)

struct ("foo", {1, 2, 3})
  ⇒ 1x3 struct array containing the fields:
    foo
```

The first case is an ordinary scalar struct—one field, one value. The second produces an empty struct array with one field and no values, since s being passed an empty cell array of struct array values. When the value is a cell array containing a single entry, this becomes a scalar struct with that single entry as the value of the field. That single entry happens to be an empty cell array.

Finally, if the value is a non-scalar cell array, then **struct** produces a struct **array**.

See also: [cell2struct], page 118, [fieldnames], page 105, [orderfields], page 107, [getfield], page 107, [setfield], page 106, [rmfield], page 107, [structfun], page 466.

The function **isstruct** can be used to test if an object is a structure or a structure array.

isstruct (*x*)   [Built-in Function]
> Return true if *x* is a structure or a structure array.

See also: [ismatrix], page 62, [iscell], page 111, [isa], page 39.

6.1.4 Manipulating Structures

Other functions that can manipulate the fields of a structure are given below.

nfields (*s*)   [Built-in Function]
> Return the number of fields of the structure *s*.

See also: [fieldnames], page 105.

names = fieldnames (*struct*)   [Function File]
names = fieldnames (*obj*)   [Function File]
names = fieldnames (*javaobj*)   [Function File]
names = fieldnames ("*jclassname*")   [Function File]
> Return a cell array of strings with the names of the fields in the specified input.

> When the input is a structure *struct*, the names are the elements of the structure.

When the input is an Octave object *obj*, the names are the public properties of the object.

When the input is a Java object *javaobj* or Java classname *jclassname* the name are the public data elements of the object or class.

See also: [nfields], page 105, [isfield], page 106, [orderfields], page 107, [struct], page 104, [methods], page 674.

isfield (*x*, "*name*")   [Built-in Function]
isfield (*x*, *name*)   [Built-in Function]
> Return true if the *x* is a structure and it includes an element named *name*. If *name* is a cell array of strings then a logical array of equal dimension is returned.

See also: [fieldnames], page 105.

s = setfield (*s*, *field*, *val*)   [Function File]
s = setfield (*s*, *idx1*, *field1*, *idx2*, *field2*, ..., *val*)   [Function File]
> Set a field member *field* in a structure *s* equal to *val*. For example:

```
s = struct ();
s = setfield (s, "foo bar", 42);
```

> This is equivalent to

```
s.("foo bar") = 42;
```

> Note that ordinary structure syntax `s.foo bar = 42` cannot be used here, as the field name is not a valid Octave identifier. Using arbitrary strings for field name is incompatible with MATLAB, so this usage will warn if the `Octave:matlab-incompatible` warning is set. See [XREFwarning_ids], page 205.

> With the second calling form, set a field on a structure array, possibly nested, with successive nested indices *idx1*, *idx2*, ... and fields *field1*, *field2*, ... The indices must be cells containing the desired index at this nesting depth.

> Thus consider instead,

```
s = struct ("baz", 42);
setfield (s, {1}, "foo", {1}, "bar", 5)
    ⇒ ans =
    scalar structure containing the fields:
      baz = 42
      foo =
        scalar structure containing the fields:
          bar = 54
```

> Here we first have an ordinary structure array with one field `baz` set to 42. Then we set another field in a nested scalar structure indexing with two single cells containing the unique desired indices.

> Finally an example with nested structure arrays,

```
sa.foo = 1;
sa = setfield (sa, {2}, "bar", {3}, "baz", 6);
sa(2).bar(3)
    ⇒ ans =
    scalar structure containing the fields:
      baz = 6
```

Here *sa* is a structure array whose field `fd` at elements 1 and 2 field is in turn another structure array whose third element is a structure

Note that the same result as in the above example could be achieved by:

```
SA.foo = 1;
SA(2).bar(3).baz = 6
```

See also: [getfield], page 107, [rmfield], page 107, [isfield], page 106, [fieldnames], page 105, [isstruct], page 105, [struct], page 104.

`[val] = getfield (s, field)`   [Function File]
`[val] = getfield (s, idx1, field1, idx2, field2, ...)`   [Function File]
   Extract a field from a structure (or a nested structure). The syntax is the same as `setfield`, except it omits the final *val* argument, returning this value instead of setting it.

   See also: [setfield], page 106, [rmfield], page 107, [isfield], page 106, [fieldnames], page 105, [isstruct], page 105, [struct], page 104.

`s = rmfield (s, "f")`   [Built-in Function]
`s = rmfield (s, f)`   [Built-in Function]
   Return a *copy* of the structure (array) *s* with the field *f* removed. If *f* is a cell array of strings or a character array, remove each of the named fields.

   See also: [orderfields], page 107, [fieldnames], page 105.

`[t, p] = orderfields (s1)`   [Function File]
`[t, p] = orderfields (s1, s2)`   [Function File]
   Return a copy of *s1* with fields arranged alphabetically or as specified by *s2*.

   Given one struct, arrange field names in *s1* alphabetically.

   If the second argument is a struct, arrange field names in *s1* as they appear in *s2*. The second argument may also specify the order in a permutation vector or a cell array of strings containing the fieldnames of *s1* in the desired order.

   The optional second output argument *p* is assigned the permutation vector which converts the original name order into the new name order.

   Examples:

```
s = struct ("d", 4, "b", 2, "a", 1, "c", 3);
t1 = orderfields (s)
    ⇒ t1 =
        {
            a =  1
            b =  2
            c =  3
            d =  4
        }
```

```
           t = struct ("d", {}, "c", {}, "b", {}, "a", {});
           t2 = orderfields (s, t)
                ⇒ t2 =
                   {
                      d =  4
                      c =  3
                      b =  2
                      a =  1
                   }
           t3 = orderfields (s, [3, 2, 4, 1])
                ⇒ t3 =
                   {
                      a =  1
                      b =  2
                      c =  3
                      d =  4
                   }
      [t4, p] = orderfields (s, {"d", "c", "b", "a"})
                ⇒ t4 =
                   {
                      d =  4
                      c =  3
                      b =  2
                      a =  1
                   }
                   p =
                      1
                      4
                      2
                      3
```

See also: [getfield], page 107, [rmfield], page 107, [isfield], page 106, [isstruct], page 105, [fieldnames], page 105, [struct], page 104.

substruct (*type*, *subs*, ...)   [Function File]
    Create a subscript structure for use with **subsref** or **subsasgn**. For example:

```
idx = substruct ("()", {3, ":"})
    ⇒
      idx =
      {
        type = ()
        subs =
        {
          [1,1] =  3
          [1,2] = :
        }
      }
x = [1, 2, 3; 4, 5, 6; 7, 8, 9];
subsref (x, idx)
    ⇒ 7  8  9
```

See also: [subsref], page 678, [subsasgn], page 680.

6.1.5 Processing Data in Structures

The simplest way to process data in a structure is within a **for** loop (see Section 10.5.1 [Looping Over Structure Elements], page 161). A similar effect can be achieved with the **structfun** function, where a user defined function is applied to each field of the structure. See [structfun], page 466.

Alternatively, to process the data in a structure, the structure might be converted to another type of container before being treated.

c = **struct2cell** (*s*)   [Built-in Function]
Create a new cell array from the objects stored in the struct object. If *f* is the number of fields in the structure, the resulting cell array will have a dimension vector corresponding to [*f* size(*s*)]. For example:

```
s = struct ("name", {"Peter", "Hannah", "Robert"},
            "age", {23, 16, 3});
c = struct2cell (s)
    ⇒ c = {2x1x3 Cell Array}
c(1,1,:)(:)
    ⇒
      {
        [1,1] = Peter
        [2,1] = Hannah
        [3,1] = Robert
      }
c(2,1,:)(:)
    ⇒
      {
        [1,1] = 23
        [2,1] = 16
        [3,1] = 3
      }
```

See also: [cell2struct], page 118, [fieldnames], page 105.

6.2 Cell Arrays

It can be both necessary and convenient to store several variables of different size or type in one variable. A cell array is a container class able to do just that. In general cell arrays work just like N-dimensional arrays with the exception of the use of '{' and '}' as allocation and indexing operators.

6.2.1 Basic Usage of Cell Arrays

As an example, the following code creates a cell array containing a string and a 2-by-2 random matrix

```
c = {"a string", rand(2, 2)};
```

To access the elements of a cell array, it can be indexed with the { and } operators. Thus, the variable created in the previous example can be indexed like this:

```
c{1}
     ⇒ ans = a string
```

As with numerical arrays several elements of a cell array can be extracted by indexing with a vector of indexes

```
c{1:2}
     ⇒ ans = a string
     ⇒ ans =

          0.593993    0.627732
          0.377037    0.033643
```

The indexing operators can also be used to insert or overwrite elements of a cell array. The following code inserts the scalar 3 on the third place of the previously created cell array

```
c{3} = 3
     ⇒ c =

        {
          [1,1] = a string
          [1,2] =

              0.593993    0.627732
              0.377037    0.033643

          [1,3] =   3
        }
```

Details on indexing cell arrays are explained in Section 6.2.3 [Indexing Cell Arrays], page 114.

In general nested cell arrays are displayed hierarchically as in the previous example. In some circumstances it makes sense to reference them by their index, and this can be performed by the `celldisp` function.

celldisp (*c*)   [Function File]
celldisp (*c*, *name*)   [Function File]
> Recursively display the contents of a cell array. By default the values are displayed with the name of the variable *c*. However, this name can be replaced with the variable *name*. For example:
>
> ```
> c = {1, 2, {31, 32}};
> celldisp (c, "b")
>     ⇒
>         b{1} =
>           1
>         b{2} =
>           2
>         b{3}{1} =
>           31
>         b{3}{2} =
>           32
> ```
>
> **See also:** [disp], page 223.

To test if an object is a cell array, use the `iscell` function. For example:

```
iscell (c)
    ⇒ ans = 1

iscell (3)
    ⇒ ans = 0
```

iscell (*x*)   [Built-in Function]
> Return true if *x* is a cell array object.
>
> **See also:** [ismatrix], page 62, [isstruct], page 105, [iscellstr], page 117, [isa], page 39.

6.2.2 Creating Cell Arrays

The introductory example (see Section 6.2.1 [Basic Usage of Cell Arrays], page 110) showed how to create a cell array containing currently available variables. In many situations, however, it is useful to create a cell array and then fill it with data.

The `cell` function returns a cell array of a given size, containing empty matrices. This function is similar to the `zeros` function for creating new numerical arrays. The following example creates a 2-by-2 cell array containing empty matrices

```
c = cell (2,2)
    ⇒ c =

        {
          [1,1] = [](0x0)
          [2,1] = [](0x0)
          [1,2] = [](0x0)
          [2,2] = [](0x0)
        }
```

Just like numerical arrays, cell arrays can be multi-dimensional. The `cell` function accepts any number of positive integers to describe the size of the returned cell array. It is also possible to set the size of the cell array through a vector of positive integers. In the following example two cell arrays of equal size are created, and the size of the first one is displayed

```
c1 = cell (3, 4, 5);
c2 = cell ( [3, 4, 5] );
size (c1)
     ⇒ ans =
        3    4    5
```

As can be seen, the [size], page 44 function also works for cell arrays. As do other functions describing the size of an object, such as [length], page 44, [numel], page 43, [rows], page 43, and [columns], page 43.

cell (*n*)   [Built-in Function]
cell (*m*, *n*)   [Built-in Function]
cell (*m*, *n*, *k*, ...)   [Built-in Function]
cell ([*m n* ...])   [Built-in Function]
    Create a new cell array object.

    If invoked with a single scalar integer argument, return a square NxN cell array. If invoked with two or more scalar integer arguments, or a vector of integer values, return an array with the given dimensions.

    See also: [cellstr], page 117, [mat2cell], page 113, [num2cell], page 112, [struct2cell], page 109.

As an alternative to creating empty cell arrays, and then filling them, it is possible to convert numerical arrays into cell arrays using the `num2cell`, `mat2cell` and `cellslices` functions.

C = num2cell (*A*)   [Built-in Function]
C = num2cell (*A*, *dim*)   [Built-in Function]
    Convert the numeric matrix *A* to a cell array. If *dim* is defined, the value *C* is of dimension 1 in this dimension and the elements of *A* are placed into *C* in slices. For example:

```
num2cell ([1,2;3,4])
   ⇒
     {
        [1,1]  =   1
        [2,1]  =   3
        [1,2]  =   2
        [2,2]  =   4
     }
num2cell ([1,2;3,4],1)
   ⇒
     {
        [1,1]  =
            1
            3
        [1,2]  =
            2
            4
     }
```

See also: [mat2cell], page 113.

C = mat2cell (*A*, *m*, *n*)   [Built-in Function]
C = mat2cell (*A*, *d1*, *d2*, ...)   [Built-in Function]
C = mat2cell (*A*, *r*)   [Built-in Function]

Convert the matrix *A* to a cell array. If *A* is 2-D, then it is required that sum (*m*) == size (*A*, 1) and sum (*n*) == size (*A*, 2). Similarly, if *A* is multi-dimensional and the number of dimensional arguments is equal to the dimensions of *A*, then it is required that sum (*di*) == size (*A*, i).

Given a single dimensional argument *r*, the other dimensional arguments are assumed to equal size (*A*,*i*).

An example of the use of mat2cell is

```
mat2cell (reshape (1:16,4,4), [3,1], [3,1])
   ⇒
{
   [1,1]  =

       1    5    9
       2    6   10
       3    7   11

   [2,1]  =

       4    8   12

   [1,2]  =

      13
```

```
              14
              15

         [2,2] = 16
       }
```

See also: [num2cell], page 112, [cell2mat], page 117.

sl = cellslices (*x*, *lb*, *ub*, *dim*)   [Built-in Function]
  Given an array *x*, this function produces a cell array of slices from the array determined by the index vectors *lb*, *ub*, for lower and upper bounds, respectively. In other
  words, it is equivalent to the following code:

```
      n = length (lb);
      sl = cell (1, n);
      for i = 1:length (lb)
        sl{i} = x(:,...,lb(i):ub(i),...,:);
      endfor
```

  The position of the index is determined by *dim*. If not specified, slicing is done along
  the first non-singleton dimension.

  See also: [cell2mat], page 117, [cellindexmat], page 116, [cellfun], page 464.

6.2.3 Indexing Cell Arrays

As shown in see Section 6.2.1 [Basic Usage of Cell Arrays], page 110 elements can be
extracted from cell arrays using the '{' and '}' operators. If you want to extract or access
subarrays which are still cell arrays, you need to use the '(' and ')' operators. The following
example illustrates the difference:

```
      c = {"1", "2", "3"; "x", "y", "z"; "4", "5", "6"};
      c{2,3}
          ⇒ ans = z

      c(2,3)
          ⇒ ans =
             {
               [1,1] = z
             }
```

So with '{}' you access elements of a cell array, while with '()' you access a sub array of a
cell array.

  Using the '(' and ')' operators, indexing works for cell arrays like for multi-dimensional
arrays. As an example, all the rows of the first and third column of a cell array can be set
to 0 with the following command:

```
c(:, [1, 3]) = {0}
    ⇒   =
        {
          [1,1] = 0
          [2,1] = 0
          [3,1] = 0
          [1,2] = 2
          [2,2] =  10
          [3,2] =  20
          [1,3] = 0
          [2,3] = 0
          [3,3] = 0
        }
```

Note, that the above can also be achieved like this:

```
c(:, [1, 3]) = 0;
```

Here, the scalar '0' is automatically promoted to cell array '{0}' and then assigned to the subarray of c.

To give another example for indexing cell arrays with '()', you can exchange the first and the second row of a cell array as in the following command:

```
c = {1, 2, 3; 4, 5, 6};
c([1, 2], :) = c([2, 1], :)
    ⇒ =
        {
          [1,1] =  4
          [2,1] =  1
          [1,2] =  5
          [2,2] =  2
          [1,3] =  6
          [2,3] =  3
        }
```

Accessing multiple elements of a cell array with the '{' and '}' operators will result in a comma-separated list of all the requested elements (see Section 6.3 [Comma Separated Lists], page 118). Using the '{' and '}' operators the first two rows in the above example can be swapped back like this:

```
[c{[1,2], :}] = deal (c{[2, 1], :})
    ⇒ =
        {
          [1,1] =  1
          [2,1] =  4
          [1,2] =  2
          [2,2] =  5
          [1,3] =  3
          [2,3] =  6
        }
```

As for struct arrays and numerical arrays, the empty matrix '[]' can be used to delete elements from a cell array:

```
x = {"1", "2"; "3", "4"};
x(1, :) = []
     ⇒ x =
        {
            [1,1] = 3
            [1,2] = 4
        }
```

The following example shows how to just remove the contents of cell array elements but not delete the space for them:

```
x = {"1", "2"; "3", "4"};
x{1, :} = []
⇒ x =
        {
            [1,1] = [](0x0)
            [2,1] = 3
            [1,2] = [](0x0)
            [2,2] = 4
        }
```

The indexing operations operate on the cell array and not on the objects within the cell array. By contrast, `cellindexmat` applies matrix indexing to the objects within each cell array entry and returns the requested values.

y = cellindexmat (*x, varargin*)   [Built-in Function]
    Given a cell array of matrices x, this function computes

```
Y = cell (size (X));
for i = 1:numel (X)
  Y{i} = X{i}(varargin{:});
endfor
```

    See also: [cellslices], page 114, [cellfun], page 464.

6.2.4 Cell Arrays of Strings

One common use of cell arrays is to store multiple strings in the same variable. It is also possible to store multiple strings in a character matrix by letting each row be a string. This, however, introduces the problem that all strings must be of equal length. Therefore, it is recommended to use cell arrays to store multiple strings. For cases, where the character matrix representation is required for an operation, there are several functions that convert a cell array of strings to a character array and back. **char** and **strvcat** convert cell arrays to a character array (see Section 5.3.1 [Concatenating Strings], page 68), while the function **cellstr** converts a character array to a cell array of strings:

```
a = ["hello"; "world"];
c = cellstr (a)
    ⇒ c =
       {
          [1,1] = hello
          [2,1] = world
       }
```

cstr = cellstr (*strmat*)   [Built-in Function]

Create a new cell array object from the elements of the string array *strmat*.

Each row of *strmat* becomes an element of *cstr*. Any trailing spaces in a row are deleted before conversion.

To convert back from a cellstr to a character array use **char**.

See also: [cell], page 112, [char], page 69.

One further advantage of using cell arrays to store multiple strings is that most functions for string manipulations included with Octave support this representation. As an example, it is possible to compare one string with many others using the **strcmp** function. If one of the arguments to this function is a string and the other is a cell array of strings, each element of the cell array will be compared to the string argument:

```
c = {"hello", "world"};
strcmp ("hello", c)
    ⇒ ans =
       1   0
```

The following string functions support cell arrays of strings: **char**, **strvcat**, **strcat** (see Section 5.3.1 [Concatenating Strings], page 68), **strcmp**, **strncmp**, **strcmpi**, **strncmpi** (see Section 5.4 [Comparing Strings], page 73), **str2double**, **deblank**, **strtrim**, **strtrunc**, **strfind**, **strmatch**, , **regexp**, **regexpi** (see Section 5.5 [Manipulating Strings], page 75) and **str2double** (see Section 5.6 [String Conversions], page 88).

The function **iscellstr** can be used to test if an object is a cell array of strings.

iscellstr (*cell*)   [Built-in Function]

Return true if every element of the cell array *cell* is a character string.

See also: [ischar], page 66.

6.2.5 Processing Data in Cell Arrays

Data that is stored in a cell array can be processed in several ways depending on the actual data. The simplest way to process that data is to iterate through it using one or more **for** loops. The same idea can be implemented more easily through the use of the **cellfun** function that calls a user-specified function on all elements of a cell array. See [cellfun], page 464.

An alternative is to convert the data to a different container, such as a matrix or a data structure. Depending on the data this is possible using the **cell2mat** and **cell2struct** functions.

m = cell2mat (*c*)   [Function File]
>   Convert the cell array *c* into a matrix by concatenating all elements of *c* into a
>   hyperrectangle. Elements of *c* must be numeric, logical, or char matrices; or cell
>   arrays; or structs; and `cat` must be able to concatenate them together.

>   **See also:** [mat2cell], page 113, [num2cell], page 112.

cell2struct (*cell*, *fields*)   [Built-in Function]
cell2struct (*cell*, *fields*, *dim*)   [Built-in Function]
>   Convert *cell* to a structure. The number of fields in *fields* must match the number of
>   elements in *cell* along dimension *dim*, that is `numel (`*fields*`) == size (`*cell*`, `*dim*`)`.
>   If *dim* is omitted, a value of 1 is assumed.

```
        A = cell2struct ({"Peter", "Hannah", "Robert";
                          185, 170, 168},
                         {"Name","Height"}, 1);
        A(1)
            ⇒
              {
                Name   = Peter
                Height = 185
              }
```

>   **See also:** [struct2cell], page 109, [cell2mat], page 117, [struct], page 104.

6.3 Comma Separated Lists

Comma separated lists[1] are the basic argument type to all Octave functions - both for input
and return arguments. In the example

```
    max (a, b)
```

'`a, b`' is a comma separated list. Comma separated lists can appear on both the right and
left hand side of an assignment. For example

```
    x = [1 0 1 0 0 1 1; 0 0 0 0 0 0 7];
    [i, j] = find (x, 2, "last");
```

Here, '`x, 2, "last"`' is a comma separated list constituting the input arguments of `find`.
`find` returns a comma separated list of output arguments which is assigned element by
element to the comma separated list '`i, j`'.

Another example of where comma separated lists are used is in the creation of a new
array with `[]` (see Section 4.1 [Matrices], page 48) or the creation of a cell array with `{}`
(see Section 6.2.1 [Basic Usage of Cell Arrays], page 110). In the expressions

```
    a = [1, 2, 3, 4];
    c = {4, 5, 6, 7};
```

both '`1, 2, 3, 4`' and '`4, 5, 6, 7`' are comma separated lists.

Comma separated lists cannot be directly manipulated by the user. However, both
structure arrays and cell arrays can be converted into comma separated lists, and thus used

---

[1] Comma-separated lists are also sometimes informally referred to as *cs-lists*.

in place of explicitly written comma separated lists. This feature is useful in many ways, as will be shown in the following subsections.

6.3.1 Comma Separated Lists Generated from Cell Arrays

As has been mentioned above (see Section 6.2.3 [Indexing Cell Arrays], page 114), elements of a cell array can be extracted into a comma separated list with the { and } operators. By surrounding this list with [and], it can be concatenated into an array. For example:

```
a = {1, [2, 3], 4, 5, 6};
b = [a{1:4}]
    ⇒ b =
          1   2   3   4   5
```

Similarly, it is possible to create a new cell array containing cell elements selected with {}. By surrounding the list with '{' and '}' a new cell array will be created, as the following example illustrates:

```
a = {1, rand(2, 2), "three"};
b = { a{ [1, 3] } }
    ⇒ b =
        {
          [1,1] =  1
          [1,2] = three
        }
```

Furthermore, cell elements (accessed by {}) can be passed directly to a function. The list of elements from the cell array will be passed as an argument list to a given function as if it is called with the elements as individual arguments. The two calls to `printf` in the following example are identical but the latter is simpler and can handle cell arrays of an arbitrary size:

```
c = {"GNU", "Octave", "is", "Free", "Software"};
printf ("%s ", c{1}, c{2}, c{3}, c{4}, c{5});
      ⊣ GNU Octave is Free Software
printf ("%s ", c{:});
      ⊣ GNU Octave is Free Software
```

If used on the left-hand side of an assignment, a comma separated list generated with {} can be assigned to. An example is

```
in{1} = [10, 20, 30, 40, 50, 60, 70, 80, 90];
in{2} = inf;
in{3} = "last";
in{4} = "first";
out = cell (4, 1);
[out{1:3}] = find (in{1 : 3});
[out{4:6}] = find (in{[1, 2, 4]})
    ⇒ out =
      {
        [1,1] = 1
        [2,1] = 9
        [3,1] = 90
        [4,1] = 1
        [3,1] = 1
        [4,1] = 10
      }
```

6.3.2 Comma Separated Lists Generated from Structure Arrays

Structure arrays can equally be used to create comma separated lists. This is done by addressing one of the fields of a structure array. For example:

```
x = ceil (randn (10, 1));
in = struct ("call1", {x, 3, "last"},
             "call2", {x, inf, "first"});
out = struct ("call1", cell (2, 1), "call2", cell (2, 1));
[out.call1] = find (in.call1);
[out.call2] = find (in.call2);
```

7 Variables

Variables let you give names to values and refer to them later. You have already seen variables in many of the examples. The name of a variable must be a sequence of letters, digits and underscores, but it may not begin with a digit. Octave does not enforce a limit on the length of variable names, but it is seldom useful to have variables with names longer than about 30 characters. The following are all valid variable names

```
x
x15
__foo_bar_baz__
fucnrdthsucngtagdjb
```

However, names like `__foo_bar_baz__` that begin and end with two underscores are understood to be reserved for internal use by Octave. You should not use them in code you write, except to access Octave's documented internal variables and built-in symbolic constants.

Case is significant in variable names. The symbols `a` and `A` are distinct variables.

A variable name is a valid expression by itself. It represents the variable's current value. Variables are given new values with *assignment operators* and *increment operators*. See Section 8.6 [Assignment Expressions], page 145.

There is one built-in variable with a special meaning. The `ans` variable always contains the result of the last computation, where the output wasn't assigned to any variable. The code `a = cos (pi)` will assign the value -1 to the variable `a`, but will not change the value of `ans`. However, the code `cos (pi)` will set the value of `ans` to -1.

Variables in Octave do not have fixed types, so it is possible to first store a numeric value in a variable and then to later use the same name to hold a string value in the same program. Variables may not be used before they have been given a value. Doing so results in an error.

`ans`   [Automatic Variable]
> The most recently computed result that was not explicitly assigned to a variable. For example, after the expression
>
> ```
> 3^2 + 4^2
> ```
>
> is evaluated, the value returned by `ans` is 25.

`isvarname (name)`   [Built-in Function]
> Return true if *name* is a valid variable name.
>
> **See also:** [iskeyword], page 855, [exist], page 128, [who], page 126.

`varname = genvarname (str)`   [Function File]
`varname = genvarname (str, exclusions)`   [Function File]
> Create unique variable(s) from *str*. If *exclusions* is given, then the variable(s) will be unique to each other and to *exclusions* (*exclusions* may be either a string or a cellstr).
>
> If *str* is a cellstr, then a unique variable is created for each cell in *str*.
>
> ```
> x = 3.141;
> genvarname ("x", who ())
>    ⇒ x1
> ```
>
> If *wanted* is a cell array, genvarname will make sure the returned strings are distinct:

```
genvarname ({"foo", "foo"})
   ⇒
      {
        [1,1] = foo
        [1,2] = foo1
      }
```

Note that the result is a char array/cell array of strings, not the variables themselves. To define a variable, eval() can be used. The following trivial example sets x to 42.

```
name = genvarname ("x");
eval ([name " = 42"]);
   ⇒ x =  42
```

Also, this can be useful for creating unique struct field names.

```
x = struct ();
for i = 1:3
  x.(genvarname ("a", fieldnames (x))) = i;
endfor
   ⇒ x =
      {
        a  = 1
        a1 = 2
        a2 = 3
      }
```

Since variable names may only contain letters, digits and underscores, genvarname replaces any sequence of disallowed characters with an underscore. Also, variables may not begin with a digit; in this case an underscore is added before the variable name.

Variable names beginning and ending with two underscores "__" are valid but they are used internally by octave and should generally be avoided, therefore genvarname will not generate such names.

genvarname will also make sure that returned names do not clash with keywords such as "for" and "if". A number will be appended if necessary. Note, however, that this does **not** include function names, such as "sin". Such names should be included in *avoid* if necessary.

See also: [isvarname], page 121, [exist], page 128, [tmpnam], page 255, [eval], page 151.

namelengthmax ()   [Function File]
Return the MATLAB compatible maximum variable name length. Octave is capable of storing strings up to $2^{31}-1$ in length. However for MATLAB compatibility all variable, function, and structure field names should be shorter than the length supplied by namelengthmax. In particular variables stored to a MATLAB file format will have their names truncated to this length.

7.1 Global Variables

A variable that has been declared *global* may be accessed from within a function body without having to pass it as a formal parameter.

A variable may be declared global using a `global` declaration statement. The following statements are all global declarations.

```
global a
global a b
global c = 2
global d = 3 e f = 5
```

A global variable may only be initialized once in a `global` statement. For example, after executing the following code

```
global gvar = 1
global gvar = 2
```

the value of the global variable `gvar` is 1, not 2. Issuing a 'clear gvar' command does not change the above behavior, but 'clear all' does.

It is necessary declare a variable as global within a function body in order to access it. For example,

```
global x
function f ()
   x = 1;
endfunction
f ()
```

does *not* set the value of the global variable `x` to 1. In order to change the value of the global variable `x`, you must also declare it to be global within the function body, like this

```
function f ()
   global x;
   x = 1;
endfunction
```

Passing a global variable in a function parameter list will make a local copy and not modify the global value. For example, given the function

```
function f (x)
   x = 0
endfunction
```

and the definition of `x` as a global variable at the top level,

```
global x = 13
```

the expression

```
f (x)
```

will display the value of `x` from inside the function as 0, but the value of `x` at the top level remains unchanged, because the function works with a *copy* of its argument.

`isglobal (name)`   [Built-in Function]
    Return true if *name* is a globally visible variable. For example:

```
global x
isglobal ("x")
    ⇒ 1
```

See also: [isvarname], page 121, [exist], page 128.

7.2 Persistent Variables

A variable that has been declared *persistent* within a function will retain its contents in memory between subsequent calls to the same function. The difference between persistent variables and global variables is that persistent variables are local in scope to a particular function and are not visible elsewhere.

The following example uses a persistent variable to create a function that prints the number of times it has been called.

```
function count_calls ()
  persistent calls = 0;
  printf ("'count_calls' has been called %d times\n",
          ++calls);
endfunction

for i = 1:3
  count_calls ();
endfor

⊣ 'count_calls' has been called 1 times
⊣ 'count_calls' has been called 2 times
⊣ 'count_calls' has been called 3 times
```

As the example shows, a variable may be declared persistent using a `persistent` declaration statement. The following statements are all persistent declarations.

```
persistent a
persistent a b
persistent c = 2
persistent d = 3 e f = 5
```

The behavior of persistent variables is equivalent to the behavior of static variables in C. The command `static` in Octave is also recognized and is equivalent to `persistent`.

Like global variables, a persistent variable may only be initialized once. For example, after executing the following code

```
persistent pvar = 1
persistent pvar = 2
```

the value of the persistent variable `pvar` is 1, not 2.

If a persistent variable is declared but not initialized to a specific value, it will contain an empty matrix. So, it is also possible to initialize a persistent variable by checking whether it is empty, as the following example illustrates.

```
function count_calls ()
  persistent calls;
  if (isempty (calls))
    calls = 0;
  endif
  printf ("'count_calls' has been called %d times\n",
          ++calls);
endfunction
```

This implementation behaves in exactly the same way as the previous implementation of
`count_calls`.

The value of a persistent variable is kept in memory until it is explicitly cleared. Assuming that the implementation of `count_calls` is saved on disk, we get the following behavior.

```
for i = 1:2
  count_calls ();
endfor
⊣ 'count_calls' has been called 1 times
⊣ 'count_calls' has been called 2 times

clear
for i = 1:2
  count_calls ();
endfor
⊣ 'count_calls' has been called 3 times
⊣ 'count_calls' has been called 4 times

clear all
for i = 1:2
  count_calls ();
endfor
⊣ 'count_calls' has been called 1 times
⊣ 'count_calls' has been called 2 times

clear count_calls
for i = 1:2
  count_calls ();
endfor
⊣ 'count_calls' has been called 1 times
⊣ 'count_calls' has been called 2 times
```

That is, the persistent variable is only removed from memory when the function containing the variable is removed. Note that if the function definition is typed directly into the Octave prompt, the persistent variable will be cleared by a simple `clear` command as the entire function definition will be removed from memory. If you do not want a persistent variable to be removed from memory even if the function is cleared, you should use the `mlock` function (see Section 11.9.6 [Function Locking], page 187).

7.3 Status of Variables

When creating simple one-shot programs it can be very convenient to see which variables are available at the prompt. The function who and its siblings whos and whos_line_format will show different information about what is in memory, as the following shows.

```
str = "A random string";
who -variables
    ⊣ *** local user variables:
    ⊣
    ⊣ __nargin__  str
```

who	[Command]
who *pattern* ...	[Command]
who *option pattern* ...	[Command]
C = who ("*pattern*", ...)	[Command]

List currently defined variables matching the given patterns. Valid pattern syntax is the same as described for the clear command. If no patterns are supplied, all variables are listed. By default, only variables visible in the local scope are displayed.

The following are valid options but may not be combined.

global   List variables in the global scope rather than the current scope.

-regexp   The patterns are considered to be regular expressions when matching the variables to display. The same pattern syntax accepted by the regexp function is used.

-file   The next argument is treated as a filename. All variables found within the specified file are listed. No patterns are accepted when reading variables from a file.

If called as a function, return a cell array of defined variable names matching the given patterns.

See also: [whos], page 126, [isglobal], page 123, [isvarname], page 121, [exist], page 128, [regexp], page 84.

whos	[Command]
whos *pattern* ...	[Command]
whos *option pattern* ...	[Command]
S = whos ("*pattern*", ...)	[Command]

Provide detailed information on currently defined variables matching the given patterns. Options and pattern syntax are the same as for the who command. Extended information about each variable is summarized in a table with the following default entries.

Attr   Attributes of the listed variable. Possible attributes are:

blank   Variable in local scope

a   Automatic variable. An automatic variable is one created by the interpreter, for example argn.

c   Variable of complex type.

f	Formal parameter (function argument).
g	Variable with global scope.
p	Persistent variable.

Name   The name of the variable.

Size   The logical size of the variable. A scalar is 1x1, a vector is 1xN or Nx1, a 2-D matrix is MxN.

Bytes   The amount of memory currently used to store the variable.

Class   The class of the variable. Examples include double, single, char, uint16, cell, and struct.

The table can be customized to display more or less information through the function `whos_line_format`.

If **whos** is called as a function, return a struct array of defined variable names matching the given patterns. Fields in the structure describing each variable are: name, size, bytes, class, global, sparse, complex, nesting, persistent.

See also: [who], page 126, [whos_line_format], page 127.

`val = whos_line_format ()`   [Built-in Function]
`old_val = whos_line_format (`*new_val*`)`   [Built-in Function]
`whos_line_format (`*new_val*`, "`*local*`")`   [Built-in Function]

Query or set the format string used by the command **whos**.

A full format string is:

```
%[modifier]<command>[:width[:left-min[:balance]]];
```

The following command sequences are available:

%a   Prints attributes of variables (g=global, p=persistent, f=formal parameter, a=automatic variable).

%b   Prints number of bytes occupied by variables.

%c   Prints class names of variables.

%e   Prints elements held by variables.

%n   Prints variable names.

%s   Prints dimensions of variables.

%t   Prints type names of variables.

Every command may also have an alignment modifier:

l   Left alignment.

r   Right alignment (default).

c   Column-aligned (only applicable to command %s).

The `width` parameter is a positive integer specifying the minimum number of columns used for printing. No maximum is needed as the field will auto-expand as required.

The parameters `left-min` and `balance` are only available when the column-aligned modifier is used with the command '`%s`'. `balance` specifies the column number within the field width which will be aligned between entries. Numbering starts from 0 which indicates the leftmost column. `left-min` specifies the minimum field width to the left of the specified balance column.

The default format is `" %a:4; %ln:6; %cs:16:6:1; %rb:12; %lc:-1;\n"`.

When called from inside a function with the `"local"` option, the variable is changed locally for the function and any subroutines it calls. The original variable value is restored when exiting the function.

See also: [whos], page 126.

Instead of displaying which variables are in memory, it is possible to determine if a given variable is available. That way it is possible to alter the behavior of a program depending on the existence of a variable. The following example illustrates this.

```
if (! exist ("meaning", "var"))
  disp ("The program has no 'meaning'");
endif
```

`exist (`*name*`, `*type*`)`   [Built-in Function]

Return 1 if the name exists as a variable, 2 if the name is an absolute file name, an ordinary file in Octave's `path`, or (after appending '`.m`') a function file in Octave's `path`, 3 if the name is a '`.oct`' or '`.mex`' file in Octave's `path`, 5 if the name is a built-in function, 7 if the name is a directory, or 103 if the name is a function not associated with a file (entered on the command line).

Otherwise, return 0.

This function also returns 2 if a regular file called *name* exists in Octave's search path. If you want information about other types of files, you should use some combination of the functions `file_in_path` and `stat` instead.

If the optional argument *type* is supplied, check only for symbols of the specified type. Valid types are

`"var"`   Check only for variables.

`"builtin"`
                Check only for built-in functions.

`"file"`   Check only for files and directories.

`"dir"`   Check only for directories.

See also: [file_in_loadpath], page 182, [file_in_path], page 714, [find_dir_in_path], page 182, [stat], page 710.

Usually Octave will manage the memory, but sometimes it can be practical to remove variables from memory manually. This is usually needed when working with large variables that fill a substantial part of the memory. On a computer that uses the IEEE floating point format, the following program allocates a matrix that requires around 128 MB memory.

```
large_matrix = zeros (4000, 4000);
```

Since having this variable in memory might slow down other computations, it can be necessary to remove it manually from memory. The `clear` function allows this.

clear [*options*] *pattern* ...   [Command]

> Delete the names matching the given patterns from the symbol table. The pattern may contain the following special characters:
>
> ?   Match any single character.
>
> *   Match zero or more characters.
>
> [*list*]   Match the list of characters specified by *list*. If the first character is ! or ^, match all characters except those specified by *list*. For example, the pattern '[a-zA-Z]' will match all lowercase and uppercase alphabetic characters.
>
> For example, the command
>
> ```
>     clear foo b*r
> ```
>
> clears the name `foo` and all names that begin with the letter `b` and end with the letter `r`.
>
> If `clear` is called without any arguments, all user-defined variables (local and global) are cleared from the symbol table. If `clear` is called with at least one argument, only the visible names matching the arguments are cleared. For example, suppose you have defined a function `foo`, and then hidden it by performing the assignment `foo = 2`. Executing the command *clear foo* once will clear the variable definition and restore the definition of `foo` as a function. Executing *clear foo* a second time will clear the function definition.
>
> The following options are available in both long and short form
>
> -all, -a   Clears all local and global user-defined variables and all functions from the symbol table.
>
> -exclusive, -x
> > Clears the variables that don't match the following pattern.
>
> -functions, -f
> > Clears the function names and the built-in symbols names.
>
> -global, -g
> > Clears the global symbol names.
>
> -variables, -v
> > Clears the local variable names.
>
> -classes, -c
> > Clears the class structure table and clears all objects.
>
> -regexp, -r
> > The arguments are treated as regular expressions as any variables that match will be cleared.
>
> With the exception of **exclusive**, all long options can be used without the dash as well.

pack ()   [Function File]
> Consolidate workspace memory in MATLAB. This function is provided for compatibility, but does nothing in Octave.

Information about a function or variable such as its location in the file system can also be acquired from within Octave. This is usually only useful during development of programs, and not within a program.

type *name* ...   [Command]
type *-q name* ...   [Command]
text = type ("*name*", ...)   [Function File]
> Display the contents of *name* which may be a file, function (m-file), variable, operator, or keyword.
>
> type normally prepends a header line describing the category of *name* such as function or variable; The '-q' option suppresses this behavior.
>
> If no output variable is used the contents are displayed on screen. Otherwise, a cell array of strings is returned, where each element corresponds to the contents of each requested function.

which *name* ...   [Command]
> Display the type of each *name*. If *name* is defined from a function file, the full name of the file is also displayed.
>
> **See also:** [help], page 20, [lookfor], page 21.

what   [Command]
what *dir*   [Command]
w = what (*dir*)   [Function File]
> List the Octave specific files in directory *dir*. If *dir* is not specified then the current directory is used. If a return argument is requested, the files found are returned in the structure *w*.
>
> **See also:** [which], page 130.

8 Expressions

Expressions are the basic building block of statements in Octave. An expression evaluates to a value, which you can print, test, store in a variable, pass to a function, or assign a new value to a variable with an assignment operator.

An expression can serve as a statement on its own. Most other kinds of statements contain one or more expressions which specify data to be operated on. As in other languages, expressions in Octave include variables, array references, constants, and function calls, as well as combinations of these with various operators.

8.1 Index Expressions

An *index expression* allows you to reference or extract selected elements of a matrix or vector.

Indices may be scalars, vectors, ranges, or the special operator ':', which may be used to select entire rows or columns.

Vectors are indexed using a single index expression. Matrices (2-D) and higher multi-dimensional arrays are indexed using either one index or N indices where N is the dimension of the array. When using a single index expression to index 2-D or higher data the elements of the array are taken in column-first order (like Fortran).

The output from indexing assumes the dimensions of the index expression. For example:

```
a(2)       # result is a scalar
a(1:2)     # result is a row vector
a([1; 2])  # result is a column vector
```

As a special case, when a colon is used as a single index, the output is a column vector containing all the elements of the vector or matrix. For example:

```
a(:)       # result is a column vector
a(:)'      # result is a row vector
```

The above two code idioms are often used in place of **reshape** when a simple vector, rather than an arbitrarily sized array, is needed.

Given the matrix

```
a = [1, 2; 3, 4]
```

all of the following expressions are equivalent and select the first row of the matrix.

```
a(1, [1, 2])  # row 1, columns 1 and 2
a(1, 1:2)     # row 1, columns in range 1-2
a(1, :)       # row 1, all columns
```

In index expressions the keyword **end** automatically refers to the last entry for a particular dimension. This magic index can also be used in ranges and typically eliminates the needs to call **size** or **length** to gather array bounds before indexing. For example:

```
a = [1, 2, 3, 4];

a(1:end/2)          # first half of a => [1, 2]
a(end + 1) = 5;     # append element
a(end) = [];        # delete element
a(1:2:end)          # odd elements of a => [1, 3]
a(2:2:end)          # even elements of a => [2, 4]
a(end:-1:1)         # reversal of a => [4, 3, 2 , 1]
```

8.1.1 Advanced Indexing

An array with 'n' dimensions can be indexed using 'm' indices. More generally, the set of index tuples determining the result is formed by the Cartesian product of the index vectors (or ranges or scalars).

For the ordinary and most common case, m == n, and each index corresponds to its respective dimension. If m < n and every index is less than the size of the array in the ith dimension, m(i) < n(i), then the index expression is padded with trailing singleton dimensions ([ones (m-n, 1)]). If m < n but one of the indices m(i) is outside the size of the current array, then the last n-m+1 dimensions are folded into a single dimension with an extent equal to the product of extents of the original dimensions. This is easiest to understand with an example.

```
a = reshape (1:8, 2, 2, 2)  # Create 3-D array
a =

ans(:,:,1) =

   1   3
   2   4

ans(:,:,2) =

   5   7
   6   8

a(2,1,2);   # Case (m == n): ans = 6
a(2,1);     # Case (m < n), idx within array:
            # equivalent to a(2,1,1), ans = 2
a(2,4);     # Case (m < n), idx outside array:
            # Dimension 2 & 3 folded into new dimension of size 2x2 = 4
            # Select 2nd row, 4th element of [2, 4, 6, 8], ans = 8
```

One advanced use of indexing is to create arrays filled with a single value. This can be done by using an index of ones on a scalar value. The result is an object with the dimensions of the index expression and every element equal to the original scalar. For example, the following statements

```
a = 13;
a(ones (1, 4))
```

produce a vector whose four elements are all equal to 13.

Similarly, by indexing a scalar with two vectors of ones it is possible to create a matrix. The following statements

```
a = 13;
a(ones (1, 2), ones (1, 3))
```

create a 2x3 matrix with all elements equal to 13.

The last example could also be written as

```
13(ones (2, 3))
```

It is more efficient to use indexing rather than the code construction `scalar * ones (N, M, ...)` because it avoids the unnecessary multiplication operation. Moreover, multiplication may not be defined for the object to be replicated whereas indexing an array is always defined. The following code shows how to create a 2x3 cell array from a base unit which is not itself a scalar.

```
{"Hello"}(ones (2, 3))
```

It should be, noted that `ones (1, n)` (a row vector of ones) results in a range (with zero increment). A range is stored internally as a starting value, increment, end value, and total number of values; hence, it is more efficient for storage than a vector or matrix of ones whenever the number of elements is greater than 4. In particular, when 'r' is a row vector, the expressions

```
r(ones (1, n), :)

r(ones (n, 1), :)
```

will produce identical results, but the first one will be significantly faster, at least for 'r' and 'n' large enough. In the first case the index is held in compressed form as a range which allows Octave to choose a more efficient algorithm to handle the expression.

A general recommendation, for a user unaware of these subtleties, is to use the function `repmat` for replicating smaller arrays into bigger ones.

A second use of indexing is to speed up code. Indexing is a fast operation and judicious use of it can reduce the requirement for looping over individual array elements which is a slow operation.

Consider the following example which creates a 10-element row vector a containing the values $a_i = \sqrt{i}$.

```
for i = 1:10
  a(i) = sqrt (i);
endfor
```

It is quite inefficient to create a vector using a loop like this. In this case, it would have been much more efficient to use the expression

```
a = sqrt (1:10);
```

which avoids the loop entirely.

In cases where a loop cannot be avoided, or a number of values must be combined to form a larger matrix, it is generally faster to set the size of the matrix first (pre-allocate storage), and then insert elements using indexing commands. For example, given a matrix `a`,

```
[nr, nc] = size (a);
x = zeros (nr, n * nc);
for i = 1:n
  x(:,(i-1)*nc+1:i*nc) = a;
endfor
```

is considerably faster than

```
x = a;
for i = 1:n-1
  x = [x, a];
endfor
```

because Octave does not have to repeatedly resize the intermediate result.

ind = sub2ind (*dims*, *i*, *j*)   [Function File]
ind = sub2ind (*dims*, *s1*, *s2*, ..., *sN*)   [Function File]
    Convert subscripts to a linear index.

    The following example shows how to convert the two-dimensional index (2,3) of a
    3-by-3 matrix to a linear index. The matrix is linearly indexed moving from one
    column to next, filling up all rows in each column.

```
            linear_index = sub2ind ([3, 3], 2, 3)
            ⇒ 8
```

    See also: [ind2sub], page 134.

[*s1*, *s2*, ..., *sN*] = ind2sub (*dims*, *ind*)   [Function File]
    Convert a linear index to subscripts.

    The following example shows how to convert the linear index 8 in a 3-by-3 matrix
    into a subscript. The matrix is linearly indexed moving from one column to next,
    filling up all rows in each column.

```
            [r, c] = ind2sub ([3, 3], 8)
                ⇒ r =  2
                ⇒ c =  3
```

    See also: [sub2ind], page 134.

isindex (*ind*)   [Built-in Function]
isindex (*ind*, *n*)   [Built-in Function]
    Return true if *ind* is a valid index. Valid indices are either positive integers (although
    possibly of real data type), or logical arrays. If present, *n* specifies the maximum
    extent of the dimension to be indexed. When possible the internal result is cached so
    that subsequent indexing using *ind* will not perform the check again.

val = allow_noninteger_range_as_index ()   [Built-in Function]
old_val = allow_noninteger_range_as_index (*new_val*)   [Built-in Function]
allow_noninteger_range_as_index (*new_val*, "*local*")   [Built-in Function]
    Query or set the internal variable that controls whether non-integer ranges are allowed
    as indices. This might be useful for MATLAB compatibility; however, it is still not
    entirely compatible because MATLAB treats the range expression differently in different
    contexts.

When called from inside a function with the `"local"` option, the variable is changed locally for the function and any subroutines it calls. The original variable value is restored when exiting the function.

8.2 Calling Functions

A *function* is a name for a particular calculation. Because it has a name, you can ask for it by name at any point in the program. For example, the function `sqrt` computes the square root of a number.

A fixed set of functions are *built-in*, which means they are available in every Octave program. The `sqrt` function is one of these. In addition, you can define your own functions. See Chapter 11 [Functions and Scripts], page 167, for information about how to do this.

The way to use a function is with a *function call* expression, which consists of the function name followed by a list of *arguments* in parentheses. The arguments are expressions which give the raw materials for the calculation that the function will do. When there is more than one argument, they are separated by commas. If there are no arguments, you can omit the parentheses, but it is a good idea to include them anyway, to clearly indicate that a function call was intended. Here are some examples:

```
sqrt (x^2 + y^2)     # One argument
ones (n, m)          # Two arguments
rand ()              # No arguments
```

Each function expects a particular number of arguments. For example, the `sqrt` function must be called with a single argument, the number to take the square root of:

```
sqrt (argument)
```

Some of the built-in functions take a variable number of arguments, depending on the particular usage, and their behavior is different depending on the number of arguments supplied.

Like every other expression, the function call has a value, which is computed by the function based on the arguments you give it. In this example, the value of `sqrt (argument)` is the square root of the argument. A function can also have side effects, such as assigning the values of certain variables or doing input or output operations.

Unlike most languages, functions in Octave may return multiple values. For example, the following statement

```
[u, s, v] = svd (a)
```

computes the singular value decomposition of the matrix `a` and assigns the three result matrices to `u`, `s`, and `v`.

The left side of a multiple assignment expression is itself a list of expressions, and is allowed to be a list of variable names or index expressions. See also Section 8.1 [Index Expressions], page 131, and Section 8.6 [Assignment Ops], page 145.

8.2.1 Call by Value

In Octave, unlike Fortran, function arguments are passed by value, which means that each argument in a function call is evaluated and assigned to a temporary location in memory before being passed to the function. There is currently no way to specify that a function parameter should be passed by reference instead of by value. This means that it is impossible to directly alter the value of a function parameter in the calling function. It can only change the local copy within the function body. For example, the function

```
function f (x, n)
  while (n-- > 0)
    disp (x);
  endwhile
endfunction
```

displays the value of the first argument n times. In this function, the variable n is used as a temporary variable without having to worry that its value might also change in the calling function. Call by value is also useful because it is always possible to pass constants for any function parameter without first having to determine that the function will not attempt to modify the parameter.

The caller may use a variable as the expression for the argument, but the called function does not know this: it only knows what value the argument had. For example, given a function called as

```
foo = "bar";
fcn (foo)
```

you should not think of the argument as being "the variable foo." Instead, think of the argument as the string value, "bar".

Even though Octave uses pass-by-value semantics for function arguments, values are not copied unnecessarily. For example,

```
x = rand (1000);
f (x);
```

does not actually force two 1000 by 1000 element matrices to exist *unless* the function f modifies the value of its argument. Then Octave must create a copy to avoid changing the value outside the scope of the function f, or attempting (and probably failing!) to modify the value of a constant or the value of a temporary result.

8.2.2 Recursion

With some restrictions[1], recursive function calls are allowed. A *recursive function* is one which calls itself, either directly or indirectly. For example, here is an inefficient[2] way to compute the factorial of a given integer:

```
function retval = fact (n)
  if (n > 0)
    retval = n * fact (n-1);
  else
    retval = 1;
  endif
endfunction
```

---

[1] Some of Octave's functions are implemented in terms of functions that cannot be called recursively. For example, the ODE solver lsode is ultimately implemented in a Fortran subroutine that cannot be called recursively, so lsode should not be called either directly or indirectly from within the user-supplied function that lsode requires. Doing so will result in an error.

[2] It would be much better to use prod (1:n), or gamma (n+1) instead, after first checking to ensure that the value n is actually a positive integer.

This function is recursive because it calls itself directly. It eventually terminates because each time it calls itself, it uses an argument that is one less than was used for the previous call. Once the argument is no longer greater than zero, it does not call itself, and the recursion ends.

The built-in variable `max_recursion_depth` specifies a limit to the recursion depth and prevents Octave from recursing infinitely.

`val = max_recursion_depth ()`	[Built-in Function]
`old_val = max_recursion_depth (new_val)`	[Built-in Function]
`max_recursion_depth (new_val, "local")`	[Built-in Function]

> Query or set the internal limit on the number of times a function may be called recursively. If the limit is exceeded, an error message is printed and control returns to the top level.
>
> When called from inside a function with the `"local"` option, the variable is changed locally for the function and any subroutines it calls. The original variable value is restored when exiting the function.

8.3 Arithmetic Operators

The following arithmetic operators are available, and work on scalars and matrices. The element-by-element operators and functions broadcast (see Section 19.2 [Broadcasting], page 459).

$x + y$ Addition. If both operands are matrices, the number of rows and columns must both agree, or they must be broadcastable to the same shape.

$x .+ y$ Element-by-element addition. This operator is equivalent to +.

$x - y$ Subtraction. If both operands are matrices, the number of rows and columns of both must agree, or they must be broadcastable to the same shape.

$x .- y$ Element-by-element subtraction. This operator is equivalent to -.

$x * y$ Matrix multiplication. The number of columns of x must agree with the number of rows of y, or they must be broadcastable to the same shape.

$x .* y$ Element-by-element multiplication. If both operands are matrices, the number of rows and columns must both agree, or they must be broadcastable to the same shape.

x / y Right division. This is conceptually equivalent to the expression

```
(inverse (y') * x')'
```

but it is computed without forming the inverse of y'.

If the system is not square, or if the coefficient matrix is singular, a minimum norm solution is computed.

$x ./ y$ Element-by-element right division.

$x \backslash y$ Left division. This is conceptually equivalent to the expression

```
inverse (x) * y
```

but it is computed without forming the inverse of x.

If the system is not square, or if the coefficient matrix is singular, a minimum norm solution is computed.

$x\, .\backslash\, y$ Element-by-element left division. Each element of y is divided by each corresponding element of x.

$x\,\hat{}\, y$
$x\, **\, y$ Power operator. If x and y are both scalars, this operator returns x raised to the power y. If x is a scalar and y is a square matrix, the result is computed using an eigenvalue expansion. If x is a square matrix, the result is computed by repeated multiplication if y is an integer, and by an eigenvalue expansion if y is not an integer. An error results if both x and y are matrices.

The implementation of this operator needs to be improved.

$x\, .\hat{}\, y$
$x\, .**\, y$ Element-by-element power operator. If both operands are matrices, the number of rows and columns must both agree, or they must be broadcastable to the same shape. If several complex results are possible, the one with smallest non-negative argument (angle) is taken. This rule may return a complex root even when a real root is also possible. Use `realpow`, `realsqrt`, `cbrt`, or `nthroot` if a real result is preferred.

$-x$ Negation.

$+x$ Unary plus. This operator has no effect on the operand.

x' Complex conjugate transpose. For real arguments, this operator is the same as the transpose operator. For complex arguments, this operator is equivalent to the expression

```
conj (x.')
```

$x.'$ Transpose.

Note that because Octave's element-by-element operators begin with a '.', there is a possible ambiguity for statements like

```
1./m
```

because the period could be interpreted either as part of the constant or as part of the operator. To resolve this conflict, Octave treats the expression as if you had typed

```
(1) ./ m
```

and not

```
(1.) / m
```

Although this is inconsistent with the normal behavior of Octave's lexer, which usually prefers to break the input into tokens by preferring the longest possible match at any given point, it is more useful in this case.

`ctranspose (x)`   [Built-in Function]
    Return the complex conjugate transpose of x. This function and `x'` are equivalent.

    See also: [transpose], page 140.

ldivide (*x*, *y*)   [Built-in Function]

    Return the element-by-element left division of *x* and *y*. This function and x .\ y are equivalent.

    See also: [rdivide], page 140, [mldivide], page 139, [times], page 140, [plus], page 139.

minus (*x*, *y*)   [Built-in Function]

    This function and x - y are equivalent.

    See also: [plus], page 139, [uminus], page 140.

mldivide (*x*, *y*)   [Built-in Function]

    Return the matrix left division of *x* and *y*. This function and x \ y are equivalent.

    See also: [mrdivide], page 139, [ldivide], page 139, [rdivide], page 140.

mpower (*x*, *y*)   [Built-in Function]

    Return the matrix power operation of *x* raised to the *y* power. This function and x ^ y are equivalent.

    See also: [power], page 139, [mtimes], page 139, [plus], page 139, [minus], page 139.

mrdivide (*x*, *y*)   [Built-in Function]

    Return the matrix right division of *x* and *y*. This function and x / y are equivalent.

    See also: [mldivide], page 139, [rdivide], page 140, [plus], page 139, [minus], page 139.

mtimes (*x*, *y*)   [Built-in Function]
mtimes (*x1*, *x2*, ...)   [Built-in Function]

    Return the matrix multiplication product of inputs. This function and x * y are equivalent. If more arguments are given, the multiplication is applied cumulatively from left to right:

        (...((x1 * x2) * x3) * ...)

    At least one argument is required.

    See also: [times], page 140, [plus], page 139, [minus], page 139, [rdivide], page 140, [mrdivide], page 139, [mldivide], page 139, [mpower], page 139.

plus (*x*, *y*)   [Built-in Function]
plus (*x1*, *x2*, ...)   [Built-in Function]

    This function and x + y are equivalent. If more arguments are given, the summation is applied cumulatively from left to right:

        (...((x1 + x2) + x3) + ...)

    At least one argument is required.

    See also: [minus], page 139, [uplus], page 140.

power (*x*, *y*)   [Built-in Function]

    Return the element-by-element operation of *x* raised to the *y* power. If several complex results are possible, returns the one with smallest non-negative argument (angle). Use realpow, realsqrt, cbrt, or nthroot if a real result is preferred.

    This function and x .^ y are equivalent.

    See also: [mpower], page 139, [realpow], page 408, [realsqrt], page 408, [cbrt], page 408, [nthroot], page 408.

`rdivide (x, y)`   [Built-in Function]
>   Return the element-by-element right division of x and y. This function and `x ./ y`
>   are equivalent.
>
>   **See also:** [ldivide], page 139, [mrdivide], page 139, [times], page 140, [plus], page 139.

`times (x, y)`   [Built-in Function]
`times (x1, x2, ...)`   [Built-in Function]
>   Return the element-by-element multiplication product of inputs. This function and
>   `x .* y` are equivalent. If more arguments are given, the multiplication is applied
>   cumulatively from left to right:
>
>   (...((x1 .* x2) .* x3) .* ...)
>
>   At least one argument is required.
>
>   **See also:** [mtimes], page 139, [rdivide], page 140.

`transpose (x)`   [Built-in Function]
>   Return the transpose of x. This function and `x.'` are equivalent.
>
>   **See also:** [ctranspose], page 138.

`uminus (x)`   [Built-in Function]
>   This function and `- x` are equivalent.
>
>   **See also:** [uplus], page 140, [minus], page 139.

`uplus (x)`   [Built-in Function]
>   This function and `+ x` are equivalent.
>
>   **See also:** [uminus], page 140, [plus], page 139, [minus], page 139.

8.4 Comparison Operators

Comparison operators compare numeric values for relationships such as equality. They are
written using *relational operators*.

All of Octave's comparison operators return a value of 1 if the comparison is true, or 0
if it is false. For matrix values, they all work on an element-by-element basis. Broadcasting
rules apply. See Section 19.2 [Broadcasting], page 459. For example:

```
[1, 2; 3, 4] == [1, 3; 2, 4]
    ⇒   1  0
        0  1
```

According to broadcasting rules, if one operand is a scalar and the other is a matrix, the
scalar is compared to each element of the matrix in turn, and the result is the same size as
the matrix.

`x < y`   True if x is less than y.

`x <= y`   True if x is less than or equal to y.

`x == y`   True if x is equal to y.

`x >= y`   True if x is greater than or equal to y.

`x > y`   True if x is greater than y.

x != y

x ~= y True if x is not equal to y.

For complex numbers, the following ordering is defined: $z1 < z2$ if and only if

```
abs (z1) < abs (z2)
|| (abs (z1) == abs (z2) && arg (z1) < arg (z2))
```

This is consistent with the ordering used by *max*, *min* and *sort*, but is not consistent with MATLAB, which only compares the real parts.

String comparisons may also be performed with the **strcmp** function, not with the comparison operators listed above. See Chapter 5 [Strings], page 65.

eq (x, y) [Built-in Function]
    Return true if the two inputs are equal. This function is equivalent to x == y.

    See also: [ne], page 141, [isequal], page 141, [le], page 141, [ge], page 141, [gt], page 141, [ne], page 141, [lt], page 141.

ge (x, y) [Built-in Function]
    This function is equivalent to x >= y.

    See also: [le], page 141, [eq], page 141, [gt], page 141, [ne], page 141, [lt], page 141.

gt (x, y) [Built-in Function]
    This function is equivalent to x > y.

    See also: [le], page 141, [eq], page 141, [ge], page 141, [ne], page 141, [lt], page 141.

isequal $(x1, x2, \ldots)$ [Function File]
    Return true if all of $x1$, $x2$, $\ldots$ are equal.

    See also: [isequaln], page 141.

isequaln $(x1, x2, \ldots)$ [Function File]
    Return true if all of $x1$, $x2$, $\ldots$ are equal under the additional assumption that NaN == NaN (no comparison of NaN placeholders in dataset).

    See also: [isequal], page 141.

le (x, y) [Built-in Function]
    This function is equivalent to x <= y.

    See also: [eq], page 141, [ge], page 141, [gt], page 141, [ne], page 141, [lt], page 141.

lt (x, y) [Built-in Function]
    This function is equivalent to x < y.

    See also: [le], page 141, [eq], page 141, [ge], page 141, [gt], page 141, [ne], page 141.

ne (x, y) [Built-in Function]
    Return true if the two inputs are not equal. This function is equivalent to x != y.

    See also: [eq], page 141, [isequal], page 141, [le], page 141, [ge], page 141, [lt], page 141.

8.5 Boolean Expressions

8.5.1 Element-by-element Boolean Operators

An *element-by-element boolean expression* is a combination of comparison expressions using the boolean operators "or" ('|'), "and" ('&'), and "not" ('!'), along with parentheses to control nesting. The truth of the boolean expression is computed by combining the truth values of the corresponding elements of the component expressions. A value is considered to be false if it is zero, and true otherwise.

Element-by-element boolean expressions can be used wherever comparison expressions can be used. They can be used in `if` and `while` statements. However, a matrix value used as the condition in an `if` or `while` statement is only true if *all* of its elements are nonzero.

Like comparison operations, each element of an element-by-element boolean expression also has a numeric value (1 if true, 0 if false) that comes into play if the result of the boolean expression is stored in a variable, or used in arithmetic.

Here are descriptions of the three element-by-element boolean operators.

boolean1 & *boolean2*

> Elements of the result are true if both corresponding elements of *boolean1* and *boolean2* are true.

boolean1 | *boolean2*

> Elements of the result are true if either of the corresponding elements of *boolean1* or *boolean2* is true.

! *boolean*
~ *boolean*

> Each element of the result is true if the corresponding element of *boolean* is false.

These operators work on an element-by-element basis. For example, the expression

```
[1, 0; 0, 1] & [1, 0; 2, 3]
```

returns a two by two identity matrix.

For the binary operators, broadcasting rules apply. See Section 19.2 [Broadcasting], page 459. In particular, if one of the operands is a scalar and the other a matrix, the operator is applied to the scalar and each element of the matrix.

For the binary element-by-element boolean operators, both subexpressions *boolean1* and *boolean2* are evaluated before computing the result. This can make a difference when the expressions have side effects. For example, in the expression

```
a & b++
```

the value of the variable *b* is incremented even if the variable *a* is zero.

This behavior is necessary for the boolean operators to work as described for matrix-valued operands.

and (*x*, *y*)   [Built-in Function]
and (*x1*, *x2*, ...)   [Built-in Function]

> Return the logical AND of *x* and *y*. This function is equivalent to `x & y`. If more arguments are given, the logical and is applied cumulatively from left to right:

        `(...((x1 & x2) & x3) & ...)`

At least one argument is required.

See also: [or], page 143, [not], page 143, [xor], page 379.

`not (x)`   [Built-in Function]

Return the logical NOT of *x*. This function is equivalent to `! x`.

See also: [and], page 142, [or], page 143, [xor], page 379.

`or (x, y)`   [Built-in Function]
`or (x1, x2, ...)`   [Built-in Function]

Return the logical OR of *x* and *y*. This function is equivalent to `x | y`. If more arguments are given, the logical or is applied cumulatively from left to right:

        `(...((x1 | x2) | x3) | ...)`

At least one argument is required.

See also: [and], page 142, [not], page 143, [xor], page 379.

8.5.2 Short-circuit Boolean Operators

Combined with the implicit conversion to scalar values in `if` and `while` conditions, Octave's element-by-element boolean operators are often sufficient for performing most logical operations. However, it is sometimes desirable to stop evaluating a boolean expression as soon as the overall truth value can be determined. Octave's *short-circuit* boolean operators work this way.

boolean1 `&&` *boolean2*

The expression *boolean1* is evaluated and converted to a scalar using the equivalent of the operation `all (boolean1(:))`. If it is false, the result of the overall expression is 0. If it is true, the expression *boolean2* is evaluated and converted to a scalar using the equivalent of the operation `all (boolean1(:))`. If it is true, the result of the overall expression is 1. Otherwise, the result of the overall expression is 0.

Warning: there is one exception to the rule of evaluating `all (boolean1(:))`, which is when `boolean1` is the empty matrix. The truth value of an empty matrix is always `false` so `[] && true` evaluates to `false` even though `all ([])` is `true`.

boolean1 `||` *boolean2*

The expression *boolean1* is evaluated and converted to a scalar using the equivalent of the operation `all (boolean1(:))`. If it is true, the result of the overall expression is 1. If it is false, the expression *boolean2* is evaluated and converted to a scalar using the equivalent of the operation `all (boolean1(:))`. If it is true, the result of the overall expression is 1. Otherwise, the result of the overall expression is 0.

Warning: the truth value of an empty matrix is always `false`, see the previous list item for details.

The fact that both operands may not be evaluated before determining the overall truth value of the expression can be important. For example, in the expression

```
     a && b++
```
the value of the variable *b* is only incremented if the variable *a* is nonzero.

This can be used to write somewhat more concise code. For example, it is possible write

```
     function f (a, b, c)
       if (nargin > 2 && ischar (c))
          ...
```
instead of having to use two `if` statements to avoid attempting to evaluate an argument that doesn't exist. For example, without the short-circuit feature, it would be necessary to write

```
     function f (a, b, c)
       if (nargin > 2)
         if (ischar (c))
            ...
```
Writing

```
     function f (a, b, c)
       if (nargin > 2 & ischar (c))
          ...
```
would result in an error if `f` were called with one or two arguments because Octave would be forced to try to evaluate both of the operands for the operator '&'.

MATLAB has special behavior that allows the operators '&' and '|' to short-circuit when used in the truth expression for `if` and `while` statements. The Octave parser may be instructed to behave in the same manner, but its use is strongly discouraged.

val = do_braindead_shortcircuit_evaluation ()   [Built-in Function]
old_val = do_braindead_shortcircuit_evaluation   [Built-in Function]
        (*new_val*)
do_braindead_shortcircuit_evaluation (*new_val*, "*local*")   [Built-in Function]
        Query or set the internal variable that controls whether Octave will do short-circuit evaluation of '|' and '&' operators inside the conditions of if or while statements.

        This feature is only provided for compatibility with MATLAB and should not be used unless you are porting old code that relies on this feature.

        To obtain short-circuit behavior for logical expressions in new programs, you should always use the '&&' and '||' operators.

        When called from inside a function with the "local" option, the variable is changed locally for the function and any subroutines it calls. The original variable value is restored when exiting the function.

Finally, the ternary operator (?:) is not supported in Octave. If short-circuiting is not important, it can be replaced by the `ifelse` function.

merge (*mask*, *tval*, *fval*)   [Built-in Function]
ifelse (*mask*, *tval*, *fval*)   [Built-in Function]
        Merge elements of *true_val* and *false_val*, depending on the value of *mask*. If *mask* is a logical scalar, the other two arguments can be arbitrary values. Otherwise, *mask* must be a logical array, and *tval*, *fval* should be arrays of matching class, or cell

arrays. In the scalar mask case, *tval* is returned if *mask* is true, otherwise *fval* is returned.

In the array mask case, both *tval* and *fval* must be either scalars or arrays with dimensions equal to *mask*. The result is constructed as follows:

```
result(mask) = tval(mask);
result(! mask) = fval(! mask);
```

mask can also be arbitrary numeric type, in which case it is first converted to logical.

See also: [logical], page 59, [diff], page 380.

8.6 Assignment Expressions

An *assignment* is an expression that stores a new value into a variable. For example, the following expression assigns the value 1 to the variable z:

```
z = 1
```

After this expression is executed, the variable z has the value 1. Whatever old value z had before the assignment is forgotten. The '=' sign is called an *assignment operator*.

Assignments can store string values also. For example, the following expression would store the value "this food is good" in the variable message:

```
thing = "food"
predicate = "good"
message = [ "this " , thing , " is " , predicate ]
```

(This also illustrates concatenation of strings.)

Most operators (addition, concatenation, and so on) have no effect except to compute a value. If you ignore the value, you might as well not use the operator. An assignment operator is different. It does produce a value, but even if you ignore the value, the assignment still makes itself felt through the alteration of the variable. We call this a *side effect*.

The left-hand operand of an assignment need not be a variable (see Chapter 7 [Variables], page 121). It can also be an element of a matrix (see Section 8.1 [Index Expressions], page 131) or a list of return values (see Section 8.2 [Calling Functions], page 135). These are all called *lvalues*, which means they can appear on the left-hand side of an assignment operator. The right-hand operand may be any expression. It produces the new value which the assignment stores in the specified variable, matrix element, or list of return values.

It is important to note that variables do *not* have permanent types. The type of a variable is simply the type of whatever value it happens to hold at the moment. In the following program fragment, the variable foo has a numeric value at first, and a string value later on:

```
octave:13> foo = 1
foo = 1
octave:13> foo = "bar"
foo = bar
```

When the second assignment gives foo a string value, the fact that it previously had a numeric value is forgotten.

Assignment of a scalar to an indexed matrix sets all of the elements that are referenced by the indices to the scalar value. For example, if a is a matrix with at least two columns,

```
a(:, 2) = 5
```
sets all the elements in the second column of a to 5.

Assigning an empty matrix '[]' works in most cases to allow you to delete rows or columns of matrices and vectors. See Section 4.1.1 [Empty Matrices], page 51. For example, given a 4 by 5 matrix A, the assignment

```
A (3, :) = []
```
deletes the third row of A, and the assignment

```
A (:, 1:2:5) = []
```
deletes the first, third, and fifth columns.

An assignment is an expression, so it has a value. Thus, z = 1 as an expression has the value 1. One consequence of this is that you can write multiple assignments together:

```
x = y = z = 0
```
stores the value 0 in all three variables. It does this because the value of z = 0, which is 0, is stored into y, and then the value of y = z = 0, which is 0, is stored into x.

This is also true of assignments to lists of values, so the following is a valid expression

```
[a, b, c] = [u, s, v] = svd (a)
```
that is exactly equivalent to

```
[u, s, v] = svd (a)
a = u
b = s
c = v
```

In expressions like this, the number of values in each part of the expression need not match. For example, the expression

```
[a, b] = [u, s, v] = svd (a)
```
is equivalent to

```
[u, s, v] = svd (a)
a = u
b = s
```
The number of values on the left side of the expression can, however, not exceed the number of values on the right side. For example, the following will produce an error.

```
[a, b, c, d] = [u, s, v] = svd (a);
⊣ error: element number 4 undefined in return list
```
The symbol ~ may be used as a placeholder in the list of lvalues, indicating that the corresponding return value should be ignored and not stored anywhere:

```
[~, s, v] = svd (a);
```
This is cleaner and more memory efficient than using a dummy variable. The **nargout** value for the right-hand side expression is not affected. If the assignment is used as an expression, the return value is a comma-separated list with the ignored values dropped.

A very common programming pattern is to increment an existing variable with a given value, like this

```
a = a + 2;
```
This can be written in a clearer and more condensed form using the += operator

```
    a += 2;
```
Similar operators also exist for subtraction (-=), multiplication (*=), and division (/=). An expression of the form

```
    expr1 op= expr2
```
is evaluated as

```
    expr1 = (expr1) op (expr2)
```
where *op* can be either +, -, *, or /, as long as *expr2* is a simple expression with no side effects. If *expr2* also contains an assignment operator, then this expression is evaluated as

```
    temp = expr2
    expr1 = (expr1) op temp
```
where *temp* is a placeholder temporary value storing the computed result of evaluating *expr2*. So, the expression

```
    a *= b+1
```
is evaluated as

```
    a = a * (b+1)
```
and *not*

```
    a = a * b + 1
```
You can use an assignment anywhere an expression is called for. For example, it is valid to write x != (y = 1) to set y to 1 and then test whether x equals 1. But this style tends to make programs hard to read. Except in a one-shot program, you should rewrite it to get rid of such nesting of assignments. This is never very hard.

8.7 Increment Operators

Increment operators increase or decrease the value of a variable by 1. The operator to increment a variable is written as '++'. It may be used to increment a variable either before or after taking its value.

For example, to pre-increment the variable x, you would write ++x. This would add one to x and then return the new value of x as the result of the expression. It is exactly the same as the expression x = x + 1.

To post-increment a variable x, you would write x++. This adds one to the variable x, but returns the value that x had prior to incrementing it. For example, if x is equal to 2, the result of the expression x++ is 2, and the new value of x is 3.

For matrix and vector arguments, the increment and decrement operators work on each element of the operand.

Here is a list of all the increment and decrement expressions.

++x   This expression increments the variable x. The value of the expression is the *new* value of x. It is equivalent to the expression x = x + 1.

--x   This expression decrements the variable x. The value of the expression is the *new* value of x. It is equivalent to the expression x = x - 1.

x++   This expression causes the variable x to be incremented. The value of the expression is the *old* value of x.

x--   This expression causes the variable x to be decremented. The value of the
           expression is the *old* value of x.

8.8 Operator Precedence

Operator precedence determines how operators are grouped, when different operators appear close by in one expression. For example, '*' has higher precedence than '+'. Thus, the expression a + b * c means to multiply b and c, and then add a to the product (i.e., a + (b * c)).

You can overrule the precedence of the operators by using parentheses. You can think of the precedence rules as saying where the parentheses are assumed if you do not write parentheses yourself. In fact, it is wise to use parentheses whenever you have an unusual combination of operators, because other people who read the program may not remember what the precedence is in this case. You might forget as well, and then you too could make a mistake. Explicit parentheses will help prevent any such mistake.

When operators of equal precedence are used together, the leftmost operator groups first, except for the assignment operators, which group in the opposite order. Thus, the expression a - b + c groups as (a - b) + c, but the expression a = b = c groups as a = (b = c).

The precedence of prefix unary operators is important when another operator follows the operand. For example, -x^2 means -(x^2), because '-' has lower precedence than '^'.

Here is a table of the operators in Octave, in order of decreasing precedence. Unless noted, all operators group left to right.

function call and array indexing, cell array indexing, and structure element indexing
          '()' '{}' '.'

postfix increment, and postfix decrement
          '++' '--'

          These operators group right to left.

transpose and exponentiation
          ''' '.'' '^' '**' '.^' '.**'

unary plus, unary minus, prefix increment, prefix decrement, and logical "not"
          '+' '-' '++' '--' '~' '!'

multiply and divide
          '*' '/' '\' '.\' '.*' './'

add, subtract
          '+' '-'

colon   ':'

relational
          '<' '<=' '==' '>=' '>' '!=' '~='

element-wise "and"
          '&'

```
element-wise "or"
```
          '|'

```
logical "and"
```
          '&&'

```
logical "or"
```
          '||'

```
assignment
```
          '=' '+=' '-=' '*=' '/=' '\=' '^=' '.*=' './=' '.\=' '.^=' '|=' '&='

      These operators group right to left.

9 Evaluation

Normally, you evaluate expressions simply by typing them at the Octave prompt, or by asking Octave to interpret commands that you have saved in a file.

Sometimes, you may find it necessary to evaluate an expression that has been computed and stored in a string, which is exactly what the `eval` function lets you do.

eval (*try*)   [Built-in Function]
eval (*try*, *catch*)   [Built-in Function]

> Parse the string *try* and evaluate it as if it were an Octave program. If that fails, evaluate the optional string *catch*. The string *try* is evaluated in the current context, so any results remain available after `eval` returns.
>
> The following example creates the variable A with the approximate value of 3.1416 in the current workspace.
>
> ```
> eval ("A = acos(-1);");
> ```
>
> If an error occurs during the evaluation of *try* then the *catch* string is evaluated, as the following example shows:
>
> ```
> eval ('error ("This is a bad example");',
>       'printf ("This error occurred:\n%s\n", lasterr ());');
>      ⊣ This error occurred:
>        This is a bad example
> ```
>
> Programming Note: if you are only using `eval` as an error-capturing mechanism, rather than for the execution of arbitrary code strings, Consider using try/catch blocks or unwind_protect/unwind_protect_cleanup blocks instead. These techniques have higher performance and don't introduce the security considerations that the evaluation of arbitrary code does.
>
> **See also:** [evalin], page 154.

9.1 Calling a Function by its Name

The `feval` function allows you to call a function from a string containing its name. This is useful when writing a function that needs to call user-supplied functions. The `feval` function takes the name of the function to call as its first argument, and the remaining arguments are given to the function.

The following example is a simple-minded function using `feval` that finds the root of a user-supplied function of one variable using Newton's method.

```
function result = newtroot (fname, x)

# usage: newtroot (fname, x)
#
#   fname : a string naming a function f(x).
#   x     : initial guess

  delta = tol = sqrt (eps);
  maxit = 200;
```

```
     fx = feval (fname, x);
     for i = 1:maxit
       if (abs (fx) < tol)
         result = x;
         return;
       else
         fx_new = feval (fname, x + delta);
         deriv = (fx_new - fx) / delta;
         x = x - fx / deriv;
         fx = fx_new;
       endif
     endfor

     result = x;

   endfunction
```

Note that this is only meant to be an example of calling user-supplied functions and should not be taken too seriously. In addition to using a more robust algorithm, any serious code would check the number and type of all the arguments, ensure that the supplied function really was a function, etc. See Section 4.8 [Predicates for Numeric Objects], page 61, for a list of predicates for numeric objects, and see Section 7.3 [Status of Variables], page 126, for a description of the `exist` function.

feval (*name*, ...)   [Built-in Function]
    Evaluate the function named *name*. Any arguments after the first are passed as inputs to the named function. For example,

```
     feval ("acos", -1)
        ⇒ 3.1416
```

calls the function `acos` with the argument '-1'.

The function `feval` can also be used with function handles of any sort (see Section 11.11.1 [Function Handles], page 190). Historically, `feval` was the only way to call user-supplied functions in strings, but function handles are now preferred due to the cleaner syntax they offer. For example,

```
     f = @exp;
     feval (f, 1)
        ⇒ 2.7183
     f (1)
        ⇒ 2.7183
```

are equivalent ways to call the function referred to by *f*. If it cannot be predicted beforehand whether *f* is a function handle, function name in a string, or inline function then `feval` can be used instead.

A similar function `run` exists for calling user script files, that are not necessarily on the user path

run *script*   [Command]
run (*"script"*)   [Function File]

> Run *script* in the current workspace.
>
> Scripts which reside in directories specified in Octave's load path, and which end with the extension '".m"', can be run simply by typing their name. For scripts not located on the load path, use **run**.
>
> The file name *script* can be a bare, fully qualified, or relative filename and with or without a file extension. If no extension is specified, Octave will first search for a script with the '".m"' extension before falling back to the script name without an extension.
>
> Implementation Note: If *script* includes a path component, then **run** first changes the directory to the directory where *script* is found. **run** then executes the script, and returns to the original directory.
>
> **See also:** [path], page 181, [addpath], page 180, [source], page 190.

9.2 Evaluation in a Different Context

Before you evaluate an expression you need to substitute the values of the variables used in the expression. These are stored in the symbol table. Whenever the interpreter starts a new function it saves the current symbol table and creates a new one, initializing it with the list of function parameters and a couple of predefined variables such as **nargin**. Expressions inside the function use the new symbol table.

Sometimes you want to write a function so that when you call it, it modifies variables in your own context. This allows you to use a pass-by-name style of function, which is similar to using a pointer in programming languages such as C.

Consider how you might write **save** and **load** as m-files. For example:

```
function create_data
  x = linspace (0, 10, 10);
  y = sin (x);
  save mydata x y
endfunction
```

With **evalin**, you could write **save** as follows:

```
function save (file, name1, name2)
  f = open_save_file (file);
  save_var (f, name1, evalin ("caller", name1));
  save_var (f, name2, evalin ("caller", name2));
endfunction
```

Here, 'caller' is the **create_data** function and **name1** is the string "x", which evaluates simply as the value of **x**.

You later want to load the values back from **mydata** in a different context:

```
function process_data
  load mydata
  ... do work ...
endfunction
```

With **assignin**, you could write **load** as follows:

```
function load (file)
  f = open_load_file (file);
  [name, val] = load_var (f);
  assignin ("caller", name, val);
  [name, val] = load_var (f);
  assignin ("caller", name, val);
endfunction
```

Here, 'caller' is the process_data function.

You can set and use variables at the command prompt using the context 'base' rather than 'caller'.

These functions are rarely used in practice. One example is the fail ('code', 'pattern') function which evaluates 'code' in the caller's context and checks that the error message it produces matches the given pattern. Other examples such as save and load are written in C++ where all Octave variables are in the 'caller' context and evalin is not needed.

evalin (*context*, *try*)   [Built-in Function]
evalin (*context*, *try*, *catch*)   [Built-in Function]
    Like eval, except that the expressions are evaluated in the context *context*, which may be either "caller" or "base".

    See also: [eval], page 151, [assignin], page 154.

assignin (*context*, *varname*, *value*)   [Built-in Function]
    Assign *value* to *varname* in context *context*, which may be either "base" or "caller".

    See also: [evalin], page 154.

10 Statements

Statements may be a simple constant expression or a complicated list of nested loops and conditional statements.

Control statements such as `if`, `while`, and so on control the flow of execution in Octave programs. All the control statements start with special keywords such as `if` and `while`, to distinguish them from simple expressions. Many control statements contain other statements; for example, the `if` statement contains another statement which may or may not be executed.

Each control statement has a corresponding *end* statement that marks the end of the control statement. For example, the keyword `endif` marks the end of an `if` statement, and `endwhile` marks the end of a `while` statement. You can use the keyword `end` anywhere a more specific end keyword is expected, but using the more specific keywords is preferred because if you use them, Octave is able to provide better diagnostics for mismatched or missing end tokens.

The list of statements contained between keywords like `if` or `while` and the corresponding end statement is called the *body* of a control statement.

10.1 The if Statement

The `if` statement is Octave's decision-making statement. There are three basic forms of an `if` statement. In its simplest form, it looks like this:

```
if (condition)
  then-body
endif
```

condition is an expression that controls what the rest of the statement will do. The *then-body* is executed only if *condition* is true.

The condition in an `if` statement is considered true if its value is non-zero, and false if its value is zero. If the value of the conditional expression in an `if` statement is a vector or a matrix, it is considered true only if it is non-empty and *all* of the elements are non-zero.

The second form of an if statement looks like this:

```
if (condition)
  then-body
else
  else-body
endif
```

If *condition* is true, *then-body* is executed; otherwise, *else-body* is executed.

Here is an example:

```
if (rem (x, 2) == 0)
  printf ("x is even\n");
else
  printf ("x is odd\n");
endif
```

In this example, if the expression `rem (x, 2) == 0` is true (that is, the value of `x` is divisible by 2), then the first `printf` statement is evaluated, otherwise the second `printf` statement is evaluated.

The third and most general form of the `if` statement allows multiple decisions to be combined in a single statement. It looks like this:

```
if (condition)
  then-body
elseif (condition)
  elseif-body
else
  else-body
endif
```

Any number of `elseif` clauses may appear. Each condition is tested in turn, and if one is found to be true, its corresponding *body* is executed. If none of the conditions are true and the `else` clause is present, its body is executed. Only one `else` clause may appear, and it must be the last part of the statement.

In the following example, if the first condition is true (that is, the value of `x` is divisible by 2), then the first `printf` statement is executed. If it is false, then the second condition is tested, and if it is true (that is, the value of `x` is divisible by 3), then the second `printf` statement is executed. Otherwise, the third `printf` statement is performed.

```
if (rem (x, 2) == 0)
  printf ("x is even\n");
elseif (rem (x, 3) == 0)
  printf ("x is odd and divisible by 3\n");
else
  printf ("x is odd\n");
endif
```

Note that the `elseif` keyword must not be spelled `else if`, as is allowed in Fortran. If it is, the space between the `else` and `if` will tell Octave to treat this as a new `if` statement within another `if` statement's `else` clause. For example, if you write

```
if (c1)
  body-1
else if (c2)
  body-2
endif
```

Octave will expect additional input to complete the first `if` statement. If you are using Octave interactively, it will continue to prompt you for additional input. If Octave is reading this input from a file, it may complain about missing or mismatched **end** statements, or, if you have not used the more specific **end** statements (`endif`, `endfor`, etc.), it may simply produce incorrect results, without producing any warning messages.

It is much easier to see the error if we rewrite the statements above like this,

```
if (c1)
  body-1
else
  if (c2)
    body-2
  endif
```

using the indentation to show how Octave groups the statements. See Chapter 11 [Functions and Scripts], page 167.

10.2 The switch Statement

It is very common to take different actions depending on the value of one variable. This is possible using the if statement in the following way

```
if (X == 1)
  do_something ();
elseif (X == 2)
  do_something_else ();
else
  do_something_completely_different ();
endif
```

This kind of code can however be very cumbersome to both write and maintain. To overcome this problem Octave supports the switch statement. Using this statement, the above example becomes

```
switch (X)
  case 1
    do_something ();
  case 2
    do_something_else ();
  otherwise
    do_something_completely_different ();
endswitch
```

This code makes the repetitive structure of the problem more explicit, making the code easier to read, and hence maintain. Also, if the variable X should change its name, only one line would need changing compared to one line per case when if statements are used.

The general form of the switch statement is

```
switch (expression)
  case label
    command_list
  case label
    command_list
  ...

  otherwise
    command_list
endswitch
```

where *label* can be any expression. However, duplicate *label* values are not detected, and only the *command_list* corresponding to the first match will be executed. For the `switch` statement to be meaningful at least one `case label command_list` clause must be present, while the `otherwise command_list` clause is optional.

If *label* is a cell array the corresponding *command_list* is executed if *any* of the elements of the cell array match *expression*. As an example, the following program will print 'Variable is either 6 or 7'.

```
A = 7;
switch (A)
  case { 6, 7 }
    printf ("variable is either 6 or 7\n");
  otherwise
    printf ("variable is neither 6 nor 7\n");
endswitch
```

As with all other specific `end` keywords, `endswitch` may be replaced by `end`, but you can get better diagnostics if you use the specific forms.

One advantage of using the `switch` statement compared to using `if` statements is that the *label*s can be strings. If an `if` statement is used it is *not* possible to write

```
if (X == "a string") # This is NOT valid
```

since a character-to-character comparison between X and the string will be made instead of evaluating if the strings are equal. This special-case is handled by the `switch` statement, and it is possible to write programs that look like this

```
switch (X)
  case "a string"
    do_something
  ...
endswitch
```

10.2.1 Notes for the C Programmer

The `switch` statement is also available in the widely used C programming language. There are, however, some differences between the statement in Octave and C

- Cases are exclusive, so they don't 'fall through' as do the cases in the `switch` statement of the C language.

- The *command_list* elements are not optional. Making the list optional would have meant requiring a separator between the label and the command list. Otherwise, things like

```
switch (foo)
  case (1) -2
  ...
```

would produce surprising results, as would

```
switch (foo)
  case (1)
  case (2)
    doit ();
  ...
```

particularly for C programmers. If `doit()` should be executed if *foo* is either 1 or 2, the above code should be written with a cell array like this

```
switch (foo)
  case { 1, 2 }
    doit ();
  ...
```

10.3 The while Statement

In programming, a *loop* means a part of a program that is (or at least can be) executed two or more times in succession.

The `while` statement is the simplest looping statement in Octave. It repeatedly executes a statement as long as a condition is true. As with the condition in an `if` statement, the condition in a `while` statement is considered true if its value is non-zero, and false if its value is zero. If the value of the conditional expression in a `while` statement is a vector or a matrix, it is considered true only if it is non-empty and *all* of the elements are non-zero.

Octave's `while` statement looks like this:

```
while (condition)
  body
endwhile
```

Here *body* is a statement or list of statements that we call the *body* of the loop, and *condition* is an expression that controls how long the loop keeps running.

The first thing the `while` statement does is test *condition*. If *condition* is true, it executes the statement *body*. After *body* has been executed, *condition* is tested again, and if it is still true, *body* is executed again. This process repeats until *condition* is no longer true. If *condition* is initially false, the body of the loop is never executed.

This example creates a variable `fib` that contains the first ten elements of the Fibonacci sequence.

```
fib = ones (1, 10);
i = 3;
while (i <= 10)
  fib (i) = fib (i-1) + fib (i-2);
  i++;
endwhile
```

Here the body of the loop contains two statements.

The loop works like this: first, the value of `i` is set to 3. Then, the `while` tests whether `i` is less than or equal to 10. This is the case when `i` equals 3, so the value of the `i`-th element of `fib` is set to the sum of the previous two values in the sequence. Then the `i++` increments the value of `i` and the loop repeats. The loop terminates when `i` reaches 11.

A newline is not required between the condition and the body; but using one makes the program clearer unless the body is very simple.

10.4 The do-until Statement

The `do-until` statement is similar to the `while` statement, except that it repeatedly executes a statement until a condition becomes true, and the test of the condition is at the end
of the loop, so the body of the loop is always executed at least once. As with the condition
in an `if` statement, the condition in a `do-until` statement is considered true if its value
is non-zero, and false if its value is zero. If the value of the conditional expression in a
`do-until` statement is a vector or a matrix, it is considered true only if it is non-empty and
all of the elements are non-zero.

Octave's `do-until` statement looks like this:

```
do
  body
until (condition)
```

Here *body* is a statement or list of statements that we call the *body* of the loop, and
condition is an expression that controls how long the loop keeps running.

This example creates a variable `fib` that contains the first ten elements of the Fibonacci
sequence.

```
fib = ones (1, 10);
i = 2;
do
  i++;
  fib (i) = fib (i-1) + fib (i-2);
until (i == 10)
```

A newline is not required between the `do` keyword and the body; but using one makes
the program clearer unless the body is very simple.

10.5 The for Statement

The `for` statement makes it more convenient to count iterations of a loop. The general
form of the `for` statement looks like this:

```
for var = expression
  body
endfor
```

where *body* stands for any statement or list of statements, *expression* is any valid expression,
and *var* may take several forms. Usually it is a simple variable name or an indexed variable.
If the value of *expression* is a structure, *var* may also be a vector with two elements. See
Section 10.5.1 [Looping Over Structure Elements], page 161, below.

The assignment expression in the `for` statement works a bit differently than Octave's
normal assignment statement. Instead of assigning the complete result of the expression, it
assigns each column of the expression to *var* in turn. If *expression* is a range, a row vector,
or a scalar, the value of *var* will be a scalar each time the loop body is executed. If *var* is a
column vector or a matrix, *var* will be a column vector each time the loop body is executed.

The following example shows another way to create a vector containing the first ten
elements of the Fibonacci sequence, this time using the `for` statement:

```
fib = ones (1, 10);
for i = 3:10
  fib (i) = fib (i-1) + fib (i-2);
endfor
```

This code works by first evaluating the expression 3:10, to produce a range of values from 3 to 10 inclusive. Then the variable i is assigned the first element of the range and the body of the loop is executed once. When the end of the loop body is reached, the next value in the range is assigned to the variable i, and the loop body is executed again. This process continues until there are no more elements to assign.

Within Octave is it also possible to iterate over matrices or cell arrays using the `for` statement. For example consider

```
disp ("Loop over a matrix")
for i = [1,3;2,4]
  i
endfor
disp ("Loop over a cell array")
for i = {1,"two";"three",4}
  i
endfor
```

In this case the variable i takes on the value of the columns of the matrix or cell matrix. So the first loop iterates twice, producing two column vectors [1;2], followed by [3;4], and likewise for the loop over the cell array. This can be extended to loops over multi-dimensional arrays. For example:

```
a = [1,3;2,4]; c = cat (3, a, 2*a);
for i = c
  i
endfor
```

In the above case, the multi-dimensional matrix c is reshaped to a two-dimensional matrix as `reshape (c, rows (c), prod (size (c)(2:end)))` and then the same behavior as a loop over a two dimensional matrix is produced.

Although it is possible to rewrite all `for` loops as `while` loops, the Octave language has both statements because often a `for` loop is both less work to type and more natural to think of. Counting the number of iterations is very common in loops and it can be easier to think of this counting as part of looping rather than as something to do inside the loop.

10.5.1 Looping Over Structure Elements

A special form of the `for` statement allows you to loop over all the elements of a structure:

```
for [ val, key ] = expression
  body
endfor
```

In this form of the `for` statement, the value of *expression* must be a structure. If it is, *key* and *val* are set to the name of the element and the corresponding value in turn, until there are no more elements. For example:

```
x.a = 1
x.b = [1, 2; 3, 4]
x.c = "string"
for [val, key] = x
  key
  val
endfor
```

```
        ⊣  key = a
        ⊣  val = 1
        ⊣  key = b
        ⊣  val =
        ⊣
        ⊣    1   2
        ⊣    3   4
        ⊣
        ⊣  key = c
        ⊣  val = string
```

The elements are not accessed in any particular order. If you need to cycle through the list in a particular way, you will have to use the function **fieldnames** and sort the list yourself.

The *key* variable may also be omitted. If it is, the brackets are also optional. This is useful for cycling through the values of all the structure elements when the names of the elements do not need to be known.

10.6 The break Statement

The **break** statement jumps out of the innermost **while**, **do-until**, or **for** loop that encloses it. The **break** statement may only be used within the body of a loop. The following example finds the smallest divisor of a given integer, and also identifies prime numbers:

```
num = 103;
div = 2;
while (div*div <= num)
  if (rem (num, div) == 0)
    break;
  endif
  div++;
endwhile
if (rem (num, div) == 0)
  printf ("Smallest divisor of %d is %d\n", num, div)
else
  printf ("%d is prime\n", num);
endif
```

When the remainder is zero in the first **while** statement, Octave immediately *breaks out* of the loop. This means that Octave proceeds immediately to the statement following the loop and continues processing. (This is very different from the **exit** statement which stops the entire Octave program.)

Here is another program equivalent to the previous one. It illustrates how the *condition* of a **while** statement could just as well be replaced with a **break** inside an **if**:

```
num = 103;
div = 2;
while (1)
  if (rem (num, div) == 0)
    printf ("Smallest divisor of %d is %d\n", num, div);
    break;
  endif
  div++;
  if (div*div > num)
    printf ("%d is prime\n", num);
    break;
  endif
endwhile
```

10.7 The continue Statement

The **continue** statement, like **break**, is used only inside **while**, **do-until**, or **for** loops. It skips over the rest of the loop body, causing the next cycle around the loop to begin immediately. Contrast this with **break**, which jumps out of the loop altogether. Here is an example:

```
# print elements of a vector of random
# integers that are even.

# first, create a row vector of 10 random
# integers with values between 0 and 100:

vec = round (rand (1, 10) * 100);

# print what we're interested in:

for x = vec
  if (rem (x, 2) != 0)
    continue;
  endif
  printf ("%d\n", x);
endfor
```

If one of the elements of *vec* is an odd number, this example skips the print statement for that element, and continues back to the first statement in the loop.

This is not a practical example of the **continue** statement, but it should give you a clear understanding of how it works. Normally, one would probably write the loop like this:

```
for x = vec
  if (rem (x, 2) == 0)
    printf ("%d\n", x);
  endif
endfor
```

10.8 The unwind_protect Statement

Octave supports a limited form of exception handling modeled after the unwind-protect form of Lisp.

The general form of an `unwind_protect` block looks like this:

```
unwind_protect
  body
unwind_protect_cleanup
  cleanup
end_unwind_protect
```

where *body* and *cleanup* are both optional and may contain any Octave expressions or commands. The statements in *cleanup* are guaranteed to be executed regardless of how control exits *body*.

This is useful to protect temporary changes to global variables from possible errors. For example, the following code will always restore the original value of the global variable `frobnosticate` even if an error occurs in the first part of the `unwind_protect` block.

```
save_frobnosticate = frobnosticate;
unwind_protect
  frobnosticate = true;
  ...
unwind_protect_cleanup
  frobnosticate = save_frobnosticate;
end_unwind_protect
```

Without `unwind_protect`, the value of *frobnosticate* would not be restored if an error occurs while evaluating the first part of the `unwind_protect` block because evaluation would stop at the point of the error and the statement to restore the value would not be executed.

In addition to unwind_protect, Octave supports another form of exception handling, the `try` block.

10.9 The try Statement

The original form of a `try` block looks like this:

```
try
  body
catch
  cleanup
end_try_catch
```

where *body* and *cleanup* are both optional and may contain any Octave expressions or commands. The statements in *cleanup* are only executed if an error occurs in *body*.

No warnings or error messages are printed while *body* is executing. If an error does occur during the execution of *body*, *cleanup* can use the functions `lasterr` or `lasterror` to access the text of the message that would have been printed, as well as its identifier. The alternative form,

```
try
  body
catch err
  cleanup
end_try_catch
```

will automatically store the output of `lasterror` in the structure *err*. See Chapter 12 [Errors and Warnings], page 197, for more information about the `lasterr` and `lasterror` functions.

10.10 Continuation Lines

In the Octave language, most statements end with a newline character and you must tell Octave to ignore the newline character in order to continue a statement from one line to the next. Lines that end with the characters ... are joined with the following line before they are divided into tokens by Octave's parser. For example, the lines

```
x = long_variable_name ...
    + longer_variable_name ...
    - 42
```

form a single statement.

Any text between the continuation marker and the newline character is ignored. For example, the statement

```
x = long_variable_name ...    # comment one
    + longer_variable_name ...comment two
    - 42                       # last comment
```

is equivalent to the one shown above.

Inside double-quoted string constants, the character \ has to be used as continuation marker. The \ must appear at the end of the line just before the newline character:

```
s = "This text starts in the first line \
and is continued in the second line."
```

Input that occurs inside parentheses can be continued to the next line without having to use a continuation marker. For example, it is possible to write statements like

```
if (fine_dining_destination == on_a_boat
    || fine_dining_destination == on_a_train)
  seuss (i, will, not, eat, them, sam, i, am, i,
         will, not, eat, green, eggs, and, ham);
endif
```

without having to add to the clutter with continuation markers.

11 Functions and Scripts

Complicated Octave programs can often be simplified by defining functions. Functions can be defined directly on the command line during interactive Octave sessions, or in external files, and can be called just like built-in functions.

11.1 Introduction to Function and Script Files

There are seven different things covered in this section.

1. Typing in a function at the command prompt.

2. Storing a group of commands in a file — called a script file.

3. Storing a function in a file—called a function file.

4. Subfunctions in function files.

5. Multiple functions in one script file.

6. Private functions.

7. Nested functions.

Both function files and script files end with an extension of .m, for MATLAB compatibility. If you want more than one independent functions in a file, it must be a script file (see Section 11.10 [Script Files], page 189), and to use these functions you must execute the script file before you can use the functions that are in the script file.

11.2 Defining Functions

In its simplest form, the definition of a function named *name* looks like this:

```
function name
  body
endfunction
```

A valid function name is like a valid variable name: a sequence of letters, digits and underscores, not starting with a digit. Functions share the same pool of names as variables.

The function *body* consists of Octave statements. It is the most important part of the definition, because it says what the function should actually *do*.

For example, here is a function that, when executed, will ring the bell on your terminal (assuming that it is possible to do so):

```
function wakeup
  printf ("\a");
endfunction
```

The **printf** statement (see Chapter 14 [Input and Output], page 223) simply tells Octave to print the string "\a". The special character '\a' stands for the alert character (ASCII 7). See Chapter 5 [Strings], page 65.

Once this function is defined, you can ask Octave to evaluate it by typing the name of the function.

Normally, you will want to pass some information to the functions you define. The syntax for passing parameters to a function in Octave is

```
function name (arg-list)
  body
endfunction
```

where *arg-list* is a comma-separated list of the function's arguments. When the function is called, the argument names are used to hold the argument values given in the call. The list of arguments may be empty, in which case this form is equivalent to the one shown above.

To print a message along with ringing the bell, you might modify the `wakeup` to look like this:

```
function wakeup (message)
  printf ("\a%s\n", message);
endfunction
```

Calling this function using a statement like this

```
wakeup ("Rise and shine!");
```

will cause Octave to ring your terminal's bell and print the message 'Rise and shine!', followed by a newline character (the '\n' in the first argument to the `printf` statement).

In most cases, you will also want to get some information back from the functions you define. Here is the syntax for writing a function that returns a single value:

```
function ret-var = name (arg-list)
  body
endfunction
```

The symbol *ret-var* is the name of the variable that will hold the value to be returned by the function. This variable must be defined before the end of the function body in order for the function to return a value.

Variables used in the body of a function are local to the function. Variables named in *arg-list* and *ret-var* are also local to the function. See Section 7.1 [Global Variables], page 123, for information about how to access global variables inside a function.

For example, here is a function that computes the average of the elements of a vector:

```
function retval = avg (v)
  retval = sum (v) / length (v);
endfunction
```

If we had written `avg` like this instead,

```
function retval = avg (v)
  if (isvector (v))
    retval = sum (v) / length (v);
  endif
endfunction
```

and then called the function with a matrix instead of a vector as the argument, Octave would have printed an error message like this:

```
error: value on right hand side of assignment is undefined
```

because the body of the `if` statement was never executed, and `retval` was never defined. To prevent obscure errors like this, it is a good idea to always make sure that the return variables will always have values, and to produce meaningful error messages when problems are encountered. For example, `avg` could have been written like this:

```
function retval = avg (v)
  retval = 0;
  if (isvector (v))
    retval = sum (v) / length (v);
  else
    error ("avg: expecting vector argument");
  endif
endfunction
```

There is still one additional problem with this function. What if it is called without an argument? Without additional error checking, Octave will probably print an error message that won't really help you track down the source of the error. To allow you to catch errors like this, Octave provides each function with an automatic variable called `nargin`. Each time a function is called, `nargin` is automatically initialized to the number of arguments that have actually been passed to the function. For example, we might rewrite the `avg` function like this:

```
function retval = avg (v)
  retval = 0;
  if (nargin != 1)
    usage ("avg (vector)");
  endif
  if (isvector (v))
    retval = sum (v) / length (v);
  else
    error ("avg: expecting vector argument");
  endif
endfunction
```

Although Octave does not automatically report an error if you call a function with more arguments than expected, doing so probably indicates that something is wrong. Octave also does not automatically report an error if a function is called with too few arguments, but any attempt to use a variable that has not been given a value will result in an error. To avoid such problems and to provide useful messages, we check for both possibilities and issue our own error message.

`nargin ()`   [Built-in Function]
`nargin (fcn)`   [Built-in Function]
> Within a function, return the number of arguments passed to the function. At the top level, return the number of command line arguments passed to Octave.
>
> If called with the optional argument *fcn*—a function name or handle— return the declared number of arguments that the function can accept.
>
> If the last argument to *fcn* is *varargin* the returned value is negative. For example, the function `union` for sets is declared as

```
function [y, ia, ib] = union (a, b, varargin)
```

and

```
nargin ("union")
    ⇒ -3
```

Programming Note: `nargin` does not work on built-in functions.

See also: [nargout], page 172, [varargin], page 174, [isargout], page 175, [varargout], page 174, [nthargout], page 171.

`inputname (n)`   [Function File]
    Return the name of the n-th argument to the calling function. If the argument is not a simple variable name, return an empty string.

`val = silent_functions ()`   [Built-in Function]
`old_val = silent_functions (new_val)`   [Built-in Function]
`silent_functions (new_val, "local")`   [Built-in Function]
    Query or set the internal variable that controls whether internal output from a function is suppressed. If this option is disabled, Octave will display the results produced by evaluating expressions within a function body that are not terminated with a semicolon.

    When called from inside a function with the `"local"` option, the variable is changed locally for the function and any subroutines it calls. The original variable value is restored when exiting the function.

11.3 Multiple Return Values

Unlike many other computer languages, Octave allows you to define functions that return more than one value. The syntax for defining functions that return multiple values is

```
function [ret-list] = name (arg-list)
  body
endfunction
```

where *name*, *arg-list*, and *body* have the same meaning as before, and *ret-list* is a comma-separated list of variable names that will hold the values returned from the function. The list of return values must have at least one element. If *ret-list* has only one element, this form of the `function` statement is equivalent to the form described in the previous section.

Here is an example of a function that returns two values, the maximum element of a vector and the index of its first occurrence in the vector.

```
function [max, idx] = vmax (v)
  idx = 1;
  max = v (idx);
  for i = 2:length (v)
    if (v (i) > max)
      max = v (i);
      idx = i;
    endif
  endfor
endfunction
```

In this particular case, the two values could have been returned as elements of a single array, but that is not always possible or convenient. The values to be returned may not have compatible dimensions, and it is often desirable to give the individual return values distinct names.

It is possible to use the **nthargout** function to obtain only some of the return values or several at once in a cell array. See Section 3.1.5 [Cell Array Objects], page 43.

nthargout (*n*, *func*, ...)   [Function File]
nthargout (*n*, *ntot*, *func*, ...)   [Function File]
    Return the *n*th output argument of function given by the function handle or string *func*. Any arguments after *func* are passed to *func*. The total number of arguments to call *func* with can be passed in *ntot*; by default *ntot* is *n*. The input *n* can also be a vector of indices of the output, in which case the output will be a cell array of the requested output arguments.

    The intended use **nthargout** is to avoid intermediate variables. For example, when finding the indices of the maximum entry of a matrix, the following two compositions of nthargout

```
m = magic (5);
cell2mat (nthargout ([1, 2], @ind2sub, size (m),
                nthargout (2, @max, m(:))))
⇒ 5   3
```

    are completely equivalent to the following lines:

```
m = magic (5);
[~, idx] = max (M(:));
[i, j] = ind2sub (size (m), idx);
[i, j]
⇒ 5   3
```

    It can also be helpful to have all output arguments in a single cell in the following manner:

```
USV = nthargout ([1:3], @svd, hilb (5));
```

    See also: [nargin], page 169, [nargout], page 172, [varargin], page 174, [varargout], page 174, [isargout], page 175.

In addition to setting **nargin** each time a function is called, Octave also automatically initializes **nargout** to the number of values that are expected to be returned. This allows you to write functions that behave differently depending on the number of values that the user of the function has requested. The implicit assignment to the built-in variable **ans** does not figure in the count of output arguments, so the value of **nargout** may be zero.

The **svd** and **lu** functions are examples of built-in functions that behave differently depending on the value of **nargout**.

It is possible to write functions that only set some return values. For example, calling the function

```
function [x, y, z] = f ()
  x = 1;
  z = 2;
endfunction
```

as

```
[a, b, c] = f ()
```

produces:

```
a = 1

b = [](0x0)

c = 2
```

along with a warning.

nargout ()   [Built-in Function]
nargout (*fcn*)   [Built-in Function]

    Within a function, return the number of values the caller expects to receive. If called with the optional argument *fcn*—a function name or handle—return the number of declared output values that the function can produce. If the final output argument is *varargout* the returned value is negative.

    For example,

```
f ()
```

    will cause **nargout** to return 0 inside the function **f** and

```
[s, t] = f ()
```

    will cause **nargout** to return 2 inside the function **f**.

    In the second usage,

```
nargout (@histc) % or nargout ("histc")
```

    will return 2, because **histc** has two outputs, whereas

```
nargout (@imread)
```

    will return -2, because **imread** has two outputs and the second is *varargout*.

    At the top level, **nargout** with no argument is undefined and will produce an error. **nargout** does not work for built-in functions and returns -1 for all anonymous functions.

    See also: [nargin], page 169, [varargin], page 174, [isargout], page 175, [varargout], page 174, [nthargout], page 171.

It is good practice at the head of a function to verify that it has been called correctly. In Octave the following idiom is seen frequently

```
if (nargin < min_#_inputs || nargin > max_#_inputs)
  print_usage ();
endif
```

which stops the function execution and prints a message about the correct way to call the function whenever the number of inputs is wrong.

For compatibility with MATLAB, `nargchk`, `narginchk` and `nargoutchk` are available which provide similar error checking.

msgstr = nargchk (*minargs*, *maxargs*, *nargs*)   [Function File]
msgstr = nargchk (*minargs*, *maxargs*, *nargs*, "*string*")   [Function File]
msgstruct = nargchk (*minargs*, *maxargs*, *nargs*, "*struct*")   [Function File]

> Return an appropriate error message string (or structure) if the number of inputs requested is invalid.
>
> This is useful for checking to see that the number of input arguments supplied to a function is within an acceptable range.
>
> **See also:** [nargoutchk], page 173, [narginchk], page 173, [error], page 197, [nargin], page 169, [nargout], page 172.

narginchk (*minargs*, *maxargs*)   [Function File]

> Check for correct number of arguments or generate an error message if the number of arguments in the calling function is outside the range *minargs* and *maxargs*. Otherwise, do nothing.
>
> Both *minargs* and *maxargs* need to be scalar numeric values. Zero, Inf and negative values are all allowed, and *minargs* and *maxargs* may be equal.
>
> Note that this function evaluates `nargin` on the caller.
>
> **See also:** [nargchk], page 173, [nargoutchk], page 173, [error], page 197, [nargout], page 172, [nargin], page 169.

nargoutchk (*minargs*, *maxargs*)   [Function File]
msgstr = nargoutchk (*minargs*, *maxargs*, *nargs*)   [Function File]
msgstr = nargoutchk (*minargs*, *maxargs*, *nargs*, "*string*")   [Function File]
msgstruct = nargoutchk (*minargs*, *maxargs*, *nargs*, "*struct*")   [Function File]

> Check for correct number of output arguments.
>
> On the first form, returns an error unless the number of arguments in its caller is between the values of *minargs* and *maxargs*. It does nothing otherwise. Note that this function evaluates the value of `nargout` on the caller so its value must have not been tampered with.
>
> Both *minargs* and *maxargs* need to be a numeric scalar. Zero, Inf and negative are all valid, and they can have the same value.
>
> For backward compatibility reasons, the other forms return an appropriate error message string (or structure) if the number of outputs requested is invalid.
>
> This is useful for checking to see that the number of output arguments supplied to a function is within an acceptable range.

See also: [nargchk], page 173, [narginchk], page 173, [error], page 197, [nargout], page 172, [nargin], page 169.

11.4 Variable-length Argument Lists

Sometimes the number of input arguments is not known when the function is defined. As an example think of a function that returns the smallest of all its input arguments. For example:

```
a = smallest (1, 2, 3);
b = smallest (1, 2, 3, 4);
```

In this example both `a` and `b` would be 1. One way to write the `smallest` function is

```
function val = smallest (arg1, arg2, arg3, arg4, arg5)
  body
endfunction
```

and then use the value of `nargin` to determine which of the input arguments should be considered. The problem with this approach is that it can only handle a limited number of input arguments.

If the special parameter name `varargin` appears at the end of a function parameter list it indicates that the function takes a variable number of input arguments. Using `varargin` the function looks like this

```
function val = smallest (varargin)
  body
endfunction
```

In the function body the input arguments can be accessed through the variable `varargin`. This variable is a cell array containing all the input arguments. See Section 6.2 [Cell Arrays], page 110, for details on working with cell arrays. The `smallest` function can now be defined like this

```
function val = smallest (varargin)
  val = min ([varargin{:}]);
endfunction
```

This implementation handles any number of input arguments, but it's also a very simple solution to the problem.

A slightly more complex example of `varargin` is a function `print_arguments` that prints all input arguments. Such a function can be defined like this

```
function print_arguments (varargin)
  for i = 1:length (varargin)
    printf ("Input argument %d: ", i);
    disp (varargin{i});
  endfor
endfunction
```

This function produces output like this

```
print_arguments (1, "two", 3);
     ⊣ Input argument 1:  1
     ⊣ Input argument 2: two
     ⊣ Input argument 3:  3
```

`[reg, prop]` = parseparams (*params*)   [Function File]
`[reg, var1, ...]` = parseparams (*params*, *name1*, *default1*,   [Function File]
    ...)

> Return in *reg* the cell elements of *param* up to the first string element and in *prop* all remaining elements beginning with the first string element. For example:

```
[reg, prop] = parseparams ({1, 2, "linewidth", 10})
reg =
{
  [1,1] = 1
  [1,2] = 2
}
prop =
{
  [1,1] = linewidth
  [1,2] = 10
}
```

> The parseparams function may be used to separate regular numeric arguments from additional arguments given as property/value pairs of the *varargin* cell array.
>
> In the second form of the call, available options are specified directly with their default values given as name-value pairs. If *params* do not form name-value pairs, or if an option occurs that does not match any of the available options, an error occurs. When called from an m-file function, the error is prefixed with the name of the caller function. The matching of options is case-insensitive.
>
> **See also:** [varargin], page 174.

11.5 Ignoring Arguments

In the formal argument list, it is possible to use the dummy placeholder ~ instead of a name. This indicates that the corresponding argument value should be ignored and not stored to any variable.

```
function val = pick2nd (~, arg2)
  val = arg2;
endfunction
```

The value of **nargin** is not affected by using this declaration.

Return arguments can also be ignored using the same syntax. Functions may take advantage of ignored outputs to reduce the number of calculations performed. To do so, use the **isargout** function to query whether the output argument is wanted. For example:

```
function [out1, out2] = long_function (x, y, z)
  if (isargout (1))
    ## Long calculation
    ...
    out1 = result;
  endif
  ...
endfunction
```

isargout (*k*)   [Built-in Function]
> Within a function, return a logical value indicating whether the argument *k* will be
> assigned to a variable on output. If the result is false, the argument has been ignored
> during the function call through the use of the tilde (~) special output argument.
> Functions can use `isargout` to avoid performing unnecessary calculations for outputs
> which are unwanted.
>
> If *k* is outside the range `1:max (nargout)`, the function returns false. *k* can also be
> an array, in which case the function works element-by-element and a logical array is
> returned. At the top level, `isargout` returns an error.
>
> **See also:** [nargout], page 172, [nargin], page 169, [varargin], page 174, [varargout],
> page 174, [nthargout], page 171.

11.6 Variable-length Return Lists

It is possible to return a variable number of output arguments from a function using a
syntax that's similar to the one used with the special `varargin` parameter name. To let a
function return a variable number of output arguments the special output parameter name
`varargout` is used. As with `varargin`, `varargout` is a cell array that will contain the
requested output arguments.

As an example the following function sets the first output argument to 1, the second to
2, and so on.

```
function varargout = one_to_n ()
  for i = 1:nargout
    varargout{i} = i;
  endfor
endfunction
```

When called this function returns values like this

```
[a, b, c] = one_to_n ()
      ⇒ a =  1
      ⇒ b =  2
      ⇒ c =  3
```

If `varargin` (`varargout`) does not appear as the last element of the input (output)
parameter list, then it is not special, and is handled the same as any other parameter name.

[*r1*, *r2*, ..., *rn*] = deal (*a*)   [Function File]
[*r1*, *r2*, ..., *rn*] = deal (*a1*, *a2*, ..., *an*)   [Function File]
> Copy the input parameters into the corresponding output parameters. If only one
> input parameter is supplied, its value is copied to each of the outputs.
>
> For example,
>
> ```
>     [a, b, c] = deal (x, y, z);
> ```
>
> is equivalent to
>
> ```
>     a = x;
>     b = y;
>     c = z;
> ```
>
> and

```
[a, b, c] = deal (x);
```

is equivalent to

```
a = b = c = x;
```

11.7 Returning from a Function

The body of a user-defined function can contain a **return** statement. This statement returns control to the rest of the Octave program. It looks like this:

```
return
```

Unlike the **return** statement in C, Octave's **return** statement cannot be used to return a value from a function. Instead, you must assign values to the list of return variables that are part of the **function** statement. The **return** statement simply makes it easier to exit a function from a deeply nested loop or conditional statement.

Here is an example of a function that checks to see if any elements of a vector are nonzero.

```
function retval = any_nonzero (v)
  retval = 0;
  for i = 1:length (v)
    if (v (i) != 0)
      retval = 1;
      return;
    endif
  endfor
  printf ("no nonzero elements found\n");
endfunction
```

Note that this function could not have been written using the **break** statement to exit the loop once a nonzero value is found without adding extra logic to avoid printing the message if the vector does contain a nonzero element.

return   [Keyword]
> When Octave encounters the keyword **return** inside a function or script, it returns control to the caller immediately. At the top level, the return statement is ignored. A **return** statement is assumed at the end of every function definition.

11.8 Default Arguments

Since Octave supports variable number of input arguments, it is very useful to assign default values to some input arguments. When an input argument is declared in the argument list it is possible to assign a default value to the argument like this

```
function name (arg1 = val1, ...)
  body
endfunction
```

If no value is assigned to *arg1* by the user, it will have the value *val1*.

As an example, the following function implements a variant of the classic "Hello, World" program.

```
function hello (who = "World")
  printf ("Hello, %s!\n", who);
endfunction
```

When called without an input argument the function prints the following

```
hello ();
        ⊣ Hello, World!
```

and when it's called with an input argument it prints the following

```
hello ("Beautiful World of Free Software");
        ⊣ Hello, Beautiful World of Free Software!
```

Sometimes it is useful to explicitly tell Octave to use the default value of an input argument. This can be done writing a ':' as the value of the input argument when calling the function.

```
hello (:);
        ⊣ Hello, World!
```

11.9 Function Files

Except for simple one-shot programs, it is not practical to have to define all the functions you need each time you need them. Instead, you will normally want to save them in a file so that you can easily edit them, and save them for use at a later time.

Octave does not require you to load function definitions from files before using them. You simply need to put the function definitions in a place where Octave can find them.

When Octave encounters an identifier that is undefined, it first looks for variables or functions that are already compiled and currently listed in its symbol table. If it fails to find a definition there, it searches a list of directories (the *path*) for files ending in '.m' that have the same base name as the undefined identifier.[1] Once Octave finds a file with a name that matches, the contents of the file are read. If it defines a *single* function, it is compiled and executed. See Section 11.10 [Script Files], page 189, for more information about how you can define more than one function in a single file.

When Octave defines a function from a function file, it saves the full name of the file it read and the time stamp on the file. If the time stamp on the file changes, Octave may reload the file. When Octave is running interactively, time stamp checking normally happens at most once each time Octave prints the prompt. Searching for new function definitions also occurs if the current working directory changes.

Checking the time stamp allows you to edit the definition of a function while Octave is running, and automatically use the new function definition without having to restart your Octave session.

To avoid degrading performance unnecessarily by checking the time stamps on functions that are not likely to change, Octave assumes that function files in the directory tree '*octave-home*/share/octave/*version*/m' will not change, so it doesn't have to check their time stamps every time the functions defined in those files are used. This is normally a very good assumption and provides a significant improvement in performance for the function files that are distributed with Octave.

---

[1] The '.m' suffix was chosen for compatibility with MATLAB.

If you know that your own function files will not change while you are running Octave, you can improve performance by calling `ignore_function_time_stamp` ("all"), so that Octave will ignore the time stamps for all function files. Passing "system" to this function resets the default behavior.

`edit` *name*   [Command]
`edit` *field value*   [Command]
value = `edit` *get field*   [Command]

     Edit the named function, or change editor settings.

     If `edit` is called with the name of a file or function as its argument it will be opened in the text editor defined by `EDITOR`.

- If the function *name* is available in a file on your path and that file is modifiable, then it will be edited in place. If it is a system function, then it will first be copied to the directory `HOME` (see below) and then edited. If no file is found, then the m-file variant, ending with ".m", will be considered. If still no file is found, then variants with a leading "@" and then with both a leading "@" and trailing ".m" will be considered.

- If *name* is the name of a function defined in the interpreter but not in an m-file, then an m-file will be created in `HOME` to contain that function along with its current definition.

- If *name*`.cc` is specified, then it will search for *name*`.cc` in the path and try to modify it, otherwise it will create a new '.cc' file in the current directory. If *name* happens to be an m-file or interpreter defined function, then the text of that function will be inserted into the .cc file as a comment.

- If '*name*`.ext`' is on your path then it will be edited, otherwise the editor will be started with '*name*`.ext`' in the current directory as the filename. If '*name*`.ext`' is not modifiable, it will be copied to `HOME` before editing.

  Warning: You may need to clear *name* before the new definition is available. If you are editing a .cc file, you will need to execute `mkoctfile` '*name*`.cc`' before the definition will be available.

     If `edit` is called with *field* and *value* variables, the value of the control field *field* will be set to *value*. If an output argument is requested and the first input argument is `get` then `edit` will return the value of the control field *field*. If the control field does not exist, edit will return a structure containing all fields and values. Thus, `edit get all` returns a complete control structure. The following control fields are used:

'home'   This is the location of user local m-files. Be sure it is in your path. The default is '`~/octave`'.

'author'   This is the name to put after the "## Author:" field of new functions. By default it guesses from the `gecos` field of the password database.

'email'   This is the e-mail address to list after the name in the author field. By default it guesses `<$LOGNAME@$HOSTNAME>`, and if `$HOSTNAME` is not defined it uses `uname -n`. You probably want to override this. Be sure to use the format `<user@host>`.

'license'

'gpl'   GNU General Public License (default).

'bsd'   BSD-style license without advertising clause.

'pd'   Public domain.

'"text"'   Your own default copyright and license.

Unless you specify 'pd', edit will prepend the copyright statement with "Copyright (C) yyyy Function Author".

'mode'   This value determines whether the editor should be started in async mode (editor is started in the background and Octave continues) or sync mode (Octave waits until the editor exits). Set it to "sync" to start the editor in sync mode. The default is "async" (see [system], page 721).

'editinplace'
             Determines whether files should be edited in place, without regard to whether they are modifiable or not. The default is false.

mfilename ()   [Built-in Function]
mfilename ("*fullpath*")   [Built-in Function]
mfilename ("*fullpathext*")   [Built-in Function]
Return the name of the currently executing file.

When called from outside an m-file return the empty string. Given the argument "fullpath", include the directory part of the file name, but not the extension. Given the argument "fullpathext", include the directory part of the file name and the extension.

val = ignore_function_time_stamp ()   [Built-in Function]
old_val = ignore_function_time_stamp (*new_val*)   [Built-in Function]
Query or set the internal variable that controls whether Octave checks the time stamp on files each time it looks up functions defined in function files. If the internal variable is set to "system", Octave will not automatically recompile function files in subdirectories of '*octave-home*/lib/*version*' if they have changed since they were last compiled, but will recompile other function files in the search path if they change. If set to "all", Octave will not recompile any function files unless their definitions are removed with clear. If set to "none", Octave will always check time stamps on files to determine whether functions defined in function files need to recompiled.

11.9.1 Manipulating the Load Path

When a function is called, Octave searches a list of directories for a file that contains the function declaration. This list of directories is known as the load path. By default the load path contains a list of directories distributed with Octave plus the current working directory. To see your current load path call the path function without any input or output arguments.

It is possible to add or remove directories to or from the load path using addpath and rmpath. As an example, the following code adds '~/Octave' to the load path.

```
addpath ("~/Octave")
```

After this the directory '~/Octave' will be searched for functions.

addpath (*dir1*, ...)   [Built-in Function]
addpath (*dir1*, ..., *option*)   [Built-in Function]
> Add named directories to the function search path. If *option* is "-begin" or 0 (the default), prepend the directory name to the current path. If *option* is "-end" or 1, append the directory name to the current path. Directories added to the path must exist.
>
> In addition to accepting individual directory arguments, lists of directory names separated by **pathsep** are also accepted. For example:
>
>   addpath ("dir1:/dir2:~/dir3")
>
> **See also:** [path], page 181, [rmpath], page 181, [genpath], page 181, [pathdef], page 182, [savepath], page 181, [pathsep], page 182.

genpath (*dir*)   [Built-in Function]
genpath (*dir*, *skip*, ...)   [Built-in Function]
> Return a path constructed from *dir* and all its subdirectories. If additional string parameters are given, the resulting path will exclude directories with those names.

rmpath (*dir1*, ...)   [Built-in Function]
> Remove *dir1*, ... from the current function search path.
>
> In addition to accepting individual directory arguments, lists of directory names separated by **pathsep** are also accepted. For example:
>
>   rmpath ("dir1:/dir2:~/dir3")
>
> **See also:** [path], page 181, [addpath], page 180, [genpath], page 181, [pathdef], page 182, [savepath], page 181, [pathsep], page 182.

savepath ()   [Function File]
savepath (*file*)   [Function File]
status = savepath (...)   [Function File]
> Save the unique portion of the current function search path that is not set during Octave's initialization process to *file*. If *file* is omitted, '~/.octaverc' is used. If successful, **savepath** returns 0.
>
> **See also:** [path], page 181, [addpath], page 180, [rmpath], page 181, [genpath], page 181, [pathdef], page 182.

path (...)   [Built-in Function]
> Modify or display Octave's load path.
>
> If *nargin* and *nargout* are zero, display the elements of Octave's load path in an easy to read format.
>
> If *nargin* is zero and nargout is greater than zero, return the current load path.
>
> If *nargin* is greater than zero, concatenate the arguments, separating them with **pathsep**. Set the internal search path to the result and return it.
>
> No checks are made for duplicate elements.
>
> **See also:** [addpath], page 180, [rmpath], page 181, [genpath], page 181, [pathdef], page 182, [savepath], page 181, [pathsep], page 182.

`val = pathdef ()`   [Function File]
> Return the default path for Octave. The path information is extracted from one of three sources. The possible sources, in order of preference, are:
>
> 1. '`~/.octaverc`'
> 2. '`<octave-home>/.../<version>/m/startup/octaverc`'
> 3. Octave's path prior to changes by any octaverc.
>
> **Scc also:** [path], page 181, [addpath], page 180, [rmpath], page 181, [genpath], page 181, [savepath], page 181.

`val = pathsep ()`   [Built-in Function]
`old_val = pathsep (new_val)`   [Built-in Function]
> Query or set the character used to separate directories in a path.
>
> **See also:** [filesep], page 714.

`rehash ()`   [Built-in Function]
> Reinitialize Octave's load path directory cache.

`file_in_loadpath (file)`   [Built-in Function]
`file_in_loadpath (file, "all")`   [Built-in Function]
> Return the absolute name of *file* if it can be found in the list of directories specified by `path`. If no file is found, return an empty character string.
>
> If the first argument is a cell array of strings, search each directory of the loadpath for element of the cell array and return the first that matches.
>
> If the second optional argument `"all"` is supplied, return a cell array containing the list of all files that have the same name in the path. If no files are found, return an empty cell array.
>
> **See also:** [file_in_path], page 714, [find_dir_in_path], page 182, [path], page 181.

`restoredefaultpath (...)`   [Built-in Function]
> Restore Octave's path to its initial state at startup.
>
> **See also:** [path], page 181, [addpath], page 180, [rmpath], page 181, [genpath], page 181, [pathdef], page 182, [savepath], page 181, [pathsep], page 182.

`command_line_path (...)`   [Built-in Function]
> Return the command line path variable.
>
> **See also:** [path], page 181, [addpath], page 180, [rmpath], page 181, [genpath], page 181, [pathdef], page 182, [savepath], page 181, [pathsep], page 182.

`find_dir_in_path (dir)`   [Built-in Function]
`find_dir_in_path (dir, "all")`   [Built-in Function]
> Return the full name of the path element matching *dir*. The match is performed at the end of each path element. For example, if *dir* is `"foo/bar"`, it matches the path element `"/some/dir/foo/bar"`, but not `"/some/dir/foo/bar/baz"` `"/some/dir/allfoo/bar"`.
>
> The second argument is optional. If it is supplied, return a cell array containing all name matches rather than just the first.
>
> **See also:** [file_in_path], page 714, [file_in_loadpath], page 182, [path], page 181.

11.9.2 Subfunctions

A function file may contain secondary functions called *subfunctions*. These secondary functions are only visible to the other functions in the same function file. For example, a file 'f.m' containing

```
function f ()
  printf ("in f, calling g\n");
  g ()
endfunction
function g ()
  printf ("in g, calling h\n");
  h ()
endfunction
function h ()
  printf ("in h\n")
endfunction
```

defines a main function f and two subfunctions. The subfunctions g and h may only be called from the main function f or from the other subfunctions, but not from outside the file 'f.m'.

11.9.3 Private Functions

In many cases one function needs to access one or more helper functions. If the helper function is limited to the scope of a single function, then subfunctions as discussed above might be used. However, if a single helper function is used by more than one function, then this is no longer possible. In this case the helper functions might be placed in a subdirectory, called "private", of the directory in which the functions needing access to this helper function are found.

As a simple example, consider a function func1, that calls a helper function func2 to do much of the work. For example:

```
function y = func1 (x)
  y = func2 (x);
endfunction
```

Then if the path to func1 is <directory>/func1.m, and if func2 is found in the directory <directory>/private/func2.m, then func2 is only available for use of the functions, like func1, that are found in <directory>.

11.9.4 Nested Functions

Nested functions are similar to subfunctions in that only the main function is visible outside the file. However, they also allow for child functions to access the local variables in their parent function. This shared access mimics using a global variable to share information — but a global variable which is not visible to the rest of Octave. As a programming strategy, sharing data this way can create code which is difficult to maintain. It is recommended to use subfunctions in place of nested functions when possible.

As a simple example, consider a parent function foo, that calls a nested child function bar, with a shared variable x.

```
function y = foo ()
  x = 10;
  bar ();
  y = x;

  function bar ()
    x = 20;
  endfunction
endfunction

foo ()
  ⇒ 20
```

Notice that there is no special syntax for sharing x. This can lead to problems with accidental variable sharing between a parent function and its child. While normally variables are inherited, child function parameters and return values are local to the child function.

Now consider the function `foobar` that uses variables x and y. `foobar` calls a nested function `foo` which takes x as a parameter and returns y. `foo` then calls `bat` which does some computation.

```
function z = foobar ()
  x = 0;
  y = 0;
  z = foo (5);
  z += x + y;

  function y = foo (x)
    y = x + bat ();

    function z = bat ()
      z = x;
    endfunction
  endfunction
endfunction

foobar ()
  ⇒ 10
```

It is important to note that the x and y in `foobar` remain zero, as in `foo` they are a return value and parameter respectively. The x in `bat` refers to the x in `foo`.

Variable inheritance leads to a problem for `eval` and scripts. If a new variable is created in a parent function, it is not clear what should happen in nested child functions. For example, consider a parent function `foo` with a nested child function `bar`:

```
function y = foo (to_eval)
  bar ();
  eval (to_eval);

  function bar ()
    eval ("x = 100;");
    eval ("y = x;");
  endfunction
endfunction

foo ("x = 5;")
    ⇒ error: can not add variable "x" to a static workspace

foo ("y = 10;")
    ⇒ 10

foo ("")
    ⇒ 100
```

The parent function foo is unable to create a new variable x, but the child function bar was successful. Furthermore, even in an eval statement y in bar is the same y as in its parent function foo. The use of eval in conjunction with nested functions is best avoided.

As with subfunctions, only the first nested function in a file may be called from the outside. Inside a function the rules are more complicated. In general a nested function may call:

0. Globally visible functions

1. Any function that the nested function's parent can call

2. Sibling functions (functions that have the same parents)

3. Direct children

As a complex example consider a parent function ex_top with two child functions, ex_a and ex_b. In addition, ex_a has two more child functions, ex_aa and ex_ab. For example:

```
function ex_top ()
  ## Can call: ex_top, ex_a, and ex_b
  ## Can NOT call: ex_aa and ex_ab

  function ex_a ()
    ## Call call everything

    function ex_aa ()
      ## Can call everything
    endfunction

    function ex_ab ()
      ## Can call everything
    endfunction
  endfunction
```

```
function ex_b ()
   ## Can call: ex_top, ex_a, and ex_b
   ## Can NOT call: ex_aa and ex_ab
 endfunction
endfunction
```

11.9.5 Overloading and Autoloading

Functions can be overloaded to work with different input arguments. For example, the operator '+' has been overloaded in Octave to work with single, double, uint8, int32, and many other arguments. The preferred way to overload functions is through classes and object oriented programming (see Section 34.4.1 [Function Overloading], page 683). Occasionally, however, one needs to undo user overloading and call the default function associated with a specific type. The builtin function exists for this purpose.

[...] = builtin (*f*, ...)   [Built-in Function]
> Call the base function *f* even if *f* is overloaded to another function for the given type signature.
>
> This is normally useful when doing object-oriented programming and there is a requirement to call one of Octave's base functions rather than the overloaded one of a new class.
>
> A trivial example which redefines the sin function to be the cos function shows how builtin works.
>
> ```
> sin (0)
>   ⇒ 0
> function y = sin (x), y = cos (x); endfunction
> sin (0)
>   ⇒ 1
> builtin ("sin", 0)
>   ⇒ 0
> ```

A single dynamically linked file might define several functions. However, as Octave searches for functions based on the functions filename, Octave needs a manner in which to find each of the functions in the dynamically linked file. On operating systems that support symbolic links, it is possible to create a symbolic link to the original file for each of the functions which it contains.

However, there is at least one well known operating system that doesn't support symbolic links. Making copies of the original file for each of the functions is undesirable as it increases the amount of disk space used by Octave. Instead Octave supplies the autoload function, that permits the user to define in which file a certain function will be found.

autoload_map = autoload ()   [Built-in Function]
autoload (*function*, *file*)   [Built-in Function]
autoload (..., "*remove*")   [Built-in Function]
> Define *function* to autoload from *file*.

The second argument, *file*, should be an absolute file name or a file name in the same directory as the function or script from which the autoload command was run. *file should not* depend on the Octave load path.

Normally, calls to `autoload` appear in PKG_ADD script files that are evaluated when a directory is added to Octave's load path. To avoid having to hardcode directory names in *file*, if *file* is in the same directory as the PKG_ADD script then

```
autoload ("foo", "bar.oct");
```

will load the function `foo` from the file `bar.oct`. The above usage when `bar.oct` is not in the same directory, or usages such as

```
autoload ("foo", file_in_loadpath ("bar.oct"))
```

are strongly discouraged, as their behavior may be unpredictable.

With no arguments, return a structure containing the current autoload map.

If a third argument `"remove"` is given, the function is cleared and not loaded anymore during the current Octave session.

See also: [PKG_ADD], page 754.

11.9.6 Function Locking

It is sometime desirable to lock a function into memory with the `mlock` function. This is typically used for dynamically linked functions in Oct-files or mex-files that contain some initialization, and it is desirable that calling `clear` does not remove this initialization.

As an example,

```
function my_function ()
  mlock ();
  ...
```

prevents `my_function` from being removed from memory after it is called, even if `clear` is called. It is possible to determine if a function is locked into memory with the `mislocked`, and to unlock a function with `munlock`, which the following illustrates.

```
my_function ();
mislocked ("my_function")
⇒ ans = 1
munlock ("my_function");
mislocked ("my_function")
⇒ ans = 0
```

A common use of `mlock` is to prevent persistent variables from being removed from memory, as the following example shows:

```
function count_calls ()
  mlock ();
  persistent calls = 0;
  printf ("'count_calls' has been called %d times\n",
          ++calls);
endfunction

count_calls ();
⊣ 'count_calls' has been called 1 times

clear count_calls
count_calls ();
⊣ 'count_calls' has been called 2 times
```

mlock might equally be used to prevent changes to a function from having effect in Octave, though a similar effect can be had with the `ignore_function_time_stamp` function.

mlock ()   [Built-in Function]
 Lock the current function into memory so that it can't be cleared.

 See also: [munlock], page 188, [mislocked], page 188, [persistent], page 124.

munlock ()   [Built-in Function]
munlock (*fcn*)   [Built-in Function]
 Unlock the named function *fcn*. If no function is named then unlock the current function.

 See also: [mlock], page 188, [mislocked], page 188, [persistent], page 124.

mislocked ()   [Built-in Function]
mislocked (*fcn*)   [Built-in Function]
 Return true if the named function *fcn* is locked. If no function is named then return true if the current function is locked.

 See also: [mlock], page 188, [munlock], page 188, [persistent], page 124.

11.9.7 Function Precedence

Given the numerous different ways that Octave can define a function, it is possible and even likely that multiple versions of a function, might be defined within a particular scope. The precedence of which function will be used within a particular scope is given by

1. Subfunction A subfunction with the required function name in the given scope.

2. Private function A function defined within a private directory of the directory which contains the current function.

3. Class constructor A function that constuctors a user class as defined in chapter Chapter 34 [Object Oriented Programming], page 673.

4. Class method An overloaded function of a class as in chapter Chapter 34 [Object Oriented Programming], page 673.

5. Command-line Function A function that has been defined on the command-line.

6. Autoload function A function that is marked as autoloaded with See [autoload], page 186.

7. A Function on the Path A function that can be found on the users load-path. There can also be Oct-file, mex-file or m-file versions of this function and the precedence between these versions are in that order.

8. Built-in function A function that is a part of core Octave such as `numel`, `size`, etc.

11.10 Script Files

A script file is a file containing (almost) any sequence of Octave commands. It is read and evaluated just as if you had typed each command at the Octave prompt, and provides a convenient way to perform a sequence of commands that do not logically belong inside a function.

Unlike a function file, a script file must *not* begin with the keyword `function`. If it does, Octave will assume that it is a function file, and that it defines a single function that should be evaluated as soon as it is defined.

A script file also differs from a function file in that the variables named in a script file are not local variables, but are in the same scope as the other variables that are visible on the command line.

Even though a script file may not begin with the `function` keyword, it is possible to define more than one function in a single script file and load (but not execute) all of them at once. To do this, the first token in the file (ignoring comments and other white space) must be something other than `function`. If you have no other statements to evaluate, you can use a statement that has no effect, like this:

```
# Prevent Octave from thinking that this
# is a function file:

1;

# Define function one:

function one ()
  ...
```

To have Octave read and compile these functions into an internal form, you need to make sure that the file is in Octave's load path (accessible through the `path` function), then simply type the base name of the file that contains the commands. (Octave uses the same rules to search for script files as it does to search for function files.)

If the first token in a file (ignoring comments) is `function`, Octave will compile the function and try to execute it, printing a message warning about any non-whitespace characters that appear after the function definition.

Note that Octave does not try to look up the definition of any identifier until it needs to evaluate it. This means that Octave will compile the following statements if they appear in a script file, or are typed at the command line,

```
# not a function file:
1;
function foo ()
  do_something ();
endfunction
function do_something ()
  do_something_else ();
endfunction
```

even though the function `do_something` is not defined before it is referenced in the function `foo`. This is not an error because Octave does not need to resolve all symbols that are referenced by a function until the function is actually evaluated.

Since Octave doesn't look for definitions until they are needed, the following code will always print 'bar = 3' whether it is typed directly on the command line, read from a script file, or is part of a function body, even if there is a function or script file called 'bar.m' in Octave's path.

```
eval ("bar = 3");
bar
```

Code like this appearing within a function body could fool Octave if definitions were resolved as the function was being compiled. It would be virtually impossible to make Octave clever enough to evaluate this code in a consistent fashion. The parser would have to be able to perform the call to `eval` at compile time, and that would be impossible unless all the references in the string to be evaluated could also be resolved, and requiring that would be too restrictive (the string might come from user input, or depend on things that are not known until the function is evaluated).

Although Octave normally executes commands from script files that have the name '*file*.m', you can use the function `source` to execute commands from any file.

source (*file*)   [Built-in Function]
    Parse and execute the contents of *file*.

    This is equivalent to executing commands from a script file, but without requiring the file to be named '*file*.m'.

    See also: [run], page 152.

11.11 Function Handles, Anonymous Functions, Inline Functions

It can be very convenient store a function in a variable so that it can be passed to a different function. For example, a function that performs numerical minimization needs access to the function that should be minimized.

11.11.1 Function Handles

A function handle is a pointer to another function and is defined with the syntax

    @*function-name*

For example,

```
f = @sin;
```

creates a function handle called f that refers to the function sin.

Function handles are used to call other functions indirectly, or to pass a function as an argument to another function like quad or fsolve. For example:

```
f = @sin;
quad (f, 0, pi)
    ⇒ 2
```

You may use feval to call a function using function handle, or simply write the name of the function handle followed by an argument list. If there are no arguments, you must use an empty argument list '()'. For example:

```
f = @sin;
feval (f, pi/4)
    ⇒ 0.70711
f (pi/4)
    ⇒ 0.70711
```

is_function_handle (*x*)   [Built-in Function]
    Return true if *x* is a function handle.

    See also: [isa], page 39, [typeinfo], page 39, [class], page 39, [functions], page 191.

***s* = functions (*fcn_handle*)**   [Built-in Function]
    Return a structure containing information about the function handle *fcn_handle*.

    The structure *s* always contains these 3 fields:

    function   The function name. For an anonymous function (no name) this will be the actual function definition.

    type   Type of the function.

        anonymous
            The function is anonymous.

        private   The function is private.

        overloaded
            The function overloads an existing function.

        simple   The function is a built-in or m-file function.

        subfunction
            The function is a subfunction within an m-file.

    file   The m-file that will be called to perform the function. This field is empty for anonymous and built-in functions.

    In addition, some function types may return more information in additional fields.

    Warning: functions is provided for debugging purposes only. It's behavior may change in the future and programs should not depend on a particular output.

func2str (*fcn_handle*)   [Built-in Function]
> Return a string containing the name of the function referenced by the function handle *fcn_handle*.

> **See also:** [str2func], page 192, [functions], page 191.

str2func (*fcn_name*)   [Built-in Function]
str2func (*fcn_name*, "*global*")   [Built-in Function]
> Return a function handle constructed from the string *fcn_name*. If the optional "**global**" argument is passed, locally visible functions are ignored in the lookup.

> **See also:** [func2str], page 192, [inline], page 192.

11.11.2 Anonymous Functions

Anonymous functions are defined using the syntax

```
@(argument-list) expression
```

Any variables that are not found in the argument list are inherited from the enclosing scope. Anonymous functions are useful for creating simple unnamed functions from expressions or for wrapping calls to other functions to adapt them for use by functions like **quad**. For example,

```
f = @(x) x.^2;
quad (f, 0, 10)
    ⇒ 333.33
```

creates a simple unnamed function from the expression `x.^2` and passes it to **quad**,

```
quad (@(x) sin (x), 0, pi)
    ⇒ 2
```

wraps another function, and

```
a = 1;
b = 2;
quad (@(x) betainc (x, a, b), 0, 0.4)
    ⇒ 0.13867
```

adapts a function with several parameters to the form required by **quad**. In this example, the values of a and b that are passed to **betainc** are inherited from the current environment.

11.11.3 Inline Functions

An inline function is created from a string containing the function body using the **inline** function. The following code defines the function $f(x) = x^2 + 2$.

```
f = inline ("x^2 + 2");
```

After this it is possible to evaluate f at any x by writing **f(x)**.

inline (*str*)   [Built-in Function]
inline (*str*, *arg1*, ...)   [Built-in Function]
inline (*str*, *n*)   [Built-in Function]
> Create an inline function from the character string *str*.

> If called with a single argument, the arguments of the generated function are extracted from the function itself. The generated function arguments will then be in alphabetical

order. It should be noted that i, and j are ignored as arguments due to the ambiguity between their use as a variable or their use as an inbuilt constant. All arguments followed by a parenthesis are considered to be functions. If no arguments are found, a function taking a single argument named x will be created.

If the second and subsequent arguments are character strings, they are the names of the arguments of the function.

If the second argument is an integer n, the arguments are "x", "P1", ..., "PN".

Programming Note: The use of inline is discouraged and it may be removed from a future version of Octave. The preferred way to create functions from strings is through the use of anonymous functions (see Section 11.11.2 [Anonymous Functions], page 192) or str2func.

See also: [argnames], page 193, [formula], page 193, [vectorize], page 458, [str2func], page 192.

argnames (*fun*)   [Built-in Function]
> Return a cell array of character strings containing the names of the arguments of the inline function *fun*.

> **See also:** [inline], page 192, [formula], page 193, [vectorize], page 458.

formula (*fun*)   [Built-in Function]
> Return a character string representing the inline function *fun*. Note that char (*fun*) is equivalent to formula (*fun*).

> **See also:** [argnames], page 193, [inline], page 192, [vectorize], page 458.

symvar (*s*)   [Function File]
> Identify the argument names in the function defined by a string. Common constant names such as pi, NaN, Inf, eps, i or j are ignored. The arguments that are found are returned in a cell array of strings. If no variables are found then the returned cell array is empty.

11.12 Commands

Commands are a special class of functions that only accept string input arguments. A command can be called as an ordinary function, but it can also be called without the parentheses. For example,

```
my_command hello world
```

is equivalent to

```
my_command ("hello", "world")
```

The general form of a command call is

```
cmdname arg1 arg2 ...
```

which translates directly to

```
cmdname ("arg1", "arg2", ...)
```

Any regular function can be used as a command if it accepts string input arguments. For example:

```
toupper lower_case_arg
    ⇒ ans = LOWER_CASE_ARG
```

One difficulty of commands occurs when one of the string input arguments is stored in a variable. Because Octave can't tell the difference between a variable name and an ordinary string, it is not possible to pass a variable as input to a command. In such a situation a command must be called as a function. For example:

```
strvar = "hello world";
toupper strvar
    ⇒ ans = STRVAR
toupper (strvar)
    ⇒ ans = HELLO WORLD
```

11.13 Organization of Functions Distributed with Octave

Many of Octave's standard functions are distributed as function files. They are loosely organized by topic, in subdirectories of '*octave-home*/lib/octave/*version*/m', to make it easier to find them.

The following is a list of all the function file subdirectories, and the types of functions you will find there.

'audio'   Functions for playing and recording sounds.

'deprecated'
            Out-of-date functions which will eventually be removed from Octave.

'elfun'   Elementary functions, principally trigonometric.

'@ftp'   Class functions for the FTP object.

'general'   Miscellaneous matrix manipulations, like flipud, rot90, and triu, as well as other basic functions, like ismatrix, nargchk, etc.

'geometry'
            Functions related to Delaunay triangulation.

'help'   Functions for Octave's built-in help system.

'image'   Image processing tools. These functions require the X Window System.

'io'   Input-output functions.

'linear-algebra'
            Functions for linear algebra.

'miscellaneous'
            Functions that don't really belong anywhere else.

'optimization'
            Functions related to minimization, optimization, and root finding.

'path'   Functions to manage the directory path Octave uses to find functions.

'pkg'   Package manager for installing external packages of functions in Octave.

'plot'   Functions for displaying and printing two- and three-dimensional graphs.

'polynomial'
: Functions for manipulating polynomials.

'prefs'   Functions implementing user-defined preferences.

'set'   Functions for creating and manipulating sets of unique values.

'signal'   Functions for signal processing applications.

'sparse'   Functions for handling sparse matrices.

'specfun'   Special functions such as `bessel` or `factor`.

'special-matrix'
: Functions that create special matrix forms such as Hilbert or Vandermonde matrices.

'startup'   Octave's system-wide startup file.

'statistics'
: Statistical functions.

'strings'   Miscellaneous string-handling functions.

'testfun'   Functions for performing unit tests on other functions.

'time'   Functions related to time and date processing.

12 Errors and Warnings

Octave includes several functions for printing error and warning messages. When you write functions that need to take special action when they encounter abnormal conditions, you should print the error messages using the functions described in this chapter.

Since many of Octave's functions use these functions, it is also useful to understand them, so that errors and warnings can be handled.

12.1 Handling Errors

An error is something that occurs when a program is in a state where it doesn't make sense to continue. An example is when a function is called with too few input arguments. In this situation the function should abort with an error message informing the user of the lacking input arguments.

Since an error can occur during the evaluation of a program, it is very convenient to be able to detect that an error occurred, so that the error can be fixed. This is possible with the **try** statement described in Section 10.9 [The try Statement], page 164.

12.1.1 Raising Errors

The most common use of errors is for checking input arguments to functions. The following example calls the **error** function if the function **f** is called without any input arguments.

```
function f (arg1)
  if (nargin == 0)
    error ("not enough input arguments");
  endif
endfunction
```

When the **error** function is called, it prints the given message and returns to the Octave prompt. This means that no code following a call to **error** will be executed.

error (*template*, ...)   [Built-in Function]
error (*id*, *template*, ...)   [Built-in Function]
> Format the optional arguments under the control of the template string *template* using the same rules as the **printf** family of functions (see Section 14.2.4 [Formatted Output], page 243) and print the resulting message on the **stderr** stream. The message is prefixed by the character string 'error: '.
>
> Calling **error** also sets Octave's internal error state such that control will return to the top level without evaluating any more commands. This is useful for aborting from functions or scripts.
>
> If the error message does not end with a new line character, Octave will print a traceback of all the function calls leading to the error. For example, given the following function definitions:
>
> ```
> function f () g (); end
> function g () h (); end
> function h () nargin == 1 || error ("nargin != 1"); end
> ```
>
> calling the function **f** will result in a list of messages that can help you to quickly locate the exact location of the error:

```
f ()
error: nargin != 1
error: called from:
error:   error at line -1, column -1
error:   h at line 1, column 27
error:   g at line 1, column 15
error:   f at line 1, column 15
```

If the error message ends in a new line character, Octave will print the message but will not display any traceback messages as it returns control to the top level. For example, modifying the error message in the previous example to end in a new line causes Octave to only print a single message:

```
function h () nargin == 1 || error ("nargin != 1\n"); end
f ()
error: nargin != 1
```

A null string ("") input to error will be ignored and the code will continue running as if the statement were a NOP. This is for compatibility with MATLAB. It also makes it possible to write code such as

```
err_msg = "";
if (CONDITION 1)
  err_msg = "CONDITION 1 found";
elseif (CONDITION2)
  err_msg = "CONDITION 2 found";
...
endif
error (err_msg);
```

which will only stop execution if an error has been found.

Implementation Note: For compatibility with MATLAB, escape sequences (e.g., "\n" => newline) are processed in *template* regardless of whether *template* has been defined within single quotes as long as there are two or more input arguments. Use a second backslash to stop interpolation of the escape sequence (e.g., "\\n") or use the regexptranslate function.

See also: [warning], page 204, [lasterror], page 201.

Since it is common to use errors when there is something wrong with the input to a function, Octave supports functions to simplify such code. When the print_usage function is called, it reads the help text of the function calling print_usage, and presents a useful error. If the help text is written in Texinfo it is possible to present an error message that only contains the function prototypes as described by the @deftypefn parts of the help text. When the help text isn't written in Texinfo, the error message contains the entire help message.

Consider the following function.

```
## -*- texinfo -*-
## @deftypefn {Function File} f (@var{arg1})
## Function help text goes here...
## @end deftypefn
function f (arg1)
  if (nargin == 0)
    print_usage ();
  endif
endfunction
```

When it is called with no input arguments it produces the following error.

```
f ()
```

```
⊣   error: Invalid call to f.  Correct usage is:
⊣
⊣     -- Function File: f (ARG1)
⊣
⊣
⊣   Additional help for built-in functions and operators is
⊣   available in the online version of the manual.  Use the command
⊣   'doc <topic>' to search the manual index.
⊣
⊣   Help and information about Octave is also available on the WWW
⊣   at http://www.octave.org and via the help@octave.org
⊣   mailing list.
```

print_usage ()   [Function File]
print_usage (_name_**)**   [Function File]
> Print the usage message for a function. When called with no input arguments the
> **print_usage** function displays the usage message of the currently executing function.

> **See also:** [help], page 20.

usage (_msg_**)**   [Built-in Function]
> Print the message _msg_, prefixed by the string 'usage: ', and set Octave's internal
> error state such that control will return to the top level without evaluating any more
> commands. This is useful for aborting from functions.

> After **usage** is evaluated, Octave will print a traceback of all the function calls leading
> to the usage message.

> You should use this function for reporting problems errors that result from an im-
> proper call to a function, such as calling a function with an incorrect number of
> arguments, or with arguments of the wrong type. For example, most functions dis-
> tributed with Octave begin with code like this

```
if (nargin != 2)
  usage ("foo (a, b)");
endif
```

> to check for the proper number of arguments.

`beep ()`   [Function File]
> Produce a beep from the speaker (or visual bell).

> **See also:** [puts], page 242, [fputs], page 242, [printf], page 243, [fprintf], page 243.

`val = beep_on_error ()`   [Built-in Function]
`old_val = beep_on_error (new_val)`   [Built-in Function]
`beep_on_error (new_val, "local")`   [Built-in Function]
> Query or set the internal variable that controls whether Octave will try to ring the terminal bell before printing an error message.

> When called from inside a function with the `"local"` option, the variable is changed locally for the function and any subroutines it calls. The original variable value is restored when exiting the function.

12.1.2 Catching Errors

When an error occurs, it can be detected and handled using the **try** statement as described in Section 10.9 [The try Statement], page 164. As an example, the following piece of code counts the number of errors that occurs during a **for** loop.

```
number_of_errors = 0;
for n = 1:100
  try
    ...
  catch
    number_of_errors++;
  end_try_catch
endfor
```

The above example treats all errors the same. In many situations it can however be necessary to discriminate between errors, and take different actions depending on the error. The **lasterror** function returns a structure containing information about the last error that occurred. As an example, the code above could be changed to count the number of errors related to the '*' operator.

```
number_of_errors = 0;
for n = 1:100
  try
    ...
  catch
    msg = lasterror.message;
    if (strfind (msg, "operator *"))
      number_of_errors++;
    endif
  end_try_catch
endfor
```

Alternatively, the output of the **lasterror** function can be found in a variable indicated immediately after the **catch** keyword, as in the example below showing how to redirect an error as a warning:

```
try
  ...
catch err
  warning(err.identifier, err.message);
  ...
end_try_catch
```

lasterr = lasterror ()   [Built-in Function]
lasterror (*err*)   [Built-in Function]
lasterror ("*reset*")   [Built-in Function]
> Query or set the last error message structure. When called without arguments, return
> a structure containing the last error message and other information related to this
> error. The elements of the structure are:

> message   The text of the last error message

> identifier
> >   The message identifier of this error message

> stack   A structure containing information on where the message occurred. This
> >   may be an empty structure if the information cannot be obtained. The
> >   fields of the structure are:

> > > file   The name of the file where the error occurred

> > > name   The name of function in which the error occurred

> > > line   The line number at which the error occurred

> > > column   An optional field with the column number at which the error
> > >   occurred

> The last error structure may be set by passing a scalar structure, *err*, as input. Any
> fields of *err* that match those above are set while any unspecified fields are initialized
> with default values.

> If **lasterror** is called with the argument **"reset"**, all fields are set to their default
> values.

> **See also:** [lasterr], page 201, [error], page 197, [lastwarn], page 205.

[*msg*, *msgid*] = lasterr ()   [Built-in Function]
lasterr (*msg*)   [Built-in Function]
lasterr (*msg*, *msgid*)   [Built-in Function]
> Query or set the last error message. When called without input arguments, return
> the last error message and message identifier. With one argument, set the last error
> message to *msg*. With two arguments, also set the last message identifier.

> **See also:** [lasterror], page 201, [error], page 197, [lastwarn], page 205.

It is also possible to assign an identification string to an error. If an error has such
an ID the user can catch this error as will be shown in the next example. To assign
an ID to an error, simply call **error** with two string arguments, where the first is the
identification string, and the second is the actual error. Note that error IDs are in the format

"NAMESPACE:ERROR-NAME". The namespace "Octave" is used for Octave's own errors. Any other string is available as a namespace for user's own errors.

The next example counts indexing errors. The errors are caught using the field identifier of the structure returned by the function lasterror.

```
number_of_errors = 0;
for n = 1:100
  try
    ...
  catch
    id = lasterror.identifier;
    if (strcmp (id, "Octave:invalid-indexing"))
      number_of_errors++;
    endif
  end_try_catch
endfor
```

The functions distributed with Octave can issue one of the following errors.

Octave:invalid-context
> Indicates the error was generated by an operation that cannot be executed in the scope from which it was called. For example, the function print_usage () when called from the Octave prompt raises this error.

Octave:invalid-input-arg
> Indicates that a function was called with invalid input arguments.

Octave:invalid-fun-call
> Indicates that a function was called in an incorrect way, e.g., wrong number of input arguments.

Octave:invalid-indexing
> Indicates that a data-type was indexed incorrectly, e.g., real-value index for arrays, non-existent field of a structure.

Octave:bad-alloc
> Indicates that memory couldn't be allocated.

Octave:undefined-function
> Indicates a call to a function that is not defined. The function may exist but Octave is unable to find it in the search path.

When an error has been handled it is possible to raise it again. This can be useful when an error needs to be detected, but the program should still abort. This is possible using the rethrow function. The previous example can now be changed to count the number of errors related to the '*' operator, but still abort if another kind of error occurs.

```
number_of_errors = 0;
for n = 1:100
  try
    ...
  catch
    msg = lasterror.message;
    if (strfind (msg, "operator *"))
      number_of_errors++;
    else
      rethrow (lasterror);
    endif
  end_try_catch
endfor
```

rethrow (*err*)   [Built-in Function]

Reissue a previous error as defined by *err*. *err* is a structure that must contain at least the `"message"` and `"identifier"` fields. *err* can also contain a field `"stack"` that gives information on the assumed location of the error. Typically *err* is returned from `lasterror`.

See also: [lasterror], page 201, [lasterr], page 201, [error], page 197.

err = **errno** ()   [Built-in Function]
err = **errno** (*val*)   [Built-in Function]
err = **errno** (*name*)   [Built-in Function]

Return the current value of the system-dependent variable errno, set its value to *val* and return the previous value, or return the named error code given *name* as a character string, or -1 if *name* is not found.

errno_list ()   [Built-in Function]

Return a structure containing the system-dependent errno values.

12.1.3 Recovering From Errors

Octave provides several ways of recovering from errors. There are `try/catch` blocks, `unwind_protect/unwind_protect_cleanup` blocks, and finally the `onCleanup` command.

The `onCleanup` command associates an ordinary Octave variable (the trigger) with an arbitrary function (the action). Whenever the Octave variable ceases to exist—whether due to a function return, an error, or simply because the variable has been removed with `clear`—then the assigned function is executed.

The function can do anything necessary for cleanup such as closing open file handles, printing an error message, or restoring global variables to their initial values. The last example is a very convenient idiom for Octave code. For example:

```
function rand42
  old_state = rand ("state");
  restore_state = onCleanup (@() rand ("state", old_state);
  rand ("state", 42);
  ...
endfunction  # rand generator state restored by onCleanup
```

obj = onCleanup (*function*)   [Built-in Function]
> Create a special object that executes a given function upon destruction. If the object
> is copied to multiple variables (or cell or struct array elements) or returned from a
> function, *function* will be executed after clearing the last copy of the object. Note
> that if multiple local onCleanup variables are created, the order in which they are
> called is unspecified. For similar functionality See Section 10.8 [The unwind_protect
> Statement], page 164.

12.2 Handling Warnings

Like an error, a warning is issued when something unexpected happens. Unlike an error,
a warning doesn't abort the currently running program. A simple example of a warning is
when a number is divided by zero. In this case Octave will issue a warning and assign the
value Inf to the result.

```
a = 1/0
     ⊣ warning: division by zero
     ⇒ a = Inf
```

12.2.1 Issuing Warnings

It is possible to issue warnings from any code using the **warning** function. In its most simple
form, the **warning** function takes a string describing the warning as its input argument. As
an example, the following code controls if the variable 'a' is non-negative, and if not issues
a warning and sets 'a' to zero.

```
a = -1;
if (a < 0)
  warning ("'a' must be non-negative.  Setting 'a' to zero.");
  a = 0;
endif
     ⊣ 'a' must be non-negative.  Setting 'a' to zero.
```

Since warnings aren't fatal to a running program, it is not possible to catch a warning
using the **try** statement or something similar. It is however possible to access the last
warning as a string using the **lastwarn** function.

It is also possible to assign an identification string to a warning. If a warning has such an
ID the user can enable and disable this warning as will be described in the next section. To
assign an ID to a warning, simply call **warning** with two string arguments, where the first
is the identification string, and the second is the actual warning. Note that warning IDs are
in the format "NAMESPACE:WARNING-NAME". The namespace "Octave" is used for Octave's
own warnings. Any other string is available as a namespace for user's own warnings.

warning (*template*, ...)   [Built-in Function]
warning (*id*, *template*, ...)   [Built-in Function]
warning ("*on*", *id*)   [Built-in Function]
warning ("*off*", *id*)   [Built-in Function]
warning ("*query*", *id*)   [Built-in Function]
warning ("*error*", *id*)   [Built-in Function]

`warning (`*`state`*`, `*`id`*`, `*`"local"`*`)`   [Built-in Function]

> Format the optional arguments under the control of the template string *template* using the same rules as the `printf` family of functions (see Section 14.2.4 [Formatted Output], page 243) and print the resulting message on the `stderr` stream. The message is prefixed by the character string 'warning: '. You should use this function when you want to notify the user of an unusual condition, but only when it makes sense for your program to go on.

> The optional message identifier allows users to enable or disable warnings tagged by *id*. A message identifier is of the form "NAMESPACE:WARNING-NAME". Octave's own warnings use the "Octave" namespace (see [XREFwarning_ids], page 205). The special identifier "all" may be used to set the state of all warnings.

> If the first argument is "on" or "off", set the state of a particular warning using the identifier *id*. If the first argument is "query", query the state of this warning instead. If the identifier is omitted, a value of "all" is assumed. If you set the state of a warning to "error", the warning named by *id* is handled as if it were an error instead. So, for example, the following handles all warnings as errors:

> ```
> warning ("error");
> ```

> If the state is "on", "off", or "error" and the third argument is "local", then the warning state will be set temporarily, until the end of the current function. Changes to warning states that are set locally affect the current function and all functions called from the current scope. The previous warning state is restored on return from the current function. The "local" option is ignored if used in the top-level workspace.

> Implementation Note: For compatibility with MATLAB, escape sequences (e.g., "\n" => newline) are processed in *template* regardless of whether *template* has been defined within single quotes as long as there are two or more input arguments. Use a second backslash to stop interpolation of the escape sequence (e.g., "\\n") or use the `regexptranslate` function.

> **See also:** [warning_ids], page 205, [lastwarn], page 205, [error], page 197.

`[`*`msg`*`, `*`msgid`*`] = lastwarn ()`   [Built-in Function]
`lastwarn (`*`msg`*`)`   [Built-in Function]
`lastwarn (`*`msg`*`, `*`msgid`*`)`   [Built-in Function]

> Query or set the last warning message. When called without input arguments, return the last warning message and message identifier. With one argument, set the last warning message to *msg*. With two arguments, also set the last message identifier.

> **See also:** [warning], page 204, [lasterror], page 201, [lasterr], page 201.

The functions distributed with Octave can issue one of the following warnings.

`Octave:abbreviated-property-match`
> By default, the `Octave:abbreviated-property-match` warning is enabled.

`Octave:array-to-scalar`
> If the `Octave:array-to-scalar` warning is enabled, Octave will warn when an implicit conversion from an array to a scalar value is attempted. By default, the `Octave:array-to-scalar` warning is disabled.

`Octave:array-to-vector`

> If the `Octave:array-to-vector` warning is enabled, Octave will warn when an implicit conversion from an array to a vector value is attempted. By default, the `Octave:array-to-vector` warning is disabled.

`Octave:assign-as-truth-value`

> If the `Octave:assign-as-truth-value` warning is enabled, a warning is issued for statements like
>
>   if (s = t)
>   ...
>
> since such statements are not common, and it is likely that the intent was to write
>
>   if (s == t)
>   ...
>
> instead.
>
> There are times when it is useful to write code that contains assignments within the condition of a `while` or `if` statement. For example, statements like
>
>   while (c = getc ())
>   ...
>
> are common in C programming.
>
> It is possible to avoid all warnings about such statements by disabling the `Octave:assign-as-truth-value` warning, but that may also let real errors like
>
>   if (x = 1)   # intended to test (x == 1)!
>   ...
>
> slip by.
>
> In such cases, it is possible suppress errors for specific statements by writing them with an extra set of parentheses. For example, writing the previous example as
>
>   while ((c = getc ()))
>   ...
>
> will prevent the warning from being printed for this statement, while allowing Octave to warn about other assignments used in conditional contexts.
>
> By default, the `Octave:assign-as-truth-value` warning is enabled.

`Octave:associativity-change`

> If the `Octave:associativity-change` warning is enabled, Octave will warn about possible changes in the meaning of some code due to changes in associativity for some operators. Associativity changes have typically been made for MATLAB compatibility. By default, the `Octave:associativity-change` warning is enabled.

`Octave:autoload-relative-file-name`

> If the `Octave:autoload-relative-file-name` is enabled, Octave will warn when parsing autoload() function calls with relative paths to function files.

This usually happens when using autoload() calls in PKG_ADD files, when the PKG_ADD file is not in the same directory as the .oct file referred to by the autoload() command. By default, the `Octave:autoload-relative-file-name` warning is enabled.

`Octave:broadcast`

Warn when performing broadcasting operations. By default, this is enabled. See Section 19.2 [Broadcasting], page 459 in the chapter Vectorization and Faster Code Execution of the manual.

`Octave:built-in-variable-assignment`

By default, the `Octave:built-in-variable-assignment` warning is enabled.

`Octave:deprecated-keyword`

If the `Octave:deprecated-keyword` warning is enabled, a warning is issued when Octave encounters a keyword that is obsolete and scheduled for removal from Octave. By default, the `Octave:deprecated-keyword` warning is enabled.

`Octave:divide-by-zero`

If the `Octave:divide-by-zero` warning is enabled, a warning is issued when Octave encounters a division by zero. By default, the `Octave:divide-by-zero` warning is enabled.

`Octave:fopen-file-in-path`

By default, the `Octave:fopen-file-in-path` warning is enabled.

`Octave:function-name-clash`

If the `Octave:function-name-clash` warning is enabled, a warning is issued when Octave finds that the name of a function defined in a function file differs from the name of the file. (If the names disagree, the name declared inside the file is ignored.) By default, the `Octave:function-name-clash` warning is enabled.

`Octave:future-time-stamp`

If the `Octave:future-time-stamp` warning is enabled, Octave will print a warning if it finds a function file with a time stamp that is in the future. By default, the `Octave:future-time-stamp` warning is enabled.

`Octave:glyph-render`

By default, the `Octave:glyph-render` warning is enabled.

`Octave:imag-to-real`

If the `Octave:imag-to-real` warning is enabled, a warning is printed for implicit conversions of complex numbers to real numbers. By default, the `Octave:imag-to-real` warning is disabled.

`Octave:load-file-in-path`

By default, the `Octave:load-file-in-path` warning is enabled.

`Octave:logical-conversion`

By default, the `Octave:logical-conversion` warning is enabled.

`Octave:matlab-incompatible`

Print warnings for Octave language features that may cause compatibility problems with MATLAB. By default, the `Octave:matlab-incompatible` warning is disabled. The –traditional or –braindead startup options for Octave may also be of use, see Section 2.1.1 [Command Line Options], page 15.

`Octave:md5sum-file-in-path`
> By default, the `Octave:md5sum-file-in-path` warning is enabled.

`Octave:missing-glyph`
> By default, the `Octave:missing-glyph` warning is enabled.

`Octave:missing-semicolon`
> If the `Octave:missing-semicolon` warning is enabled, Octave will warn when statements in function definitions don't end in semicolons. By default the `Octave:missing-semicolon` warning is disabled.

`Octave:mixed-string-concat`
> If the `Octave:mixed-string-concat` warning is enabled, print a warning when concatenating a mixture of double and single quoted strings. By default, the `Octave:mixed-string-concat` warning is disabled.

`Octave:neg-dim-as-zero`
> If the `Octave:neg-dim-as-zero` warning is enabled, print a warning for expressions like
>
>   eye (-1)
>
> By default, the `Octave:neg-dim-as-zero` warning is disabled.

`Octave:nested-functions-coerced`
> By default, the `Octave:nested-functions-coerced` warning is enabled.

`Octave:noninteger-range-as-index`
> By default, the `Octave:noninteger-range-as-index` warning is enabled.

`Octave:num-to-str`
> If the `Octave:num-to-str` warning is enable, a warning is printed for implicit conversions of numbers to their ASCII character equivalents when strings are constructed using a mixture of strings and numbers in matrix notation. For example,
>
>   ["f", 111, 111]
>   ⇒ "foo"
>
> elicits a warning if the `Octave:num-to-str` warning is enabled. By default, the `Octave:num-to-str` warning is enabled.

`Octave:possible-matlab-short-circuit-operator`
> If the `Octave:possible-matlab-short-circuit-operator` warning is enabled, Octave will warn about using the not short circuiting operators & and | inside `if` or `while` conditions. They normally never short circuit, but MATLAB always short circuits if any logical operators are used in a condition. You can turn on the option
>
>   do_braindead_shortcircuit_evaluation (1)
>
> if you would like to enable this short-circuit evaluation in Octave. Note that the && and || operators always short circuit in both Octave and MATLAB, so it's only necessary to enable MATLAB-style short-circuiting if it's too arduous to modify

existing code that relies on this behavior. By default, the `Octave:possible-matlab-short-circuit-operator` warning is enabled.

`Octave:precedence-change`

If the `Octave:precedence-change` warning is enabled, Octave will warn about possible changes in the meaning of some code due to changes in precedence for some operators. Precedence changes have typically been made for MATLAB compatibility. By default, the `Octave:precedence-change` warning is enabled.

`Octave:recursive-path-search`

By default, the `Octave:recursive-path-search` warning is enabled.

`Octave:remove-init-dir`

The `path` function changes the search path that Octave uses to find functions. It is possible to set the path to a value which excludes Octave's own built-in functions. If the `Octave:remove-init-dir` warning is enabled then Octave will warn when the `path` function has been used in a way that may render Octave unworkable. By default, the `Octave:remove-init-dir` warning is enabled.

`Octave:reload-forces-clear`

If several functions have been loaded from the same file, Octave must clear all the functions before any one of them can be reloaded. If the `Octave:reload-forces-clear` warning is enabled, Octave will warn you when this happens, and print a list of the additional functions that it is forced to clear. By default, the `Octave:reload-forces-clear` warning is enabled.

`Octave:resize-on-range-error`

If the `Octave:resize-on-range-error` warning is enabled, print a warning when a matrix is resized by an indexed assignment with indices outside the current bounds. By default, the ## `Octave:resize-on-range-error` warning is disabled.

`Octave:separator-insert`

Print warning if commas or semicolons might be inserted automatically in literal matrices. By default, the `Octave:separator-insert` warning is disabled.

`Octave:shadowed-function`

By default, the `Octave:shadowed-function` warning is enabled.

`Octave:single-quote-string`

Print warning if a single quote character is used to introduce a string constant. By default, the `Octave:single-quote-string` warning is disabled.

`Octave:singular-matrix-div`

By default, the `Octave:singular-matrix-div` warning is enabled.

`Octave:sqrtm:SingularMatrix`

By default, the `Octave:sqrtm:SingularMatrix` warning is enabled.

`Octave:str-to-num`

If the `Octave:str-to-num` warning is enabled, a warning is printed for implicit conversions of strings to their numeric ASCII equivalents. For example,

```
"abc" + 0
  ⇒ 97 98 99
```

elicits a warning if the `Octave:str-to-num` warning is enabled. By default, the
`Octave:str-to-num` warning is disabled.

`Octave:undefined-return-values`

> If the `Octave:undefined-return-values` warning is disabled, print a warning
> if a function does not define all the values in the return list which are expected.
> By default, the `Octave:undefined-return-values` warning is enabled.

`Octave:variable-switch-label`

> If the `Octave:variable-switch-label` warning is enabled, Octave will print a
> warning if a switch label is not a constant or constant expression. By default,
> the `Octave:variable-switch-label` warning is disabled.

12.2.2 Enabling and Disabling Warnings

The `warning` function also allows you to control which warnings are actually printed to
the screen. If the `warning` function is called with a string argument that is either `"on"` or
`"off"` all warnings will be enabled or disabled.

It is also possible to enable and disable individual warnings through their string identi-
fications. The following code will issue a warning

```
warning ("example:non-negative-variable",
         "'a' must be non-negative.  Setting 'a' to zero.");
```

while the following won't issue a warning

```
warning ("off", "example:non-negative-variable");
warning ("example:non-negative-variable",
         "'a' must be non-negative.  Setting 'a' to zero.");
```

13 Debugging

Octave includes a built-in debugger to aid in the development of scripts. This can be used to interrupt the execution of an Octave script at a certain point, or when certain conditions are met. Once execution has stopped, and debug mode is entered, the symbol table at the point where execution has stopped can be examined and modified to check for errors.

The normal command-line editing and history functions are available in debug mode.

13.1 Entering Debug Mode

There are two basic means of interrupting the execution of an Octave script. These are breakpoints (see Section 13.3 [Breakpoints], page 212), discussed in the next section, and interruption based on some condition.

Octave supports three means to stop execution based on the values set in the functions `debug_on_interrupt`, `debug_on_warning`, and `debug_on_error`.

val = debug_on_interrupt ()   [Built-in Function]
old_val = debug_on_interrupt (*new_val*)   [Built-in Function]
debug_on_interrupt (*new_val*, "*local*")   [Built-in Function]
> Query or set the internal variable that controls whether Octave will try to enter debugging mode when it receives an interrupt signal (typically generated with `C-c`). If a second interrupt signal is received before reaching the debugging mode, a normal interrupt will occur.
>
> When called from inside a function with the `"local"` option, the variable is changed locally for the function and any subroutines it calls. The original variable value is restored when exiting the function.
>
> **See also:** [debug_on_error], page 211, [debug_on_warning], page 211.

val = debug_on_warning ()   [Built-in Function]
old_val = debug_on_warning (*new_val*)   [Built-in Function]
debug_on_warning (*new_val*, "*local*")   [Built-in Function]
> Query or set the internal variable that controls whether Octave will try to enter the debugger when a warning is encountered.
>
> When called from inside a function with the `"local"` option, the variable is changed locally for the function and any subroutines it calls. The original variable value is restored when exiting the function.
>
> **See also:** [debug_on_error], page 211, [debug_on_interrupt], page 211.

val = debug_on_error ()   [Built-in Function]
old_val = debug_on_error (*new_val*)   [Built-in Function]
debug_on_error (*new_val*, "*local*")   [Built-in Function]
> Query or set the internal variable that controls whether Octave will try to enter the debugger when an error is encountered. This will also inhibit printing of the normal traceback message (you will only see the top-level error message).
>
> When called from inside a function with the `"local"` option, the variable is changed locally for the function and any subroutines it calls. The original variable value is restored when exiting the function.

See also: [debug_on_warning], page 211, [debug_on_interrupt], page 211.

13.2 Leaving Debug Mode

Use either `dbcont` or `return` to leave the debug mode and continue the normal execution of the script.

`dbcont`   [Command]
>   Leave command-line debugging mode and continue code execution normally.
>
>   **See also:** [dbstep], page 215, [dbquit], page 212.

To quit debug mode and return directly to the prompt without executing any additional code use `dbquit`.

`dbquit`   [Command]
>   Quit debugging mode immediately without further code execution and return to the Octave prompt.
>
>   **See also:** [dbcont], page 212, [dbstep], page 215.

Finally, typing `exit` or `quit` at the debug prompt will result in Octave terminating normally.

13.3 Breakpoints

Breakpoints can be set in any m-file function by using the `dbstop` function.

`rline = dbstop ("func")`   [Built-in Function]
`rline = dbstop ("func", line)`   [Built-in Function]
`rline = dbstop ("func", line1, line2, ...)`   [Built-in Function]
>   Set a breakpoint in function *func*.
>
>   Arguments are
>
>   *func*   Function name as a string variable. When already in debug mode this
>   should be left out and only the line should be given.
>
>   *line*   Line number where the breakpoint should be set. Multiple lines may be
>   given as separate arguments or as a vector.
>
>   When called with a single argument *func*, the breakpoint is set at the first executable line in the named function.
>
>   The optional output *rline* is the real line number where the breakpoint was set. This can differ from specified line if the line is not executable. For example, if a breakpoint attempted on a blank line then Octave will set the real breakpoint at the next executable line.
>
>   **See also:** [dbclear], page 213, [dbstatus], page 213, [dbstep], page 215, [debug_on_error], page 211, [debug_on_warning], page 211, [debug_on_interrupt], page 211.

Breakpoints in class methods are also supported (e.g., dbstop ("@class/method")). How-ever, breakpoints cannot be set in built-in functions (e.g., sin, etc.) or dynamically loaded functions (i.e., oct-files).

To set a breakpoint immediately upon entering a function use line number 1, or omit the line number entirely and just give the function name. When setting the breakpoint Octave will ignore the leading comment block, and the breakpoint will be set on the first executable statement in the function. For example:

```
dbstop ("asind", 1)
⇒ 29
```

Note that the return value of 29 means that the breakpoint was effectively set to line 29. The status of breakpoints in a function can be queried with dbstatus.

dbstatus ()   [Built-in Function]
brk_list = dbstatus ()   [Built-in Function]
brk_list = dbstatus ("*func*")   [Built-in Function]
    Report the location of active breakpoints.

    When called with no input or output arguments, print the list of all functions with breakpoints and the line numbers where those breakpoints are set. If a function name *func* is specified then only report breakpoints for the named function.

    The optional return argument *brk_list* is a struct array with the following fields.

    name   The name of the function with a breakpoint.

    file   The name of the m-file where the function code is located.

    line   A line number, or vector of line numbers, with a breakpoint.

    See also: [dbclear], page 213, [dbwhere], page 214.

Reusing the previous example, dbstatus ("asind") will return 29. The breakpoints listed can then be cleared with the dbclear function.

dbclear ("*func*")   [Built-in Function]
dbclear ("*func*", *line*, ...)   [Built-in Function]
dbclear (*line*, ...)   [Built-in Function]
    Delete a breakpoint in the function *func*.

    Arguments are

    func   Function name as a string variable. When already in debug mode this argument should be omitted and only the line number should be given.

    line   Line number from which to remove a breakpoint. Multiple lines may be given as separate arguments or as a vector.

    When called without a line number specification all breakpoints in the named function are cleared.

    If the requested line is not a breakpoint no action is performed.

    See also: [dbstop], page 212, [dbstatus], page 213, [dbwhere], page 214.

A breakpoint may also be set in a subfunction. For example, if a file contains the functions

```
function y = func1 (x)
  y = func2 (x);
endfunction
function y = func2 (x)
  y = x + 1;
endfunction
```

then a breakpoint can be set at the start of the subfunction directly with

```
dbstop (["func1", filemarker(), "func2"])
⇒ 5
```

Note that `filemarker` returns the character that marks subfunctions from the file containing them. Unless the default has been changed this character is '>'. Thus, a quicker and more normal way to set the breakpoint would be

```
dbstop func1>func2
```

Another simple way of setting a breakpoint in an Octave script is the use of the `keyboard` function.

`keyboard ()`   [Built-in Function]
`keyboard ("prompt")`   [Built-in Function]

> This function is normally used for simple debugging. When the `keyboard` function is executed, Octave prints a prompt and waits for user input. The input strings are then evaluated and the results are printed. This makes it possible to examine the values of variables within a function, and to assign new values if necessary. To leave the prompt and return to normal execution type '`return`' or '`dbcont`'. The `keyboard` function does not return an exit status.
>
> If `keyboard` is invoked without arguments, a default prompt of '`debug> `' is used.
>
> **See also:** [dbstop], page 212, [dbcont], page 212, [dbquit], page 212.

The `keyboard` function is placed in a script at the point where the user desires that the execution be stopped. It automatically sets the running script into the debug mode.

13.4 Debug Mode

There are three additional support functions that allow the user to find out where in the execution of a script Octave entered the debug mode, and to print the code in the script surrounding the point where Octave entered debug mode.

`dbwhere`   [Command]

> In debugging mode, report the current file and line number where execution is stopped.
>
> **See also:** [dbstatus], page 213, [dbcont], page 212, [dbstep], page 215, [dbup], page 216.

`dbtype`   [Command]
`dbtype lineno`   [Command]

```
dbtype startl:endl                                                              [Command]
dbtype startl:end                                                               [Command]
dbtype func                                                                     [Command]
dbtype func lineno                                                              [Command]
dbtype func startl:endl                                                         [Command]
dbtype func startl:end                                                          [Command]
```
Display a script file with line numbers.

When called with no arguments in debugging mode, display the script file currently being debugged. An optional range specification can be used to list only a portion of the file. The special keyword "end" is a valid line number specification for the last line of the file.

When called with the name of a function, list that script file with line numbers.

See also: [dbwhere], page 214, [dbstatus], page 213, [dbstop], page 212.

```
dblist                                                                          [Command]
dblist n                                                                        [Command]
```
In debugging mode, list *n* lines of the function being debugged centered around the current line to be executed. If unspecified *n* defaults to 10 (+/- 5 lines)

See also: [dbwhere], page 214, [dbtype], page 214.

You may also use **isdebugmode** to determine whether the debugger is currently active.

```
isdebugmode ()                                                           [Built-in Function]
```
Return true if in debugging mode, otherwise false.

See also: [dbwhere], page 214, [dbstack], page 216, [dbstatus], page 213.

Debug mode also allows single line stepping through a function using the command **dbstep**.

```
dbstep                                                                          [Command]
dbstep n                                                                        [Command]
dbstep in                                                                       [Command]
dbstep out                                                                      [Command]
dbnext ...                                                                      [Command]
```
In debugging mode, execute the next *n* lines of code. If *n* is omitted, execute the next single line of code. If the next line of code is itself defined in terms of an m-file remain in the existing function.

Using **dbstep in** will cause execution of the next line to step into any m-files defined on the next line. Using **dbstep out** will cause execution to continue until the current function returns.

dbnext is an alias for **dbstep**.

See also: [dbcont], page 212, [dbquit], page 212.

When in debug mode the RETURN key will execute the last entered command. This is useful, for example, after hitting a breakpoint and entering **dbstep** once. After that, one can advance line by line through the code with only a single key stroke.

13.5 Call Stack

The function being debugged may be the leaf node of a series of function calls. After examining values in the current subroutine it may turn out that the problem occurred in earlier pieces of code. Use dbup and dbdown to move up and down through the series of function calls to locate where variables first took on the wrong values. dbstack shows the entire series of function calls and at what level debugging is currently taking place.

dbstack   [Command]
dbstack *n*   [Command]
dbstack *-completenames*   [Command]
[*stack, idx*] = dbstack (...)   [Built-in Function]
>   Display or return current debugging function stack information. With optional argu-
>   ment *n*, omit the *n* innermost stack frames.
>
>   Although accepted, the argument *-completenames* is silently ignored. Octave always
>   returns absolute file names. The arguments *n* and *-completenames* can be both spec-
>   ified in any order.
>
>   The optional return argument *stack* is a struct array with the following fields:
>
>   file   The name of the m-file where the function code is located.
>
>   name   The name of the function with a breakpoint.
>
>   line   The line number of an active breakpoint.
>
>   column   The column number of the line where the breakpoint begins.
>
>   scope   Undocumented.
>
>   context   Undocumented.
>
>   The return argument *idx* specifies which element of the *stack* struct array is currently
>   active.
>
>   **See also:** [dbup], page 216, [dbdown], page 216, [dbwhere], page 214, [dbstatus],
>   page 213.

dbup   [Built-in Function]
dbup (*n*)   [Built-in Function]
>   In debugging mode, move up the execution stack *n* frames. If *n* is omitted, move up
>   one frame.
>
>   **See also:** [dbstack], page 216, [dbdown], page 216.

dbdown   [Built-in Function]
dbdown (*n*)   [Built-in Function]
>   In debugging mode, move down the execution stack *n* frames. If *n* is omitted, move
>   down one frame.
>
>   **See also:** [dbstack], page 216, [dbup], page 216.

13.6 Profiling

Octave supports profiling of code execution on a per-function level. If profiling is enabled, each call to a function (supporting built-ins, operators, functions in oct- and mex-files, user-defined functions in Octave code and anonymous functions) is recorded while running Octave code. After that, this data can aid in analyzing the code behavior, and is in particular helpful for finding "hot spots" in the code which use up a lot of computation time and are the best targets to spend optimization efforts on.

The main command for profiling is `profile`, which can be used to start or stop the profiler and also to query collected data afterwards. The data is returned in an Octave data structure which can then be examined or further processed by other routines or tools.

`profile` *on*	[Command]
`profile` *off*	[Command]
`profile` *resume*	[Command]
`profile` *clear*	[Command]
S = `profile` ("*status*")	[Function File]
T = `profile` ("*info*")	[Function File]

Control the built-in profiler.

`profile on`

Start the profiler, clearing all previously collected data if there is any.

`profile off`

Stop profiling. The collected data can later be retrieved and examined with calls like S = `profile` ("info").

`profile clear`

Clear all collected profiler data.

`profile resume`

Restart profiling without cleaning up the old data and instead all newly collected statistics are added to the already existing ones.

S = `profile` ("status")

Return a structure filled with certain information about the current status of the profiler. At the moment, the only field is `ProfilerStatus` which is either `"on"` or `"off"`.

T = `profile` ("info")

Return the collected profiling statistics in the structure *T*. The flat profile is returned in the field `FunctionTable` which is an array of structures, each entry corresponding to a function which was called and for which profiling statistics are present. Furthermore, the field `Hierarchical` contains the hierarchical call-tree. Each node has an index into the `FunctionTable` identifying the function it corresponds to as well as data fields for number of calls and time spent at this level in the call-tree.

See also: [profshow], page 218, [profexplore], page 218.

An easy way to get an overview over the collected data is `profshow`. This function takes the profiler data returned by `profile` as input and prints a flat profile, for instance:

Function Attr	Time (s)	Calls
>myfib R	2.195	13529
binary <=	0.061	13529
binary -	0.050	13528
binary +	0.026	6764

This shows that most of the run time was spent executing the function 'myfib', and some minor proportion evaluating the listed binary operators. Furthermore, it is shown how often the function was called and the profiler also records that it is recursive.

profshow (*data*)   [Function File]
profshow (*data*, *n*)   [Function File]
> Show flat profiler results.
>
> This command prints out profiler data as a flat profile. *data* is the structure returned by profile ("info"). If *n* is given, it specifies the number of functions to show in the profile; functions are sorted in descending order by total time spent in them. If there are more than *n* included in the profile, those will not be shown. *n* defaults to 20.
>
> The attribute column shows 'R' for recursive functions and nothing otherwise.
>
> **See also:** [profexplore], page 218, [profile], page 217.

profexplore ()   [Function File]
profexplore (*data*)   [Function File]
> Interactively explore hierarchical profiler output.
>
> Assuming *data* is the structure with profile data returned by profile ("info"), this command opens an interactive prompt that can be used to explore the call-tree. Type *help* to get a list of possible commands. If *data* is omitted, profile ("info") is called and used in its place.
>
> **See also:** [profile], page 217, [profshow], page 218.

13.7 Profiler Example

Below, we will give a short example of a profiler session. See Section 13.6 [Profiling], page 217, for the documentation of the profiler functions in detail. Consider the code:

```
global N A;

N = 300;
A = rand (N, N);

function xt = timesteps (steps, x0, expM)
  global N;

  if (steps == 0)
    xt = NA (N, 0);
  else
    xt = NA (N, steps);
```

```
    x1 = expM * x0;
    xt(:, 1) = x1;
    xt(:, 2 : end) = timesteps (steps - 1, x1, expM);
  endif
endfunction

function foo ()
  global N A;

  initial = @(x) sin (x);
  x0 = (initial (linspace (0, 2 * pi, N)))';

  expA = expm (A);
  xt = timesteps (100, x0, expA);
endfunction

function fib = bar (N)
  if (N <= 2)
    fib = 1;
  else
    fib = bar (N - 1) + bar (N - 2);
  endif
endfunction
```

If we execute the two main functions, we get:

```
tic; foo; toc;
⇒ Elapsed time is 2.37338 seconds.

tic; bar (20); toc;
⇒ Elapsed time is 2.04952 seconds.
```

But this does not give much information about where this time is spent; for instance, whether the single call to expm is more expensive or the recursive time-stepping itself. To get a more detailed picture, we can use the profiler.

```
profile on;
foo;
profile off;

data = profile ("info");
profshow (data, 10);
```

This prints a table like:

#	Function	Attr	Time (s)	Calls
7	expm		1.034	1
3	binary *		0.823	117
41	binary \		0.188	1
38	binary ^		0.126	2
43	timesteps	R	0.111	101
44	NA		0.029	101
39	binary +		0.024	8
34	norm		0.011	1
40	binary -		0.004	101
33	balance		0.003	1

The entries are the individual functions which have been executed (only the 10 most important ones), together with some information for each of them. The entries like 'binary *' denote operators, while other entries are ordinary functions. They include both built-ins like expm and our own routines (for instance timesteps). From this profile, we can immediately deduce that expm uses up the largest proportion of the processing time, even though it is only called once. The second expensive operation is the matrix-vector product in the routine timesteps.[1]

Timing, however, is not the only information available from the profile. The attribute column shows us that timesteps calls itself recursively. This may not be that remarkable in this example (since it's clear anyway), but could be helpful in a more complex setting. As to the question of why is there a 'binary \' in the output, we can easily shed some light on that too. Note that data is a structure array (Section 6.1.2 [Structure Arrays], page 101) which contains the field FunctionTable. This stores the raw data for the profile shown. The number in the first column of the table gives the index under which the shown function can be found there. Looking up data.FunctionTable(41) gives:

```
scalar structure containing the fields:

    FunctionName = binary \
    TotalTime =  0.18765
    NumCalls =  1
    IsRecursive = 0
    Parents =  7
    Children = [](1x0)
```

Here we see the information from the table again, but have additional fields Parents and Children. Those are both arrays, which contain the indices of functions which have directly called the function in question (which is entry 7, expm, in this case) or been called by it (no functions). Hence, the backslash operator has been used internally by expm.

Now let's take a look at bar. For this, we start a fresh profiling session (profile on does this; the old data is removed before the profiler is restarted):

---

[1] We only know it is the binary multiplication operator, but fortunately this operator appears only at one place in the code and thus we know which occurrence takes so much time. If there were multiple places, we would have to use the hierarchical profile to find out the exact place which uses up the time which is not covered in this example.

```
profile on;
bar (20);
profile off;

profshow (profile ("info"));
```

This gives:

#	Function	Attr	Time (s)	Calls
1	bar	R	2.091	13529
2	binary <=		0.062	13529
3	binary -		0.042	13528
4	binary +		0.023	6764
5	profile		0.000	1
8	false		0.000	1
6	nargin		0.000	1
7	binary !=		0.000	1
9	__profiler_enable__		0.000	1

Unsurprisingly, `bar` is also recursive. It has been called 13,529 times in the course of recursively calculating the Fibonacci number in a suboptimal way, and most of the time was spent in `bar` itself.

Finally, let's say we want to profile the execution of both `foo` and `bar` together. Since we already have the run-time data collected for `bar`, we can restart the profiler without clearing the existing data and collect the missing statistics about `foo`. This is done by:

```
profile resume;
foo;
profile off;

profshow (profile ("info"), 10);
```

As you can see in the table below, now we have both profiles mixed together.

#	Function	Attr	Time (s)	Calls
1	bar	R	2.091	13529
16	expm		1.122	1
12	binary *		0.798	117
46	binary \		0.185	1
45	binary ^		0.124	2
48	timesteps	R	0.115	101
2	binary <=		0.062	13529
3	binary -		0.045	13629
4	binary +		0.041	6772
49	NA		0.036	101

14 Input and Output

Octave supports several ways of reading and writing data to or from the prompt or a file. The simplest functions for data Input and Output (I/O) are easy to use, but only provide limited control of how data is processed. For more control, a set of functions modeled after the C standard library are also provided by Octave.

14.1 Basic Input and Output

14.1.1 Terminal Output

Since Octave normally prints the value of an expression as soon as it has been evaluated, the simplest of all I/O functions is a simple expression. For example, the following expression will display the value of 'pi'

```
pi
     ⊣ pi = 3.1416
```

This works well as long as it is acceptable to have the name of the variable (or 'ans') printed along with the value. To print the value of a variable without printing its name, use the function disp.

The format command offers some control over the way Octave prints values with disp and through the normal echoing mechanism.

disp (*x*)   [Built-in Function]
  Display the value of *x*. For example:

```
disp ("The value of pi is:"), disp (pi)

     ⊣ the value of pi is:
     ⊣ 3.1416
```

  Note that the output from disp always ends with a newline.

  If an output value is requested, disp prints nothing and returns the formatted output in a string.

  See also: [fdisp], page 234.

list_in_columns (*arg*, *width*, *prefix*)   [Built-in Function]
  Return a string containing the elements of *arg* listed in columns with an overall maximum width of *width* and optional prefix *prefix*. The argument *arg* must be a cell array of character strings or a character array. If *width* is not specified or is an empty matrix, or less than or equal to zero, the width of the terminal screen is used. Newline characters are used to break the lines in the output string. For example:

```
list_in_columns ({"abc", "def", "ghijkl", "mnop", "qrs", "tuv"}, 20)
    ⇒ abc    mnop
      def    qrs
      ghijkl tuv

whos ans
    ⇒
    Variables in the current scope:

    Attr Name       Size              Bytes Class
    ==== ====       ====              ===== =====
         ans        1x37                 37 char

    Total is 37 elements using 37 bytes
```

See also: [terminal_size], page 224.

terminal_size ()   [Built-in Function]
    Return a two-element row vector containing the current size of the terminal window
    in characters (rows and columns).

    See also: [list_in_columns], page 223.

format   [Command]
format *options*   [Command]
    Reset or specify the format of the output produced by `disp` and Octave's normal
    echoing mechanism. This command only affects the display of numbers but not how
    they are stored or computed. To change the internal representation from the default
    double use one of the conversion functions such as `single`, `uint8`, `int64`, etc.

    By default, Octave displays 5 significant digits in a human readable form (option
    'short' paired with 'loose' format for matrices). If `format` is invoked without any
    options, this default format is restored.

    Valid formats for floating point numbers are listed in the following table.

short   Fixed point format with 5 significant figures in a field that is a maximum
           of 10 characters wide. (default).

           If Octave is unable to format a matrix so that columns line up on the
           decimal point and all numbers fit within the maximum field width then
           it switches to an exponential 'e' format.

long   Fixed point format with 15 significant figures in a field that is a maximum
           of 20 characters wide.

           As with the 'short' format, Octave will switch to an exponential 'e'
           format if it is unable to format a matrix properly using the current format.

short e

long e   Exponential format. The number to be represented is split between a
           mantissa and an exponent (power of 10). The mantissa has 5 significant
           digits in the short format and 15 digits in the long format. For example,
           with the 'short e' format, `pi` is displayed as `3.1416e+00`.

short E

long E   Identical to 'short e' or 'long e' but displays an uppercase 'E' to indicate
              the exponent. For example, with the 'long E' format, pi is displayed as
              3.14159265358979E+00.

short g

long g   Optimally choose between fixed point and exponential format based on
              the magnitude of the number. For example, with the 'short g' format,
              pi .^ [2; 4; 8; 16; 32] is displayed as

                  ans =

                          9.8696
                          97.409
                          9488.5
                      9.0032e+07
                      8.1058e+15

short eng

long eng   Identical to 'short e' or 'long e' but displays the value using an engi-
              neering format, where the exponent is divisible by 3. For example, with
              the 'short eng' format, 10 * pi is displayed as 31.4159e+00.

long G

short G   Identical to 'short g' or 'long g' but displays an uppercase 'E' to indicate
              the exponent.

free

none   Print output in free format, without trying to line up columns of matrices
              on the decimal point. This also causes complex numbers to be format-
              ted as numeric pairs like this '(0.60419, 0.60709)' instead of like this
              '0.60419 + 0.60709i'.

The following formats affect all numeric output (floating point and integer types).

+

+ *chars*

plus

plus *chars*

              Print a '+' symbol for nonzero matrix elements and a space for zero matrix
              elements. This format can be very useful for examining the structure of
              a large sparse matrix.

              The optional argument *chars* specifies a list of 3 characters to use for
              printing values greater than zero, less than zero and equal to zero. For
              example, with the '+ "+-."' format, [1, 0, -1; -1, 0, 1] is displayed
              as

                  ans =

                  +.-
                  -.+

bank   Print in a fixed format with two digits to the right of the decimal point.

`native-hex`
> Print the hexadecimal representation of numbers as they are stored in memory. For example, on a workstation which stores 8 byte real values in IEEE format with the least significant byte first, the value of `pi` when printed in `native-hex` format is `400921fb54442d18`.

`hex`   The same as `native-hex`, but always print the most significant byte first.

`native-bit`
> Print the bit representation of numbers as stored in memory. For example, the value of `pi` is
>
> > 01000000000010010010000111111011
> > 01010100010001000010110100011000
>
> (shown here in two 32 bit sections for typesetting purposes) when printed in native-bit format on a workstation which stores 8 byte real values in IEEE format with the least significant byte first.

`bit`   The same as `native-bit`, but always print the most significant bits first.

`rat`   Print a rational approximation, i.e., values are approximated as the ratio of small integers. For example, with the 'rat' format, `pi` is displayed as `355/113`.

The following two options affect the display of all matrices.

`compact`   Remove blank lines around column number labels and between matrices producing more compact output with more data per page.

`loose`   Insert blank lines above and below column number labels and between matrices to produce a more readable output with less data per page. (default).

See also: [fixed_point_format], page 50, [output_max_field_width], page 49, [output_precision], page 50, [split_long_rows], page 50, [rats], page 428.

14.1.1.1 Paging Screen Output

When running interactively, Octave normally sends any output intended for your terminal that is more than one screen long to a paging program, such as `less` or `more`. This avoids the problem of having a large volume of output stream by before you can read it. With `less` (and some versions of `more`) you can also scan forward and backward, and search for specific items.

Normally, no output is displayed by the pager until just before Octave is ready to print the top level prompt, or read from the standard input (for example, by using the `fscanf` or `scanf` functions). This means that there may be some delay before any output appears on your screen if you have asked Octave to perform a significant amount of work with a single command statement. The function `fflush` may be used to force output to be sent to the pager (or any other stream) immediately.

You can select the program to run as the pager using the `PAGER` function, and you can turn paging off by using the function `more`.

more   [Command]
more *on*   [Command]
more *off*   [Command]

> Turn output pagination on or off. Without an argument, `more` toggles the current state. The current state can be determined via `page_screen_output`.
>
> **See also:** [page_screen_output], page 227, [page_output_immediately], page 227, [PAGER], page 227, [PAGER_FLAGS], page 227.

val = PAGER ()   [Built-in Function]
old_val = PAGER (*new_val*)   [Built-in Function]
PAGER (*new_val*, "*local*")   [Built-in Function]

> Query or set the internal variable that specifies the program to use to display terminal output on your system. The default value is normally `"less"`, `"more"`, or `"pg"`, depending on what programs are installed on your system. See Appendix G [Installation], page 833.
>
> When called from inside a function with the `"local"` option, the variable is changed locally for the function and any subroutines it calls. The original variable value is restored when exiting the function.
>
> **See also:** [PAGER_FLAGS], page 227, [page_output_immediately], page 227, [more], page 226, [page_screen_output], page 227.

val = PAGER_FLAGS ()   [Built-in Function]
old_val = PAGER_FLAGS (*new_val*)   [Built-in Function]
PAGER_FLAGS (*new_val*, "*local*")   [Built-in Function]

> Query or set the internal variable that specifies the options to pass to the pager.
>
> When called from inside a function with the `"local"` option, the variable is changed locally for the function and any subroutines it calls. The original variable value is restored when exiting the function.
>
> **See also:** [PAGER], page 227, [more], page 226, [page_screen_output], page 227, [page_output_immediately], page 227.

val = page_screen_output ()   [Built-in Function]
old_val = page_screen_output (*new_val*)   [Built-in Function]
page_screen_output (*new_val*, "*local*")   [Built-in Function]

> Query or set the internal variable that controls whether output intended for the terminal window that is longer than one page is sent through a pager. This allows you to view one screenful at a time. Some pagers (such as `less`—see Appendix G [Installation], page 833) are also capable of moving backward on the output.
>
> When called from inside a function with the `"local"` option, the variable is changed locally for the function and any subroutines it calls. The original variable value is restored when exiting the function.
>
> **See also:** [more], page 226, [page_output_immediately], page 227, [PAGER], page 227, [PAGER_FLAGS], page 227.

val = page_output_immediately ()   [Built-in Function]
old_val = page_output_immediately (*new_val*)   [Built-in Function]

page_output_immediately (*new_val*, "*local*")   [Built-in Function]
>   Query or set the internal variable that controls whether Octave sends output to the
>   pager as soon as it is available. Otherwise, Octave buffers its output and waits until
>   just before the prompt is printed to flush it to the pager.
>
>   When called from inside a function with the "`local`" option, the variable is changed
>   locally for the function and any subroutines it calls. The original variable value is
>   restored when exiting the function.
>
>   **See also:** [page_screen_output], page 227, [more], page 226, [PAGER], page 227,
>   [PAGER_FLAGS], page 227.

fflush (*fid*)   [Built-in Function]
>   Flush output to *fid*. This is useful for ensuring that all pending output makes it to
>   the screen before some other event occurs. For example, it is always a good idea to
>   flush the standard output stream before calling `input`.
>
>   `fflush` returns 0 on success and an OS dependent error value (-1 on Unix) on error.
>
>   **See also:** [fopen], page 240, [fclose], page 242.

14.1.2 Terminal Input

Octave has three functions that make it easy to prompt users for input. The `input` and
`menu` functions are normally used for managing an interactive dialog with a user, and the
`keyboard` function is normally used for doing simple debugging.

ans = input (*prompt*)   [Built-in Function]
ans = input (*prompt*, "*s*")   [Built-in Function]
>   Print *prompt* and wait for user input.
>
>   For example,
>
>   input ("Pick a number, any number! ")
>
>   prints the prompt
>
>   Pick a number, any number!
>
>   and waits for the user to enter a value. The string entered by the user is evaluated
>   as an expression, so it may be a literal constant, a variable name, or any other valid
>   expression.
>
>   Currently, `input` only returns one value, regardless of the number of values produced
>   by the evaluation of the expression.
>
>   If you are only interested in getting a literal string value, you can call `input` with the
>   character string "`s`" as the second argument. This tells Octave to return the string
>   entered by the user directly, without evaluating it first.
>
>   Because there may be output waiting to be displayed by the pager, it is a good idea to
>   always call `fflush (stdout)` before calling `input`. This will ensure that all pending
>   output is written to the screen before your prompt.
>
>   **See also:** [yes_or_no], page 229, [kbhit], page 229, [pause], page 702, [menu], page 229,
>   [listdlg], page 743.

menu (*title*, *opt1*, ...)   [Function File]

> Print a title string followed by a series of options. Each option will be printed along with a number. The return value is the number of the option selected by the user. This function is useful for interactive programs. There is no limit to the number of options that may be passed in, but it may be confusing to present more than will fit easily on one screen.
>
> **See also:** [input], page 228, [listdlg], page 743.

ans = **yes_or_no** ("*prompt*")   [Built-in Function]

> Ask the user a yes-or-no question.
>
> Return logical true if the answer is yes or false if the answer is no. Takes one argument, *prompt*, which is the string to display when asking the question. *prompt* should end in a space; **yes-or-no** adds the string '(yes or no) ' to it. The user must confirm the answer with RET and can edit it until it has been confirmed.
>
> **See also:** [input], page 228.

For **input**, the normal command line history and editing functions are available at the prompt.

Octave also has a function that makes it possible to get a single character from the keyboard without requiring the user to type a carriage return.

kbhit ()   [Built-in Function]
kbhit (*1*)   [Built-in Function]

> Read a single keystroke from the keyboard. If called with an argument, don't wait for a keypress. For example,
>
>   x = kbhit ();
>
> will set *x* to the next character typed at the keyboard as soon as it is typed.
>
>   x = kbhit (1);
>
> is identical to the above example, but doesn't wait for a keypress, returning the empty string if no key is available.
>
> **See also:** [input], page 228, [pause], page 702.

14.1.3 Simple File I/O

The **save** and **load** commands allow data to be written to and read from disk files in various formats. The default format of files written by the **save** command can be controlled using the functions **save_default_options** and **save_precision**.

As an example the following code creates a 3-by-3 matrix and saves it to the file 'myfile.mat'.

    A = [1:3; 4:6; 7:9];
    save myfile.mat A

Once one or more variables have been saved to a file, they can be read into memory using the **load** command.

```
load myfile.mat
A
      ⊣ A =
      ⊣
      ⊣   1   2   3
      ⊣   4   5   6
      ⊣   7   8   9
```

save *file*   [Command]
save *options file*   [Command]
save *options file v1 v2 ...*   [Command]
save *options file -struct STRUCT f1 f2 ...*   [Command]

Save the named variables *v1*, *v2*, ..., in the file *file*. The special filename '-' may be used to write output to the terminal. If no variable names are listed, Octave saves all the variables in the current scope. Otherwise, full variable names or pattern syntax can be used to specify the variables to save. If the '-struct' modifier is used, fields *f1 f2* ... of the scalar structure *STRUCT* are saved as if they were variables with corresponding names. Valid options for the save command are listed in the following table. Options that modify the output format override the format specified by save_default_options.

If save is invoked using the functional form

```
save ("-option1", ..., "file", "v1", ...)
```

then the *options*, *file*, and variable name arguments (*v1*, ...) must be specified as character strings.

-append   Append to the destination instead of overwriting.

-ascii   Save a single matrix in a text file without header or any other information.

-binary   Save the data in Octave's binary data format.

-float-binary
           Save the data in Octave's binary data format but only using single precision. Only use this format if you know that all the values to be saved can be represented in single precision.

-hdf5   Save the data in HDF5 format. (HDF5 is a free, portable binary format developed by the National Center for Supercomputing Applications at the University of Illinois.) This format is only available if Octave was built with a link to the HDF5 libraries.

-float-hdf5
           Save the data in HDF5 format but only using single precision. Only use this format if you know that all the values to be saved can be represented in single precision.

-V7
-v7
-7
-mat7-binary
           Save the data in MATLAB's v7 binary data format.

```
-V6
-v6
-6
-mat
-mat-binary
```
    Save the data in MATLAB's v6 binary data format.

```
-V4
-v4
-4
-mat4-binary
```
    Save the data in the binary format written by MATLAB version 4.

`-text`   Save the data in Octave's text data format. (default).

```
-zip
-z
```
    Use the gzip algorithm to compress the file. This works equally on files that are compressed with gzip outside of octave, and gzip can equally be used to convert the files for backward compatibility. This option is only available if Octave was built with a link to the zlib libraries.

The list of variables to save may use wildcard patterns containing the following special characters:

`?`   Match any single character.

`*`   Match zero or more characters.

`[ list ]` Match the list of characters specified by *list*. If the first character is ! or ^, match all characters except those specified by *list*. For example, the pattern `[a-zA-Z]` will match all lower and uppercase alphabetic characters.

      Wildcards may also be used in the field name specifications when using the '`-struct`' modifier (but not in the struct name itself).

Except when using the MATLAB binary data file format or the '`-ascii`' format, saving global variables also saves the global status of the variable. If the variable is restored at a later time using '`load`', it will be restored as a global variable.

The command

    `save -binary data a b*`

saves the variable '`a`' and all variables beginning with '`b`' to the file '`data`' in Octave's binary format.

See also: [load], page 232, [save_default_options], page 231, [save_header_format_string], page 232, [dlmread], page 235, [csvread], page 236, [fread], page 252.

There are three functions that modify the behavior of `save`.

`val = save_default_options ()`   [Built-in Function]
`old_val = save_default_options (new_val)`   [Built-in Function]

save_default_options (*new_val*, "*local*")   [Built-in Function]
> Query or set the internal variable that specifies the default options for the `save` command, and defines the default format. Typical values include "`-ascii`", "`-text -zip`". The default value is '`-text`'.
>
> When called from inside a function with the "`local`" option, the variable is changed locally for the function and any subroutines it calls. The original variable value is restored when exiting the function.
>
> **See also:** [save], page 230.

val = save_precision ()   [Built-in Function]
old_val = save_precision (*new_val*)   [Built-in Function]
save_precision (*new_val*, "*local*")   [Built-in Function]
> Query or set the internal variable that specifies the number of digits to keep when saving data in text format.
>
> When called from inside a function with the "`local`" option, the variable is changed locally for the function and any subroutines it calls. The original variable value is restored when exiting the function.

val = save_header_format_string ()   [Built-in Function]
old_val = save_header_format_string (*new_val*)   [Built-in Function]
save_header_format_string (*new_val*, "*local*")   [Built-in Function]
> Query or set the internal variable that specifies the format string used for the comment line written at the beginning of text-format data files saved by Octave. The format string is passed to `strftime` and should begin with the character '#' and contain no newline characters. If the value of `save_header_format_string` is the empty string, the header comment is omitted from text-format data files. The default value is
>
> ```
> "# Created by Octave VERSION, %a %b %d %H:%M:%S %Y %Z <USER@HOST>"
> ```
>
> When called from inside a function with the "`local`" option, the variable is changed locally for the function and any subroutines it calls. The original variable value is restored when exiting the function.
>
> **See also:** [strftime], page 699, [save], page 230.

load *file*   [Command]
load *options file*   [Command]
load *options file v1 v2 . . .*   [Command]
S = load ("*options*", "*file*", "*v1*", "*v2*", . . .)   [Command]
load *file options*   [Command]
load *file options v1 v2 . . .*   [Command]
S = load ("*file*", "*options*", "*v1*", "*v2*", . . .)   [Command]
> Load the named variables *v1*, *v2*, . . ., from the file *file*. If no variables are specified then all variables found in the file will be loaded. As with `save`, the list of variables to extract can be full names or use a pattern syntax. The format of the file is automatically detected but may be overridden by supplying the appropriate option.
>
> If load is invoked using the functional form
>
> ```
> load ("-option1", ..., "file", "v1", ...)
> ```

then the *options*, *file*, and variable name arguments (*v1*, ...) must be specified as character strings.

If a variable that is not marked as global is loaded from a file when a global symbol with the same name already exists, it is loaded in the global symbol table. Also, if a variable is marked as global in a file and a local symbol exists, the local symbol is moved to the global symbol table and given the value from the file.

If invoked with a single output argument, Octave returns data instead of inserting variables in the symbol table. If the data file contains only numbers (TAB- or space-delimited columns), a matrix of values is returned. Otherwise, `load` returns a structure with members corresponding to the names of the variables in the file.

The `load` command can read data stored in Octave's text and binary formats, and MATLAB's binary format. If compiled with zlib support, it can also load gzip-compressed files. It will automatically detect the type of file and do conversion from different floating point formats (currently only IEEE big and little endian, though other formats may be added in the future).

Valid options for `load` are listed in the following table.

-force   This option is accepted for backward compatibility but is ignored. Octave now overwrites variables currently in memory with those of the same name found in the file.

-ascii   Force Octave to assume the file contains columns of numbers in text format without any header or other information. Data in the file will be loaded as a single numeric matrix with the name of the variable derived from the name of the file.

-binary   Force Octave to assume the file is in Octave's binary format.

-hdf5   Force Octave to assume the file is in HDF5 format. (HDF5 is a free, portable binary format developed by the National Center for Supercomputing Applications at the University of Illinois.) Note that Octave can read HDF5 files not created by itself, but may skip some datasets in formats that it cannot support. This format is only available if Octave was built with a link to the HDF5 libraries.

-import   This option is accepted for backward compatibility but is ignored. Octave can now support multi-dimensional HDF data and automatically modifies variable names if they are invalid Octave identifiers.

-mat
-mat-binary
-6
-v6
-7
-v7   Force Octave to assume the file is in MATLAB's version 6 or 7 binary format.

```
-mat4-binary
-4
-v4
-V4
```
        Force Octave to assume the file is in the binary format written by MATLAB version 4.

`-text`   Force Octave to assume the file is in Octave's text format.

See also: [save], page 230, [dlmwrite], page 234, [csvwrite], page 236, [fwrite], page 254.

str = fileread (*filename*)   [Function File]
    Read the contents of *filename* and return it as a string.

    See also: [fread], page 252, [textread], page 236, [sscanf], page 249.

native_float_format ()   [Built-in Function]
    Return the native floating point format as a string

It is possible to write data to a file in a similar way to the `disp` function for writing data to the screen. The `fdisp` works just like `disp` except its first argument is a file pointer as created by `fopen`. As an example, the following code writes to data 'myfile.txt'.

```
fid = fopen ("myfile.txt", "w");
fdisp (fid, "3/8 is ");
fdisp (fid, 3/8);
fclose (fid);
```

See Section 14.2.1 [Opening and Closing Files], page 240, for details on how to use `fopen` and `fclose`.

fdisp (*fid*, *x*)   [Built-in Function]
    Display the value of *x* on the stream *fid*. For example:

```
        fdisp (stdout, "The value of pi is:"), fdisp (stdout, pi)

           ⊣ the value of pi is:
           ⊣ 3.1416
```

    Note that the output from `fdisp` always ends with a newline.

    See also: [disp], page 223.

Octave can also read and write matrices text files such as comma separated lists.

dlmwrite (*file*, *M*)   [Function File]
dlmwrite (*file*, *M*, *delim*, *r*, *c*)   [Function File]
dlmwrite (*file*, *M*, *key*, *val* ...)   [Function File]
dlmwrite (*file*, *M*, "-append", ...)   [Function File]
dlmwrite (*fid*, ...)   [Function File]
    Write the matrix *M* to the named file using delimiters.

    file should be a file name or writable file ID given by `fopen`.

    The parameter *delim* specifies the delimiter to use to separate values on a row.

The value of r specifies the number of delimiter-only lines to add to the start of the file.

The value of c specifies the number of delimiters to prepend to each line of data.

If the argument `"-append"` is given, append to the end of *file*.

In addition, the following keyword value pairs may appear at the end of the argument list:

`"append"`   Either `"on"` or `"off"`. See `"-append"` above.

`"delimiter"`

> See *delim* above.

`"newline"`

> The character(s) to use to separate each row. Three special cases exist for this option. `"unix"` is changed into `"\n"`, `"pc"` is changed into `"\r\n"`, and `"mac"` is changed into `"\r"`. Other values for this option are kept as is.

`"roffset"`

> See r above.

`"coffset"`

> See c above.

`"precision"`

> The precision to use when writing the file. It can either be a format string (as used by fprintf) or a number of significant digits.

> ```
> dlmwrite ("file.csv", reshape (1:16, 4, 4));
> ```

> ```
> dlmwrite ("file.tex", a, "delimiter", "&", "newline", "\\n")
> ```

See also: [dlmread], page 235, [csvread], page 236, [csvwrite], page 236.

`data = dlmread (file)`	[Built-in Function]
`data = dlmread (file, sep)`	[Built-in Function]
`data = dlmread (file, sep, r0, c0)`	[Built-in Function]
`data = dlmread (file, sep, range)`	[Built-in Function]
`data = dlmread (..., "emptyvalue", EMPTYVAL)`	[Built-in Function]

Read the matrix *data* from a text file. If not defined the separator between fields is determined from the file itself. Otherwise the separation character is defined by *sep*.

Given two scalar arguments *r0* and *c0*, these define the starting row and column of the data to be read. These values are indexed from zero, such that the first row corresponds to an index of zero.

The *range* parameter may be a 4-element vector containing the upper left and lower right corner `[R0,C0,R1,C1]` where the lowest index value is zero. Alternatively, a spreadsheet style range such as `"A2..Q15"` or `"T1:AA5"` can be used. The lowest alphabetical index 'A' refers to the first column. The lowest row index is 1.

file should be a file name or file id given by **fopen**. In the latter case, the file is read until end of file is reached.

The `"emptyvalue"` option may be used to specify the value used to fill empty fields. The default is zero.

See also: [csvread], page 236, [textscan], page 236, [textread], page 236, [dlmwrite], page 234.

<pre>
csvwrite (<i>filename</i>, x)   [Function File]
csvwrite (<i>filename</i>, x, <i>dlm_opts</i>)   [Function File]
</pre>
>   Write the matrix x to the file *filename* in comma-separated-value format.
>
>   This function is equivalent to
>
> ```
> dlmwrite (filename, x, ",", ...)
> ```
>
>   **See also:** [csvread], page 236, [dlmwrite], page 234, [dlmread], page 235.

<pre>
x = csvread (<i>filename</i>)   [Function File]
x = csvread (<i>filename</i>, <i>dlm_opts</i>)   [Function File]
</pre>
>   Read the comma-separated-value file *filename* into the matrix x.
>
>   This function is equivalent to
>
> ```
> x = dlmread (filename, "," , ...)
> ```
>
>   **See also:** [csvwrite], page 236, [dlmread], page 235, [dlmwrite], page 234.

Formatted data from can be read from, or written to, text files as well.

<pre>
[a, ...] = textread (<i>filename</i>)   [Function File]
[a, ...] = textread (<i>filename</i>, <i>format</i>)   [Function File]
[a, ...] = textread (<i>filename</i>, <i>format</i>, n)   [Function File]
[a, ...] = textread (<i>filename</i>, <i>format</i>, <i>prop1</i>, <i>value1</i>, ...)   [Function File]
[a, ...] = textread (<i>filename</i>, <i>format</i>, n, <i>prop1</i>, <i>value1</i>, ...)   [Function File]
</pre>
>   Read data from a text file.
>
>   The file *filename* is read and parsed according to *format*. The function behaves like
>   **strread** except it works by parsing a file instead of a string. See the documentation
>   of **strread** for details.
>
>   In addition to the options supported by **strread**, this function supports two more:
>
>   - "headerlines": The first *value* number of lines of *filename* are skipped.
>
>   - "endofline": Specify a single character or "\r\n". If no value is given, it will be
>   inferred from the file. If set to "" (empty string) EOLs are ignored as delimiters.
>
>   The optional input n specifies the number of data lines to read; in this sense it differs
>   slightly from the format repeat count in strread.
>
>   If the format string is empty (not: omitted) and the file contains only numeric data
>   (excluding headerlines), textread will return a rectangular matrix with the number
>   of columns matching the number of numeric fields on the first data line of the file.
>   Empty fields are returned as zero values.
>
>   **See also:** [strread], page 81, [load], page 232, [dlmread], page 235, [fscanf], page 248,
>   [textscan], page 236.

<pre>
C = textscan (<i>fid</i>, <i>format</i>)   [Function File]
C = textscan (<i>fid</i>, <i>format</i>, n)   [Function File]
C = textscan (<i>fid</i>, <i>format</i>, <i>param</i>, <i>value</i>, ...)   [Function File]
C = textscan (<i>fid</i>, <i>format</i>, n, <i>param</i>, <i>value</i>, ...)   [Function File]
</pre>

C = textscan (*str*, ...)   [Function File]
[C, *position*] = textscan (*fid*, ...)   [Function File]
> Read data from a text file or string.
>
> The string *str* or file associated with *fid* is read from and parsed according to *format*. The function behaves like `strread` except it can also read from file instead of a string. See the documentation of `strread` for details.
>
> In addition to the options supported by `strread`, this function supports a few more:
>
> - "collectoutput": A value of 1 or true instructs textscan to concatenate consecutive columns of the same class in the output cell array. A value of 0 or false (default) leaves output in distinct columns.
>
> - "endofline": Specify "\r", "\n" or "\r\n" (for CR, LF, or CRLF). If no value is given, it will be inferred from the file. If set to "" (empty string) EOLs are ignored as delimiters and added to whitespace.
>
> - "headerlines": The first *value* number of lines of *fid* are skipped.
>
> - "returnonerror": If set to numerical 1 or true (default), return normally when read errors have been encountered. If set to 0 or false, return an error and no data. As the string or file is read by columns rather than by rows, and because textscan is fairly forgiving as regards read errors, setting this option may have little or no actual effect.
>
> When reading from a character string, optional input argument *n* specifies the number of times *format* should be used (i.e., to limit the amount of data read). When reading from file, *n* specifies the number of data lines to read; in this sense it differs slightly from the format repeat count in strread.
>
> The output *C* is a cell array whose second dimension is determined by the number of format specifiers.
>
> The second output, *position*, provides the position, in characters, from the beginning of the file.
>
> If the format string is empty (not: omitted) and the file contains only numeric data (excluding headerlines), textscan will return data in a number of columns matching the number of numeric fields on the first data line of the file.
>
> **See also:** [dlmread], page 235, [fscanf], page 248, [load], page 232, [strread], page 81, [textread], page 236.

The `importdata` function has the ability to work with a wide variety of data.

A = importdata (*fname*)   [Function File]
A = importdata (*fname*, *delimiter*)   [Function File]
A = importdata (*fname*, *delimiter*, *header_rows*)   [Function File]
[A, *delimiter*] = importdata (...)   [Function File]
[A, *delimiter*, *header_rows*] = importdata (...)   [Function File]
> Import data from the file *fname*.
>
> Input parameters:
>
> - *fname* The name of the file containing data.

- *delimiter* The character separating columns of data. Use `\t` for tab. (Only valid for ASCII files)

- *header_rows* The number of header rows before the data begins. (Only valid for ASCII files)

Different file types are supported:

- ASCII table

  Import ASCII table using the specified number of header rows and the specified delimiter.

- Image file

- MATLAB file

- Spreadsheet files (depending on external software)

- WAV file

See also: [textscan], page 236, [dlmread], page 235, [csvread], page 236, [load], page 232.

14.1.3.1 Saving Data on Unexpected Exits

If Octave for some reason exits unexpectedly it will by default save the variables available in the workspace to a file in the current directory. By default this file is named 'octave-workspace' and can be loaded into memory with the `load` command. While the default behavior most often is reasonable it can be changed through the following functions.

val = `crash_dumps_octave_core` ()	[Built-in Function]
old_val = `crash_dumps_octave_core` (*new_val*)	[Built-in Function]
`crash_dumps_octave_core` (*new_val*, "*local*")	[Built-in Function]

> Query or set the internal variable that controls whether Octave tries to save all current variables to the file 'octave-workspace' if it crashes or receives a hangup, terminate or similar signal.

> When called from inside a function with the `"local"` option, the variable is changed locally for the function and any subroutines it calls. The original variable value is restored when exiting the function.

> **See also:** [octave_core_file_limit], page 239, [octave_core_file_name], page 239, [octave_core_file_options], page 239.

val = `sighup_dumps_octave_core` ()	[Built-in Function]
old_val = `sighup_dumps_octave_core` (*new_val*)	[Built-in Function]
`sighup_dumps_octave_core` (*new_val*, "*local*")	[Built-in Function]

> Query or set the internal variable that controls whether Octave tries to save all current variables to the file 'octave-workspace' if it receives a hangup signal.

> When called from inside a function with the `"local"` option, the variable is changed locally for the function and any subroutines it calls. The original variable value is restored when exiting the function.

val = `sigterm_dumps_octave_core` ()	[Built-in Function]
old_val = `sigterm_dumps_octave_core` (*new_val*)	[Built-in Function]

sigterm_dumps_octave_core (*new_val*, "*local*")   [Built-in Function]
> Query or set the internal variable that controls whether Octave tries to save all current variables to the file 'octave-workspace' if it receives a terminate signal.
>
> When called from inside a function with the "local" option, the variable is changed locally for the function and any subroutines it calls. The original variable value is restored when exiting the function.

val = octave_core_file_options ()   [Built-in Function]
old_val = octave_core_file_options (*new_val*)   [Built-in Function]
octave_core_file_options (*new_val*, "*local*")   [Built-in Function]
> Query or set the internal variable that specifies the options used for saving the workspace data if Octave aborts. The value of octave_core_file_options should follow the same format as the options for the save function. The default value is Octave's binary format.
>
> When called from inside a function with the "local" option, the variable is changed locally for the function and any subroutines it calls. The original variable value is restored when exiting the function.
>
> **See also:** [crash_dumps_octave_core], page 238, [octave_core_file_name], page 239, [octave_core_file_limit], page 239.

val = octave_core_file_limit ()   [Built-in Function]
old_val = octave_core_file_limit (*new_val*)   [Built-in Function]
octave_core_file_limit (*new_val*, "*local*")   [Built-in Function]
> Query or set the internal variable that specifies the maximum amount of memory (in kilobytes) of the top-level workspace that Octave will attempt to save when writing data to the crash dump file (the name of the file is specified by *octave_core_file_name*). If *octave_core_file_options* flags specify a binary format, then *octave_core_file_limit* will be approximately the maximum size of the file. If a text file format is used, then the file could be much larger than the limit. The default value is -1 (unlimited)
>
> When called from inside a function with the "local" option, the variable is changed locally for the function and any subroutines it calls. The original variable value is restored when exiting the function.
>
> **See also:** [crash_dumps_octave_core], page 238, [octave_core_file_name], page 239, [octave_core_file_options], page 239.

val = octave_core_file_name ()   [Built-in Function]
old_val = octave_core_file_name (*new_val*)   [Built-in Function]
octave_core_file_name (*new_val*, "*local*")   [Built-in Function]
> Query or set the internal variable that specifies the name of the file used for saving data from the top-level workspace if Octave aborts. The default value is "octave-workspace"
>
> When called from inside a function with the "local" option, the variable is changed locally for the function and any subroutines it calls. The original variable value is restored when exiting the function.
>
> **See also:** [crash_dumps_octave_core], page 238, [octave_core_file_name], page 239, [octave_core_file_options], page 239.

14.2 C-Style I/O Functions

Octave's C-style input and output functions provide most of the functionality of the C programming language's standard I/O library. The argument lists for some of the input functions are slightly different, however, because Octave has no way of passing arguments by reference.

In the following, *file* refers to a file name and `fid` refers to an integer file number, as returned by `fopen`.

There are three files that are always available. Although these files can be accessed using their corresponding numeric file ids, you should always use the symbolic names given in the table below, since it will make your programs easier to understand.

`stdin ()`   [Built-in Function]
> Return the numeric value corresponding to the standard input stream. When Octave is used interactively, this is filtered through the command line editing functions.
>
> **See also:** [stdout], page 240, [stderr], page 240.

`stdout ()`   [Built-in Function]
> Return the numeric value corresponding to the standard output stream. Data written to the standard output is normally filtered through the pager.
>
> **See also:** [stdin], page 240, [stderr], page 240.

`stderr ()`   [Built-in Function]
> Return the numeric value corresponding to the standard error stream. Even if paging is turned on, the standard error is not sent to the pager. It is useful for error messages and prompts.
>
> **See also:** [stdin], page 240, [stdout], page 240.

14.2.1 Opening and Closing Files

When reading data from a file it must be opened for reading first, and likewise when writing to a file. The `fopen` function returns a pointer to an open file that is ready to be read or written. Once all data has been read from or written to the opened file it should be closed. The `fclose` function does this. The following code illustrates the basic pattern for writing to a file, but a very similar pattern is used when reading a file.

```
filename = "myfile.txt";
fid = fopen (filename, "w");
# Do the actual I/O here...
fclose (fid);
```

[fid, msg] = fopen (*name, mode, arch*)   [Built-in Function]
fid_list = fopen ("*all*")   [Built-in Function]
[file, mode, arch] = fopen (*fid*)   [Built-in Function]
> The first form of the `fopen` function opens the named file with the specified mode (read-write, read-only, etc.) and architecture interpretation (IEEE big endian, IEEE little endian, etc.), and returns an integer value that may be used to refer to the file later. If an error occurs, *fid* is set to -1 and *msg* contains the corresponding system

error message. The *mode* is a one or two character string that specifies whether the file is to be opened for reading, writing, or both.

The second form of the `fopen` function returns a vector of file ids corresponding to all the currently open files, excluding the `stdin`, `stdout`, and `stderr` streams.

The third form of the `fopen` function returns information about the open file given its file id.

For example,

```
myfile = fopen ("splat.dat", "r", "ieee-le");
```

opens the file 'splat.dat' for reading. If necessary, binary numeric values will be read assuming they are stored in IEEE format with the least significant bit first, and then converted to the native representation.

Opening a file that is already open simply opens it again and returns a separate file id. It is not an error to open a file several times, though writing to the same file through several different file ids may produce unexpected results.

The possible values 'mode' may have are

'r'   Open a file for reading.

'w'   Open a file for writing. The previous contents are discarded.

'a'   Open or create a file for writing at the end of the file.

'r+'   Open an existing file for reading and writing.

'w+'   Open a file for reading or writing. The previous contents are discarded.

'a+'   Open or create a file for reading or writing at the end of the file.

Append a `"t"` to the mode string to open the file in text mode or a `"b"` to open in binary mode. On Windows and Macintosh systems, text mode reading and writing automatically converts linefeeds to the appropriate line end character for the system (carriage-return linefeed on Windows, carriage-return on Macintosh). The default if no mode is specified is binary mode.

Additionally, you may append a `"z"` to the mode string to open a gzipped file for reading or writing. For this to be successful, you must also open the file in binary mode.

The parameter *arch* is a string specifying the default data format for the file. Valid values for *arch* are:

'native'   The format of the current machine (this is the default).

'ieee-be' IEEE big endian format.

'ieee-le' IEEE little endian format.

however, conversions are currently only supported for 'native' 'ieee-be', and 'ieee-le' formats.

See also: [fclose], page 242, [fgets], page 243, [fgetl], page 242, [fscanf], page 248, [fread], page 252, [fputs], page 242, [fdisp], page 234, [fprintf], page 243, [fwrite], page 254, [fskipl], page 243, [fseek], page 257, [frewind], page 257, [ftell], page 256, [feof], page 256, [ferror], page 256, [fclear], page 256, [fflush], page 228, [freport], page 256.

`fclose (fid)`   [Built-in Function]
`fclose ("all")`   [Built-in Function]
> Close the specified file. If successful, `fclose` returns 0, otherwise, it returns -1. The second form of the `fclose` call closes all open files except `stdout`, `stderr`, and `stdin`.
>
> **See also:** [fopen], page 240, [freport], page 256.

`is_valid_file_id (fid)`   [Function File]
> Return true if *fid* refers to an open file.
>
> **See also:** [fopen], page 240.

14.2.2 Simple Output

Once a file has been opened for writing a string can be written to the file using the `fputs` function. The following example shows how to write the string 'Free Software is needed for Free Science' to the file 'free.txt'.

```
filename = "free.txt";
fid = fopen (filename, "w");
fputs (fid, "Free Software is needed for Free Science");
fclose (fid);
```

`fputs (fid, string)`   [Built-in Function]
> Write a string to a file with no formatting.
>
> Return a non-negative number on success and EOF on error.
>
> **See also:** [fdisp], page 234, [fprintf], page 243, [fwrite], page 254, [fopen], page 240.

A function much similar to `fputs` is available for writing data to the screen. The `puts` function works just like `fputs` except it doesn't take a file pointer as its input.

`puts (string)`   [Built-in Function]
> Write a string to the standard output with no formatting.
>
> Return a non-negative number on success and EOF on error.
>
> **See also:** [fputs], page 242, [disp], page 223.

14.2.3 Line-Oriented Input

To read from a file it must be opened for reading using `fopen`. Then a line can be read from the file using `fgetl` as the following code illustrates

```
fid = fopen ("free.txt");
txt = fgetl (fid)
      ⊣ Free Software is needed for Free Science
fclose (fid);
```

This of course assumes that the file 'free.txt' exists and contains the line 'Free Software is needed for Free Science'.

`str = fgetl (fid)`   [Built-in Function]
`str = fgetl (fid, len)`   [Built-in Function]
> Read characters from a file, stopping after a newline, or EOF, or *len* characters have been read. The characters read, excluding the possible trailing newline, are returned as a string.

If *len* is omitted, `fgetl` reads until the next newline character.

If there are no more characters to read, `fgetl` returns −1.

To read a line and return the terminating newline see `fgets`.

See also: [fgets], page 243, [fscanf], page 248, [fread], page 252, [fopen], page 240.

`str = fgets (fid)`   [Built-in Function]
`str = fgets (fid, len)`   [Built-in Function]
    Read characters from a file, stopping after a newline, or EOF, or *len* characters have been read. The characters read, including the possible trailing newline, are returned as a string.

    If *len* is omitted, `fgets` reads until the next newline character.

    If there are no more characters to read, `fgets` returns −1.

    To read a line and discard the terminating newline see `fgetl`.

    See also: [fputs], page 242, [fgetl], page 242, [fscanf], page 248, [fread], page 252, [fopen], page 240.

`nlines = fskipl (fid)`   [Built-in Function]
`nlines = fskipl (fid, count)`   [Built-in Function]
`nlines = fskipl (fid, Inf)`   [Built-in Function]
    Read and skip *count* lines from the file descriptor *fid*. `fskipl` discards characters until an end-of-line is encountered exactly *count*-times, or until the end-of-file marker is found.

    If *count* is omitted, it defaults to 1. *count* may also be `Inf`, in which case lines are skipped until the end of the file. This form is suitable for counting the number of lines in a file.

    Returns the number of lines skipped (end-of-line sequences encountered).

    See also: [fgetl], page 242, [fgets], page 243, [fscanf], page 248, [fopen], page 240.

14.2.4 Formatted Output

This section describes how to call `printf` and related functions.

    The following functions are available for formatted output. They are modeled after the C language functions of the same name, but they interpret the format template differently in order to improve the performance of printing vector and matrix values.

`printf (template, ...)`   [Built-in Function]
    Print optional arguments under the control of the template string *template* to the stream `stdout` and return the number of characters printed.

    See the Formatted Output section of the GNU Octave manual for a complete description of the syntax of the template string.

    See also: [fprintf], page 243, [sprintf], page 244, [scanf], page 249.

`fprintf (fid, template, ...)`   [Built-in Function]
    This function is just like `printf`, except that the output is written to the stream *fid* instead of `stdout`. If *fid* is omitted, the output is written to `stdout`.

    See also: [fputs], page 242, [fdisp], page 234, [fwrite], page 254, [fscanf], page 248, [printf], page 243, [sprintf], page 244, [fopen], page 240.

sprintf (*template*, ...)   [Built-in Function]
>   This is like printf, except that the output is returned as a string. Unlike the C
>   library function, which requires you to provide a suitably sized string as an argument,
>   Octave's sprintf function returns the string, automatically sized to hold all of the
>   items converted.
>
>   **See also:** [printf], page 243, [fprintf], page 243, [sscanf], page 249.

The printf function can be used to print any number of arguments. The template
string argument you supply in a call provides information not only about the number of
additional arguments, but also about their types and what style should be used for printing
them.

Ordinary characters in the template string are simply written to the output stream
as-is, while *conversion specifications* introduced by a '%' character in the template cause
subsequent arguments to be formatted and written to the output stream. For example,

```
pct = 37;
filename = "foo.txt";
printf ("Processed %d%% of '%s'.\nPlease be patient.\n",
        pct, filename);
```

produces output like

```
Processed 37% of 'foo.txt'.
Please be patient.
```

This example shows the use of the '%d' conversion to specify that a scalar argument
should be printed in decimal notation, the '%s' conversion to specify printing of a string
argument, and the '%%' conversion to print a literal '%' character.

There are also conversions for printing an integer argument as an unsigned value in
octal, decimal, or hexadecimal radix ('%o', '%u', or '%x', respectively); or as a character
value ('%c').

Floating-point numbers can be printed in normal, fixed-point notation using the '%f'
conversion or in exponential notation using the '%e' conversion. The '%g' conversion uses
either '%e' or '%f' format, depending on what is more appropriate for the magnitude of the
particular number.

You can control formatting more precisely by writing *modifiers* between the '%' and
the character that indicates which conversion to apply. These slightly alter the ordinary
behavior of the conversion. For example, most conversion specifications permit you to
specify a minimum field width and a flag indicating whether you want the result left- or
right-justified within the field.

The specific flags and modifiers that are permitted and their interpretation vary depending on the particular conversion. They're all described in more detail in the following
sections.

14.2.5 Output Conversion for Matrices

When given a matrix value, Octave's formatted output functions cycle through the format
template until all the values in the matrix have been printed. For example:

```
printf ("%4.2f %10.2e %8.4g\n", hilb (3));
```

```
⊣ 1.00    5.00e-01    0.3333
⊣ 0.50    3.33e-01     0.25
⊣ 0.33    2.50e-01      0.2
```

If more than one value is to be printed in a single call, the output functions do not return to the beginning of the format template when moving on from one value to the next. This can lead to confusing output if the number of elements in the matrices are not exact multiples of the number of conversions in the format template. For example:

```
printf ("%4.2f %10.2e %8.4g\n", [1, 2], [3, 4]);
```

```
⊣ 1.00    2.00e+00         3
⊣ 4.00
```

If this is not what you want, use a series of calls instead of just one.

14.2.6 Output Conversion Syntax

This section provides details about the precise syntax of conversion specifications that can appear in a `printf` template string.

Characters in the template string that are not part of a conversion specification are printed as-is to the output stream.

The conversion specifications in a `printf` template string have the general form:

> % *flags width* [. *precision*] *type conversion*

For example, in the conversion specifier '%-10.8ld', the '-' is a flag, '10' specifies the field width, the precision is '8', the letter 'l' is a type modifier, and 'd' specifies the conversion style. (This particular type specifier says to print a numeric argument in decimal notation, with a minimum of 8 digits left-justified in a field at least 10 characters wide.)

In more detail, output conversion specifications consist of an initial '%' character followed in sequence by:

- Zero or more *flag characters* that modify the normal behavior of the conversion specification.

- An optional decimal integer specifying the *minimum field width*. If the normal conversion produces fewer characters than this, the field is padded with spaces to the specified width. This is a *minimum* value; if the normal conversion produces more characters than this, the field is *not* truncated. Normally, the output is right-justified within the field.

  You can also specify a field width of '*'. This means that the next argument in the argument list (before the actual value to be printed) is used as the field width. The value is rounded to the nearest integer. If the value is negative, this means to set the '-' flag (see below) and to use the absolute value as the field width.

- An optional *precision* to specify the number of digits to be written for the numeric conversions. If the precision is specified, it consists of a period ('.') followed optionally by a decimal integer (which defaults to zero if omitted).

  You can also specify a precision of '*'. This means that the next argument in the argument list (before the actual value to be printed) is used as the precision. The value must be an integer, and is ignored if it is negative.

- An optional *type modifier character*. This character is ignored by Octave's `printf` function, but is recognized to provide compatibility with the C language `printf`.

- A character that specifies the conversion to be applied.

The exact options that are permitted and how they are interpreted vary between the different conversion specifiers. See the descriptions of the individual conversions for information about the particular options that they use.

14.2.7 Table of Output Conversions

Here is a table summarizing what all the different conversions do:

'%d', '%i'   Print an integer as a signed decimal number. See Section 14.2.8 [Integer Conversions], page 247, for details. '%d' and '%i' are synonymous for output, but are different when used with `scanf` for input (see Section 14.2.13 [Table of Input Conversions], page 250).

'%o'   Print an integer as an unsigned octal number. See Section 14.2.8 [Integer Conversions], page 247, for details.

'%u'   Print an integer as an unsigned decimal number. See Section 14.2.8 [Integer Conversions], page 247, for details.

'%x', '%X'   Print an integer as an unsigned hexadecimal number. '%x' uses lowercase letters and '%X' uses uppercase. See Section 14.2.8 [Integer Conversions], page 247, for details.

'%f'   Print a floating-point number in normal (fixed-point) notation. See Section 14.2.9 [Floating-Point Conversions], page 247, for details.

'%e', '%E'   Print a floating-point number in exponential notation. '%e' uses lowercase letters and '%E' uses uppercase. See Section 14.2.9 [Floating-Point Conversions], page 247, for details.

'%g', '%G'   Print a floating-point number in either normal (fixed-point) or exponential notation, whichever is more appropriate for its magnitude. '%g' uses lowercase letters and '%G' uses uppercase. See Section 14.2.9 [Floating-Point Conversions], page 247, for details.

'%c'   Print a single character. See Section 14.2.10 [Other Output Conversions], page 248.

'%s'   Print a string. See Section 14.2.10 [Other Output Conversions], page 248.

'%%'   Print a literal '%' character. See Section 14.2.10 [Other Output Conversions], page 248.

If the syntax of a conversion specification is invalid, unpredictable things will happen, so don't do this. If there aren't enough function arguments provided to supply values for all the conversion specifications in the template string, or if the arguments are not of the correct types, the results are unpredictable. If you supply more arguments than conversion specifications, the extra argument values are simply ignored; this is sometimes useful.

14.2.8 Integer Conversions

This section describes the options for the '%d', '%i', '%o', '%u', '%x', and '%X' conversion specifications. These conversions print integers in various formats.

The '%d' and '%i' conversion specifications both print an numeric argument as a signed decimal number; while '%o', '%u', and '%x' print the argument as an unsigned octal, decimal, or hexadecimal number (respectively). The '%X' conversion specification is just like '%x' except that it uses the characters 'ABCDEF' as digits instead of 'abcdef'.

The following flags are meaningful:

'-'   Left-justify the result in the field (instead of the normal right-justification).

'+'   For the signed '%d' and '%i' conversions, print a plus sign if the value is positive.

' '   For the signed '%d' and '%i' conversions, if the result doesn't start with a plus or minus sign, prefix it with a space character instead. Since the '+' flag ensures that the result includes a sign, this flag is ignored if you supply both of them.

'#'   For the '%o' conversion, this forces the leading digit to be '0', as if by increasing the precision. For '%x' or '%X', this prefixes a leading '0x' or '0X' (respectively) to the result. This doesn't do anything useful for the '%d', '%i', or '%u' conversions.

'0'   Pad the field with zeros instead of spaces. The zeros are placed after any indication of sign or base. This flag is ignored if the '-' flag is also specified, or if a precision is specified.

If a precision is supplied, it specifies the minimum number of digits to appear; leading zeros are produced if necessary. If you don't specify a precision, the number is printed with as many digits as it needs. If you convert a value of zero with an explicit precision of zero, then no characters at all are produced.

14.2.9 Floating-Point Conversions

This section discusses the conversion specifications for floating-point numbers: the '%f', '%e', '%E', '%g', and '%G' conversions.

The '%f' conversion prints its argument in fixed-point notation, producing output of the form [-]ddd.ddd, where the number of digits following the decimal point is controlled by the precision you specify.

The '%e' conversion prints its argument in exponential notation, producing output of the form [-]d.ddde[+|-]dd. Again, the number of digits following the decimal point is controlled by the precision. The exponent always contains at least two digits. The '%E' conversion is similar but the exponent is marked with the letter 'E' instead of 'e'.

The '%g' and '%G' conversions print the argument in the style of '%e' or '%E' (respectively) if the exponent would be less than -4 or greater than or equal to the precision; otherwise they use the '%f' style. Trailing zeros are removed from the fractional portion of the result and a decimal-point character appears only if it is followed by a digit.

The following flags can be used to modify the behavior:

'-'   Left-justify the result in the field. Normally the result is right-justified.

'+'   Always include a plus or minus sign in the result.

' '   If the result doesn't start with a plus or minus sign, prefix it with a space
           instead. Since the '+' flag ensures that the result includes a sign, this flag is
           ignored if you supply both of them.

'#'   Specifies that the result should always include a decimal point, even if no digits
           follow it. For the '%g' and '%G' conversions, this also forces trailing zeros after
           the decimal point to be left in place where they would otherwise be removed.

'0'   Pad the field with zeros instead of spaces; the zeros are placed after any sign.
           This flag is ignored if the '-' flag is also specified.

The precision specifies how many digits follow the decimal-point character for the '%f',
'%e', and '%E' conversions. For these conversions, the default precision is 6. If the precision
is explicitly 0, this suppresses the decimal point character entirely. For the '%g' and '%G'
conversions, the precision specifies how many significant digits to print. Significant digits
are the first digit before the decimal point, and all the digits after it. If the precision is 0
or not specified for '%g' or '%G', it is treated like a value of 1. If the value being printed
cannot be expressed precisely in the specified number of digits, the value is rounded to the
nearest number that fits.

14.2.10 Other Output Conversions

This section describes miscellaneous conversions for `printf`.

The '%c' conversion prints a single character. The '-' flag can be used to specify left-
justification in the field, but no other flags are defined, and no precision or type modifier
can be given. For example:

```
printf ("%c%c%c%c%c", "h", "e", "l", "l", "o");
```

prints 'hello'.

The '%s' conversion prints a string. The corresponding argument must be a string. A
precision can be specified to indicate the maximum number of characters to write; otherwise
characters in the string up to but not including the terminating null character are written
to the output stream. The '-' flag can be used to specify left-justification in the field, but
no other flags or type modifiers are defined for this conversion. For example:

```
printf ("%3s%-6s", "no", "where");
```

prints ' nowhere ' (note the leading and trailing spaces).

14.2.11 Formatted Input

Octave provides the `scanf`, `fscanf`, and `sscanf` functions to read formatted input. There
are two forms of each of these functions. One can be used to extract vectors of data from
a file, and the other is more 'C-like'.

[*val*, *count*, *errmsg*] = fscanf (*fid*, *template*, *size*)   [Built-in Function]
[*v1*, *v2*, ..., *count*, *errmsg*] = fscanf (*fid*, *template*,   [Built-in Function]
     "*C*")
           In the first form, read from *fid* according to *template*, returning the result in the
           matrix *val*.

           The optional argument *size* specifies the amount of data to read and may be one of

           Inf   Read as much as possible, returning a column vector.

nr   Read up to *nr* elements, returning a column vector.

[nr, Inf]   Read as much as possible, returning a matrix with *nr* rows. If the number of elements read is not an exact multiple of *nr*, the last column is padded with zeros.

[nr, nc]   Read up to *nr* * *nc* elements, returning a matrix with *nr* rows. If the number of elements read is not an exact multiple of *nr*, the last column is padded with zeros.

If *size* is omitted, a value of Inf is assumed.

A string is returned if *template* specifies only character conversions.

The number of items successfully read is returned in *count*.

If an error occurs, *errmsg* contains a system-dependent error message.

In the second form, read from *fid* according to *template*, with each conversion specifier in *template* corresponding to a single scalar return value. This form is more "C-like", and also compatible with previous versions of Octave. The number of successful conversions is returned in *count*

See the Formatted Input section of the GNU Octave manual for a complete description of the syntax of the template string.

See also: [fgets], page 243, [fgetl], page 242, [fread], page 252, [scanf], page 249, [sscanf], page 249, [fopen], page 240.

[val, count, errmsg] = scanf (*template*, *size*)   [Built-in Function]
[v1, v2, ..., count, errmsg]] = scanf (*template*, "*C*")   [Built-in Function]
This is equivalent to calling fscanf with *fid* = stdin.

It is currently not useful to call scanf in interactive programs.

See also: [fscanf], page 248, [sscanf], page 249, [printf], page 243.

[val, count, errmsg, pos] = sscanf (*string*, *template*,   [Built-in Function]
       size)
[v1, v2, ..., count, errmsg] = sscanf (*string*,   [Built-in Function]
       template, "*C*")
This is like fscanf, except that the characters are taken from the string *string* instead of from a stream. Reaching the end of the string is treated as an end-of-file condition. In addition to the values returned by fscanf, the index of the next character to be read is returned in *pos*.

See also: [fscanf], page 248, [scanf], page 249, [sprintf], page 244.

Calls to scanf are superficially similar to calls to printf in that arbitrary arguments are read under the control of a template string. While the syntax of the conversion specifications in the template is very similar to that for printf, the interpretation of the template is oriented more towards free-format input and simple pattern matching, rather than fixed-field formatting. For example, most scanf conversions skip over any amount of "white space" (including spaces, tabs, and newlines) in the input file, and there is no concept of precision for the numeric input conversions as there is for the corresponding output conversions. Ordinarily, non-whitespace characters in the template are expected to match characters in the input stream exactly.

When a *matching failure* occurs, `scanf` returns immediately, leaving the first non-matching character as the next character to be read from the stream, and `scanf` returns all the items that were successfully converted.

The formatted input functions are not used as frequently as the formatted output functions. Partly, this is because it takes some care to use them properly. Another reason is that it is difficult to recover from a matching error.

14.2.12 Input Conversion Syntax

A `scanf` template string is a string that contains ordinary multibyte characters interspersed with conversion specifications that start with '%'.

Any whitespace character in the template causes any number of whitespace characters in the input stream to be read and discarded. The whitespace characters that are matched need not be exactly the same whitespace characters that appear in the template string. For example, write ' , ' in the template to recognize a comma with optional whitespace before and after.

Other characters in the template string that are not part of conversion specifications must match characters in the input stream exactly; if this is not the case, a matching failure occurs.

The conversion specifications in a `scanf` template string have the general form:

```
% flags width type conversion
```

In more detail, an input conversion specification consists of an initial '%' character followed in sequence by:

- An optional *flag character* '*', which says to ignore the text read for this specification. When `scanf` finds a conversion specification that uses this flag, it reads input as directed by the rest of the conversion specification, but it discards this input, does not return any value, and does not increment the count of successful assignments.

- An optional decimal integer that specifies the *maximum field width*. Reading of characters from the input stream stops either when this maximum is reached or when a non-matching character is found, whichever happens first. Most conversions discard initial whitespace characters, and these discarded characters don't count towards the maximum field width. Conversions that do not discard initial whitespace are explicitly documented.

- An optional type modifier character. This character is ignored by Octave's `scanf` function, but is recognized to provide compatibility with the C language `scanf`.

- A character that specifies the conversion to be applied.

The exact options that are permitted and how they are interpreted vary between the different conversion specifiers. See the descriptions of the individual conversions for information about the particular options that they allow.

14.2.13 Table of Input Conversions

Here is a table that summarizes the various conversion specifications:

'%d'   Matches an optionally signed integer written in decimal. See Section 14.2.14 [Numeric Input Conversions], page 251.

'%i'   Matches an optionally signed integer in any of the formats that the C language defines for specifying an integer constant. See Section 14.2.14 [Numeric Input Conversions], page 251.

'%o'   Matches an unsigned integer written in octal radix. See Section 14.2.14 [Numeric Input Conversions], page 251.

'%u'   Matches an unsigned integer written in decimal radix. See Section 14.2.14 [Numeric Input Conversions], page 251.

'%x', '%X' Matches an unsigned integer written in hexadecimal radix. See Section 14.2.14 [Numeric Input Conversions], page 251.

'%e', '%f', '%g', '%E', '%G'
            Matches an optionally signed floating-point number. See Section 14.2.14 [Numeric Input Conversions], page 251.

'%s'   Matches a string containing only non-whitespace characters. See Section 14.2.15 [String Input Conversions], page 252.

'%c'   Matches a string of one or more characters; the number of characters read is controlled by the maximum field width given for the conversion. See Section 14.2.15 [String Input Conversions], page 252.

'%%'   This matches a literal '%' character in the input stream. No corresponding argument is used.

If the syntax of a conversion specification is invalid, the behavior is undefined. If there aren't enough function arguments provided to supply addresses for all the conversion specifications in the template strings that perform assignments, or if the arguments are not of the correct types, the behavior is also undefined. On the other hand, extra arguments are simply ignored.

14.2.14 Numeric Input Conversions

This section describes the scanf conversions for reading numeric values.

The '%d' conversion matches an optionally signed integer in decimal radix.

The '%i' conversion matches an optionally signed integer in any of the formats that the C language defines for specifying an integer constant.

For example, any of the strings '10', '0xa', or '012' could be read in as integers under the '%i' conversion. Each of these specifies a number with decimal value 10.

The '%o', '%u', and '%x' conversions match unsigned integers in octal, decimal, and hexadecimal radices, respectively.

The '%X' conversion is identical to the '%x' conversion. They both permit either uppercase or lowercase letters to be used as digits.

Unlike the C language scanf, Octave ignores the 'h', 'l', and 'L' modifiers.

14.2.15 String Input Conversions

This section describes the `scanf` input conversions for reading string and character values: '`%s`' and '`%c`'.

The '`%c`' conversion is the simplest: it matches a fixed number of characters, always. The maximum field with says how many characters to read; if you don't specify the maximum, the default is 1. This conversion does not skip over initial whitespace characters. It reads precisely the next *n* characters, and fails if it cannot get that many.

The '`%s`' conversion matches a string of non-whitespace characters. It skips and discards initial whitespace, but stops when it encounters more whitespace after having read something.

For example, reading the input:

```
hello, world
```

with the conversion '`%10c`' produces `" hello, wo"`, but reading the same input with the conversion '`%10s`' produces `"hello,"`.

14.2.16 Binary I/O

Octave can read and write binary data using the functions `fread` and `fwrite`, which are patterned after the standard C functions with the same names. They are able to automatically swap the byte order of integer data and convert among the supported floating point formats as the data are read.

[*val*, *count*] = fread (*fid*, *size*, *precision*, *skip*, *arch*)   [Built-in Function]
    Read binary data of type *precision* from the specified file ID *fid*.

    The optional argument *size* specifies the amount of data to read and may be one of

    Inf   Read as much as possible, returning a column vector.

    nr   Read up to *nr* elements, returning a column vector.

    [*nr*, Inf] Read as much as possible, returning a matrix with *nr* rows. If the number of elements read is not an exact multiple of *nr*, the last column is padded with zeros.

    [*nr*, *nc*] Read up to *nr* `*` *nc* elements, returning a matrix with *nr* rows. If the number of elements read is not an exact multiple of *nr*, the last column is padded with zeros.

    If *size* is omitted, a value of `Inf` is assumed.

    The optional argument *precision* is a string specifying the type of data to read and may be one of

    `"schar"`
    `"signed char"`
            Signed character.

    `"uchar"`
    `"unsigned char"`
            Unsigned character.

`"int8"`
`"integer*1"`
> 8-bit signed integer.

`"int16"`
`"integer*2"`
> 16-bit signed integer.

`"int32"`
`"integer*4"`
> 32-bit signed integer.

`"int64"`
`"integer*8"`
> 64-bit signed integer.

`"uint8"`   8-bit unsigned integer.

`"uint16"`   16-bit unsigned integer.

`"uint32"`   32-bit unsigned integer.

`"uint64"`   64-bit unsigned integer.

`"single"`
`"float32"`
`"real*4"`   32-bit floating point number.

`"double"`
`"float64"`
`"real*8"`   64-bit floating point number.

`"char"`
`"char*1"`   Single character.

`"short"`   Short integer (size is platform dependent).

`"int"`   Integer (size is platform dependent).

`"long"`   Long integer (size is platform dependent).

`"ushort"`
`"unsigned short"`
> Unsigned short integer (size is platform dependent).

`"uint"`
`"unsigned int"`
> Unsigned integer (size is platform dependent).

`"ulong"`
`"unsigned long"`
> Unsigned long integer (size is platform dependent).

`"float"`   Single precision floating point number (size is platform dependent).

The default precision is `"uchar"`.

The *precision* argument may also specify an optional repeat count. For example, '32*single' causes `fread` to read a block of 32 single precision floating point numbers. Reading in blocks is useful in combination with the *skip* argument.

The *precision* argument may also specify a type conversion. For example, 'int16=>int32' causes `fread` to read 16-bit integer values and return an array of 32-bit integer values. By default, `fread` returns a double precision array. The special form '*TYPE' is shorthand for 'TYPE=>TYPE'.

The conversion and repeat counts may be combined. For example, the specification '32*single=>single' causes `fread` to read blocks of single precision floating point values and return an array of single precision values instead of the default array of double precision values.

The optional argument *skip* specifies the number of bytes to skip after each element (or block of elements) is read. If it is not specified, a value of 0 is assumed. If the final block read is not complete, the final skip is omitted. For example,

```
fread (f, 10, "3*single=>single", 8)
```

will omit the final 8-byte skip because the last read will not be a complete block of 3 values.

The optional argument *arch* is a string specifying the data format for the file. Valid values are

`"native"`   The format of the current machine.

`"ieee-be"`
        IEEE big endian.

`"ieee-le"`
        IEEE little endian.

The data read from the file is returned in *val*, and the number of values read is returned in `count`

See also: [fwrite], page 254, [fgets], page 243, [fgetl], page 242, [fscanf], page 248, [fopen], page 240.

count = fwrite (*fid*, *data*, *precision*, *skip*, *arch*)   [Built-in Function]
    Write data in binary form of type *precision* to the specified file ID *fid*, returning the number of values successfully written to the file.

The argument *data* is a matrix of values that are to be written to the file. The values are extracted in column-major order.

The remaining arguments *precision*, *skip*, and *arch* are optional, and are interpreted as described for `fread`.

The behavior of `fwrite` is undefined if the values in *data* are too large to fit in the specified precision.

See also: [fread], page 252, [fputs], page 242, [fprintf], page 243, [fopen], page 240.

14.2.17 Temporary Files

Sometimes one needs to write data to a file that is only temporary. This is most commonly used when an external program launched from within Octave needs to access data. When Octave exits all temporary files will be deleted, so this step need not be executed manually.

[*fid*, *name*, *msg*] = mkstemp (*template*, *delete*)   [Built-in Function]
    Return the file ID corresponding to a new temporary file with a unique name created from *template*. The last six characters of *template* must be **XXXXXX** and these are replaced with a string that makes the filename unique. The file is then created with mode read/write and permissions that are system dependent (on GNU/Linux systems, the permissions will be 0600 for versions of glibc 2.0.7 and later). The file is opened in binary mode and with the **O_EXCL** flag.

    If the optional argument *delete* is supplied and is true, the file will be deleted automatically when Octave exits.

    If successful, *fid* is a valid file ID, *name* is the name of the file, and *msg* is an empty string. Otherwise, *fid* is -1, *name* is empty, and *msg* contains a system-dependent error message.

    See also: [tmpfile], page 255, [tmpnam], page 255, [P_tmpdir], page 715.

[*fid*, *msg*] = tmpfile ()   [Built-in Function]
    Return the file ID corresponding to a new temporary file with a unique name. The file is opened in binary read/write ("w+b") mode. The file will be deleted automatically when it is closed or when Octave exits.

    If successful, *fid* is a valid file ID and *msg* is an empty string. Otherwise, *fid* is -1 and *msg* contains a system-dependent error message.

    See also: [tmpnam], page 255, [mkstemp], page 255, [P_tmpdir], page 715.

tmpnam ()   [Built-in Function]
tmpnam (*dir*)   [Built-in Function]
tmpnam (*dir*, *prefix*)   [Built-in Function]
    Return a unique temporary file name as a string.

    If *prefix* is omitted, a value of "oct-" is used. If *dir* is also omitted, the default directory for temporary files is used. If *dir* is provided, it must exist, otherwise the default directory for temporary files is used. Since the named file is not opened, by **tmpnam**, it is possible (though relatively unlikely) that it will not be available by the time your program attempts to open it.

    See also: [tmpfile], page 255, [mkstemp], page 255, [P_tmpdir], page 715.

14.2.18 End of File and Errors

Once a file has been opened its status can be acquired. As an example the **feof** functions determines if the end of the file has been reached. This can be very useful when reading small parts of a file at a time. The following example shows how to read one line at a time from a file until the end has been reached.

```
filename = "myfile.txt";
fid = fopen (filename, "r");
while (! feof (fid) )
  text_line = fgetl (fid);
endwhile
fclose (fid);
```

Note that in some situations it is more efficient to read the entire contents of a file and then process it, than it is to read it line by line. This has the potential advantage of removing the loop in the above code.

feof (*fid*)   [Built-in Function]
    Return 1 if an end-of-file condition has been encountered for a given file and 0 otherwise. Note that it will only return 1 if the end of the file has already been encountered, not if the next read operation will result in an end-of-file condition.

    See also: [fread], page 252, [fopen], page 240.

[err, msg] = ferror (*fid*)   [Built-in Function]
[err, msg] = ferror (*fid*, *"clear"*)   [Built-in Function]
    Return 1 if an error condition has been encountered for the file ID *fid* and 0 otherwise. Note that it will only return 1 if an error has already been encountered, not if the next operation will result in an error condition.

    The second argument is optional. If it is supplied, also clear the error condition.

    See also: [fclear], page 256, [fopen], page 240.

fclear (*fid*)   [Built-in Function]
    Clear the stream state for the specified file.

    See also: [fopen], page 240.

freport ()   [Built-in Function]
    Print a list of which files have been opened, and whether they are open for reading, writing, or both. For example:

```
freport ()
```

⊣	number	mode	name
⊣			
⊣	0	r	stdin
⊣	1	w	stdout
⊣	2	w	stderr
⊣	3	r	myfile

    See also: [fopen], page 240, [fclose], page 242.

14.2.19 File Positioning

Three functions are available for setting and determining the position of the file pointer for a given file.

ftell (*fid*)   [Built-in Function]
>   Return the position of the file pointer as the number of characters from the beginning of the file *fid*.

>   **See also:** [fseek], page 257, [feof], page 256, [fopen], page 240.

fseek (*fid*, *offset*)   [Built-in Function]
fseek (*fid*, *offset*, *origin*)   [Built-in Function]
status = fseek (...)   [Built-in Function]
>   Set the file pointer to any location within the file *fid*.

>   The pointer is positioned *offset* characters from the *origin*, which may be one of the predefined variables SEEK_CUR (current position), SEEK_SET (beginning), or SEEK_END (end of file) or strings "cof", "bof" or "eof". If *origin* is omitted, SEEK_SET is assumed. *offset* may be positive, negative, or zero but not all combinations of *origin* and *offset* can be realized.

>   Return 0 on success and -1 on error.

>   **See also:** [fskipl], page 243, [frewind], page 257, [ftell], page 256, [fopen], page 240.

SEEK_SET ()   [Built-in Function]
SEEK_CUR ()   [Built-in Function]
SEEK_END ()   [Built-in Function]
>   Return the numerical value to pass to **fseek** to perform one of the following actions:

>   SEEK_SET   Position file relative to the beginning.

>   SEEK_CUR   Position file relative to the current position.

>   SEEK_END   Position file relative to the end.

>   **See also:** [fseek], page 257.

frewind (*fid*)   [Built-in Function]
>   Move the file pointer to the beginning of the file *fid*, returning 0 for success, and -1 if an error was encountered. It is equivalent to **fseek** (*fid*, 0, SEEK_SET).

>   **See also:** [fseek], page 257, [ftell], page 256, [fopen], page 240.

The following example stores the current file position in the variable **marker**, moves the pointer to the beginning of the file, reads four characters, and then returns to the original position.

```
marker = ftell (myfile);
frewind (myfile);
fourch = fgets (myfile, 4);
fseek (myfile, marker, SEEK_SET);
```

15 Plotting

15.1 Introduction to Plotting

Earlier versions of Octave provided plotting through the use of gnuplot. This capability is still available. But, a newer plotting capability is provided by access to OpenGL. Which plotting system is used is controlled by the **graphics_toolkit** function. See Section 15.4.7 [Graphics Toolkits], page 375.

The function call **graphics_toolkit ("fltk")** selects the FLTK/OpenGL system, and **graphics_toolkit ("gnuplot")** selects the gnuplot system. The two systems may be used selectively through the use of the **graphics_toolkit** property of the graphics handle for each figure. This is explained in Section 15.3 [Graphics Data Structures], page 334. **Caution:** The FLTK toolkit uses single precision variables internally which limits the maximum value that can be displayed to approximately 10^{38}. If your data contains larger values you must use the gnuplot toolkit which supports values up to 10^{308}.

15.2 High-Level Plotting

Octave provides simple means to create many different types of two- and three-dimensional plots using high-level functions.

If you need more detailed control, see Section 15.3 [Graphics Data Structures], page 334 and Section 15.4 [Advanced Plotting], page 361.

15.2.1 Two-Dimensional Plots

The **plot** function allows you to create simple x-y plots with linear axes. For example,

```
x = -10:0.1:10;
plot (x, sin (x));
```

displays a sine wave shown in Figure 15.1. On most systems, this command will open a separate plot window to display the graph.

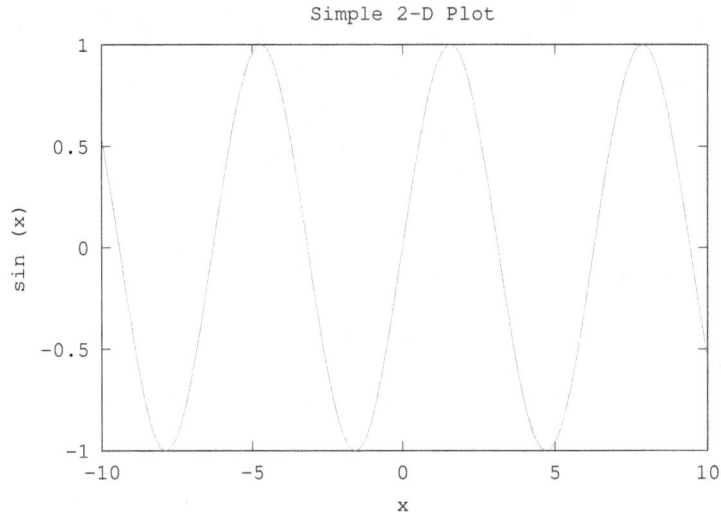

Figure 15.1: Simple Two-Dimensional Plot.

plot (y)	[Function File]
plot (x, y)	[Function File]
plot (x, y, fmt)	[Function File]
plot (..., property, value, ...)	[Function File]
plot (x1, y1, ..., xn, yn)	[Function File]
plot (hax, ...)	[Function File]
h = plot (...)	[Function File]

Produce 2-D plots.

Many different combinations of arguments are possible. The simplest form is

        plot (y)

where the argument is taken as the set of y coordinates and the x coordinates are taken to be the range 1:numel (y).

If more than one argument is given, they are interpreted as

        plot (y, property, value, ...)

or

        plot (x, y, property, value, ...)

or

        plot (x, y, fmt, ...)

and so on. Any number of argument sets may appear. The x and y values are interpreted as follows:

- If a single data argument is supplied, it is taken as the set of y coordinates and the x coordinates are taken to be the indices of the elements, starting with 1.

- If x and y are scalars, a single point is plotted.

- squeeze() is applied to arguments with more than two dimensions, but no more than two singleton dimensions.

- If both arguments are vectors, the elements of *y* are plotted versus the elements of *x*.

- If *x* is a vector and *y* is a matrix, then the columns (or rows) of *y* are plotted versus *x*. (using whichever combination matches, with columns tried first.)

- If the *x* is a matrix and *y* is a vector, *y* is plotted versus the columns (or rows) of *x*. (using whichever combination matches, with columns tried first.)

- If both arguments are matrices, the columns of *y* are plotted versus the columns of *x*. In this case, both matrices must have the same number of rows and columns and no attempt is made to transpose the arguments to make the number of rows match.

Multiple property-value pairs may be specified, but they must appear in pairs. These arguments are applied to the line objects drawn by `plot`. Useful properties to modify are `"linestyle"`, `"linewidth"`, `"color"`, `"marker"`, `"markersize"`, `"markeredgecolor"`, `"markerfacecolor"`.

The *fmt* format argument can also be used to control the plot style. The format is composed of three parts: linestyle, markerstyle, color. When a markerstyle is specified, but no linestyle, only the markers are plotted. Similarly, if a linestyle is specified, but no markerstyle, then only lines are drawn. If both are specified then lines and markers will be plotted. If no *fmt* and no *property/value* pairs are given, then the default plot style is solid lines with no markers and the color determined by the `"colororder"` property of the current axes.

Format arguments:

linestyle

'-'	Use solid lines (default).
'--'	Use dashed lines.
':'	Use dotted lines.
'-.'	Use dash-dotted lines.

markerstyle

'+'	crosshair
'o'	circle
'*'	star
'.'	point
'x'	cross
's'	square
'd'	diamond
'^'	upward-facing triangle
'v'	downward-facing triangle
'>'	right-facing triangle
'<'	left-facing triangle
'p'	pentagram
'h'	hexagram

color

'k'	blacK
'r'	Red

'g'	Green
'b'	Blue
'm'	Magenta
'c'	Cyan
'w'	White

";key;"   Here "key" is the label to use for the plot legend.

The *fmt* argument may also be used to assign legend keys. To do so, include the desired label between semicolons after the formatting sequence described above, e.g., "+b;Key Title;". Note that the last semicolon is required and Octave will generate an error if it is left out.

Here are some plot examples:

```
plot (x, y, "or", x, y2, x, y3, "m", x, y4, "+")
```

This command will plot y with red circles, y2 with solid lines, y3 with solid magenta lines, and y4 with points displayed as '+'.

```
plot (b, "*", "markersize", 10)
```

This command will plot the data in the variable b, with points displayed as '*' and a marker size of 10.

```
t = 0:0.1:6.3;
plot (t, cos(t), "-;cos(t);", t, sin(t), "-b;sin(t);");
```

This will plot the cosine and sine functions and label them accordingly in the legend.

If the first argument *hax* is an axes handle, then plot into this axis, rather than the current axes returned by gca.

The optional return value *h* is a vector of graphics handles to the created line objects.

To save a plot, in one of several image formats such as PostScript or PNG, use the print command.

See also: [axis], page 285, [box], page 316, [grid], page 316, [hold], page 321, [legend], page 313, [title], page 313, [xlabel], page 315, [ylabel], page 315, [xlim], page 287, [ylim], page 287, [ezplot], page 288, [errorbar], page 276, [fplot], page 288, [line], page 337, [plot3], page 303, [polar], page 280, [loglog], page 264, [semilogx], page 263, [semilogy], page 263, [subplot], page 318.

The plotyy function may be used to create a plot with two independent y axes.

plotyy (*x1, y1, x2, y2*)	[Function File]
plotyy (..., *fun*)	[Function File]
plotyy (..., *fun1, fun2*)	[Function File]
plotyy (*hax*, ...)	[Function File]
[*ax, h1, h2*] = plotyy (...)	[Function File]

Plot two sets of data with independent y-axes.

The arguments *x1* and *y1* define the arguments for the first plot and *x1* and *y2* for the second.

By default the arguments are evaluated with feval (@plot, x, y). However the type of plot can be modified with the *fun* argument, in which case the plots are generated by `feval` (*fun*, *x*, *y*). *fun* can be a function handle, an inline function, or a string of a function name.

The function to use for each of the plots can be independently defined with *fun1* and *fun2*.

If the first argument *hax* is an axes handle, then it defines the principal axis in which to plot the *x1* and *y1* data.

The return value *ax* is a vector with the axis handles of the two y axes. *h1* and *h2* are handles to the objects generated by the plot commands.

```
x = 0:0.1:2*pi;
y1 = sin (x);
y2 = exp (x - 1);
ax = plotyy (x, y1, x - 1, y2, @plot, @semilogy);
xlabel ("X");
ylabel (ax(1), "Axis 1");
ylabel (ax(2), "Axis 2");
```

See also: [plot], page 260.

The functions `semilogx`, `semilogy`, and `loglog` are similar to the `plot` function, but produce plots in which one or both of the axes use log scales.

`semilogx` (*y*)	[Function File]
`semilogx` (*x*, *y*)	[Function File]
`semilogx` (*x*, *y*, *property*, *value*, ...)	[Function File]
`semilogx` (*x*, *y*, *fmt*)	[Function File]
`semilogx` (*hax*, ...)	[Function File]
h = `semilogx` (...)	[Function File]

    Produce a 2-D plot using a logarithmic scale for the x-axis.

    See the documentation of `plot` for a description of the arguments that `semilogx` will accept.

    If the first argument *hax* is an axes handle, then plot into this axis, rather than the current axes returned by `gca`.

    The optional return value *h* is a graphics handle to the created plot.

    See also: [plot], page 260, [semilogy], page 263, [loglog], page 264.

`semilogy` (*y*)	[Function File]
`semilogy` (*x*, *y*)	[Function File]
`semilogy` (*x*, *y*, *property*, *value*, ...)	[Function File]
`semilogy` (*x*, *y*, *fmt*)	[Function File]
`semilogy` (*h*, ...)	[Function File]
h = `semilogy` (...)	[Function File]

    Produce a 2-D plot using a logarithmic scale for the y-axis.

    See the documentation of `plot` for a description of the arguments that `semilogy` will accept.

    If the first argument *hax* is an axes handle, then plot into this axis, rather than the current axes returned by `gca`.

The optional return value h is a graphics handle to the created plot.

See also: [plot], page 260, [semilogx], page 263, [loglog], page 264.

```
loglog (y)                                                          [Function File]
loglog (x, y)                                                       [Function File]
loglog (x, y, prop, value, ...)                                     [Function File]
loglog (x, y, fmt)                                                  [Function File]
loglog (hax, ...)                                                   [Function File]
h = loglog (...)                                                    [Function File]
```
Produce a 2-D plot using logarithmic scales for both axes.

See the documentation of `plot` for a description of the arguments that `loglog` will accept.

If the first argument hax is an axes handle, then plot into this axis, rather than the current axes returned by `gca`.

The optional return value h is a graphics handle to the created plot.

See also: [plot], page 260, [semilogx], page 263, [semilogy], page 263.

The functions `bar`, `barh`, `stairs`, and `stem` are useful for displaying discrete data. For example,

```
hist (randn (10000, 1), 30);
```

produces the histogram of 10,000 normally distributed random numbers shown in Figure 15.2.

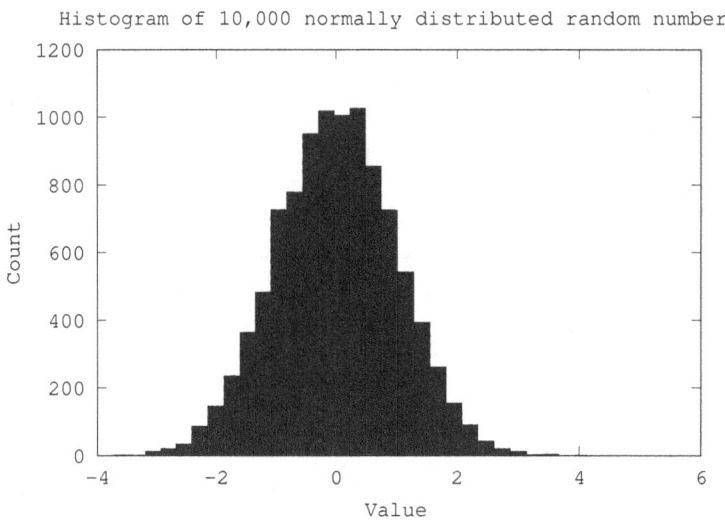

Figure 15.2: Histogram.

```
bar (y)                                                             [Function File]
bar (x, y)                                                          [Function File]
bar (..., w)                                                        [Function File]
bar (..., style)                                                    [Function File]
bar (..., prop, val, ...)                                           [Function File]
```

`bar (`*hax*`, ...)`   [Function File]
`h = bar (..., `*prop*`, `*val*`, ...)`   [Function File]

> Produce a bar graph from two vectors of X-Y data.
>
> If only one argument is given, *y*, it is taken as a vector of Y values and the X coordinates are the range `1:numel (`*y*`)`.
>
> The optional input *w* controls the width of the bars. A value of 1.0 will cause each bar to exactly touch any adjacent bars. The default width is 0.8.
>
> If *y* is a matrix, then each column of *y* is taken to be a separate bar graph plotted on the same graph. By default the columns are plotted side-by-side. This behavior can be changed by the *style* argument which can take the following values:

> `"grouped"` (default)
>> Side-by-side bars with a gap between bars and centered over the X-coordinate.

> `"stacked"`
>> Bars are stacked so that each X value has a single bar composed of multiple segments.

> `"hist"`   Side-by-side bars with no gap between bars and centered over the X-coordinate.

> `"histc"` Side-by-side bars with no gap between bars and left-aligned to the X-coordinate.

> Optional property/value pairs are passed directly to the underlying patch objects.
>
> If the first argument *hax* is an axes handle, then plot into this axis, rather than the current axes returned by `gca`.
>
> The optional return value *h* is a vector of handles to the created "bar series" hggroups with one handle per column of the variable *y*. This series makes it possible to change a common element in one bar series object and have the change reflected in the other "bar series". For example,

```
h = bar (rand (5, 10));
set (h(1), "basevalue", 0.5);
```

> changes the position on the base of all of the bar series.
>
> The following example modifies the face and edge colors using property/value pairs.

```
bar (randn (1, 100), "facecolor", "r", "edgecolor", "b");
```

> The color of the bars is taken from the figure's colormap, such that

```
bar (rand (10, 3));
colormap (summer (64));
```

> will change the colors used for the bars. The color of bars can also be set manually using the `"facecolor"` property as shown below.

```
h = bar (rand (10, 3));
set (h(1), "facecolor", "r")
set (h(2), "facecolor", "g")
set (h(3), "facecolor", "b")
```

> **See also:** [barh], page 266, [hist], page 266, [pie], page 280, [plot], page 260, [patch], page 337.

barh (*y*)	[Function File]
barh (*x*, *y*)	[Function File]
barh (..., *w*)	[Function File]
barh (..., *style*)	[Function File]
barh (..., *prop*, *val*, ...)	[Function File]
barh (*hax*, ...)	[Function File]
h = barh (..., *prop*, *val*, ...)	[Function File]

Produce a horizontal bar graph from two vectors of X-Y data.

If only one argument is given, it is taken as a vector of Y values and the X coordinates are the range 1:numel (*y*).

The optional input *w* controls the width of the bars. A value of 1.0 will cause each bar to exactly touch any adjacent bars. The default width is 0.8.

If *y* is a matrix, then each column of *y* is taken to be a separate bar graph plotted on the same graph. By default the columns are plotted side-by-side. This behavior can be changed by the *style* argument which can take the following values:

"grouped" (default)
  Side-by-side bars with a gap between bars and centered over the Y-coordinate.

"stacked"
  Bars are stacked so that each Y value has a single bar composed of multiple segments.

"hist"   Side-by-side bars with no gap between bars and centered over the Y-coordinate.

"histc"   Side-by-side bars with no gap between bars and left-aligned to the Y-coordinate.

Optional property/value pairs are passed directly to the underlying patch objects.

If the first argument *hax* is an axes handle, then plot into this axis, rather than the current axes returned by gca.

The optional return value *h* is a graphics handle to the created bar series hggroup. For a description of the use of the bar series, see [bar], page 264.

See also: [bar], page 264, [hist], page 266, [pie], page 280, [plot], page 260, [patch], page 337.

hist (*y*)	[Function File]
hist (*y*, *x*)	[Function File]
hist (*y*, *nbins*)	[Function File]
hist (*y*, *x*, *norm*)	[Function File]
hist (..., *prop*, *val*, ...)	[Function File]
hist (*hax*, ...)	[Function File]
[*nn*, *xx*] = hist (...)	[Function File]

Produce histogram counts or plots.

With one vector input argument, *y*, plot a histogram of the values with 10 bins. The range of the histogram bins is determined by the range of the data. With one matrix input argument, *y*, plot a histogram where each bin contains a bar per input column.

Given a second vector argument, x, use that as the centers of the bins, with the width of the bins determined from the adjacent values in the vector.

If scalar, the second argument, *nbins*, defines the number of bins.

If a third argument is provided, the histogram is normalized such that the sum of the bars is equal to *norm*.

Extreme values are lumped into the first and last bins.

The histogram's appearance may be modified by specifying property/value pairs. For example the face and edge color may be modified.

```
hist (randn (1, 100), 25, "facecolor", "r", "edgecolor", "b");
```

The histogram's colors also depend upon the current colormap.

```
hist (rand (10, 3));
colormap (summer ());
```

If the first argument *hax* is an axes handle, then plot into this axis, rather than the current axes returned by gca.

With two output arguments, produce the values *nn* (numbers of elements) and *xx* (bin centers) such that bar (*xx, nn*) will plot the histogram.

See also: [histc], page 567, [bar], page 264, [pie], page 280, [rose], page 273.

stemleaf (*x, caption*)   [Function File]
stemleaf (*x, caption, stem_sz*)   [Function File]
plotstr = stemleaf (...)   [Function File]
    Compute and display a stem and leaf plot of the vector *x*.

The input *x* should be a vector of integers. Any non-integer values will be converted to integer by *x* = fix (*x*). By default each element of *x* will be plotted with the last digit of the element as a leaf value and the remaining digits as the stem. For example, 123 will be plotted with the stem '12' and the leaf '3'. The second argument, *caption*, should be a character array which provides a description of the data. It is included as a heading for the output.

The optional input *stem_sz* sets the width of each stem. The stem width is determined by 10^(*stem_sz* + 1). The default stem width is 10.

The output of stemleaf is composed of two parts: a "Fenced Letter Display," followed by the stem-and-leaf plot itself. The Fenced Letter Display is described in *Exploratory Data Analysis*. Briefly, the entries are as shown:

```
        Fenced Letter Display
#% nx|_____        nx = numel (x)
M% mi|        md        |        mi median index, md median
H% hi|hl                hu| hs   hi lower hinge index, hl,hu hinges,
1    |x(1)        x(nx)|         hs h_spreadx(1), x(nx) first
        _____                  and last data value.
     _____|step |_____        step 1.5*h_spread
        f|ifl            ifh|     inner fence, lower and higher
         |nfl            nfh|     no.\ of data points within fences
        F|ofl            ofh|     outer fence, lower and higher
         |nFl            nFh|     no.\ of data points outside outer
                                  fences
```

The stem-and-leaf plot shows on each line the stem value followed by the string made up of the leaf digits. If the *stem_sz* is not 1 the successive leaf values are separated by ",".

With no return argument, the plot is immediately displayed. If an output argument is provided, the plot is returned as an array of strings.

The leaf digits are not sorted. If sorted leaf values are desired, use **xs = sort (x)** before calling **stemleaf (xs)**.

The stem and leaf plot and associated displays are described in: Ch. 3, *Exploratory Data Analysis* by J. W. Tukey, Addison-Wesley, 1977.

See also: [hist], page 266, [printd], page 268.

printd (*obj*, *filename*)	[Function File]
out_file = printd (...)	[Function File]

Convert any object acceptable to **disp** into the format selected by the suffix of *filename*. If the return argument *out_file* is given, the name of the created file is returned.

This function is intended to facilitate manipulation of the output of functions such as **stemleaf**.

See also: [stemleaf], page 267.

stairs (*y*)	[Function File]
stairs (*x*, *y*)	[Function File]
stairs (..., *style*)	[Function File]
stairs (..., *prop*, *val*, ...)	[Function File]
stairs (*hax*, ...)	[Function File]
h = stairs (...)	[Function File]
[*xstep*, *ystep*] = stairs (...)	[Function File]

Produce a stairstep plot.

The arguments *x* and *y* may be vectors or matrices. If only one argument is given, it is taken as a vector of Y values and the X coordinates are taken to be the indices of the elements.

The style to use for the plot can be defined with a line style *style* of the same format as the **plot** command.

Multiple property/value pairs may be specified, but they must appear in pairs.

If the first argument *hax* is an axis handle, then plot into this axis, rather than the current axis handle returned by `gca`.

If one output argument is requested, return a graphics handle to the created plot. If two output arguments are specified, the data are generated but not plotted. For example,

```
stairs (x, y);
```

and

```
[xs, ys] = stairs (x, y);
plot (xs, ys);
```

are equivalent.

See also: [bar], page 264, [hist], page 266, [plot], page 260, [stem], page 269.

`stem (y)`	[Function File]
`stem (x, y)`	[Function File]
`stem (..., linespec)`	[Function File]
`stem (..., "filled")`	[Function File]
`stem (..., prop, val, ...)`	[Function File]
`stem (hax, ...)`	[Function File]
`h = stem (...)`	[Function File]

Plot a 2-D stem graph.

If only one argument is given, it is taken as the y-values and the x-coordinates are taken from the indices of the elements.

If *y* is a matrix, then each column of the matrix is plotted as a separate stem graph. In this case *x* can either be a vector, the same length as the number of rows in *y*, or it can be a matrix of the same size as *y*.

The default color is "b" (blue), the default line style is "-", and the default marker is "o". The line style can be altered by the `linespec` argument in the same manner as the `plot` command. If the "filled" argument is present the markers at the top of the stems will be filled in. For example,

```
x = 1:10;
y = 2*x;
stem (x, y, "r");
```

plots 10 stems with heights from 2 to 20 in red;

Optional property/value pairs may be specified to control the appearance of the plot.

If the first argument *hax* is an axes handle, then plot into this axis, rather than the current axes returned by `gca`.

The optional return value *h* is a handle to a "stem series" hggroup. The single hggroup handle has all of the graphical elements comprising the plot as its children; This allows the properties of multiple graphics objects to be changed by modifying just a single property of the "stem series" hggroup.

For example,

```
x = [0:10]';
y = [sin(x), cos(x)]
h = stem (x, y);
set (h(2), "color", "g");
set (h(1), "basevalue", -1)
```

changes the color of the second "stem series" and moves the base line of the first.

Stem Series Properties

linestyle   The linestyle of the stem. (Default: "-")

linewidth   The width of the stem. (Default: 0.5)

color   The color of the stem, and if not separately specified, the marker. (De-
             fault: "b" [blue])

marker   The marker symbol to use at the top of each stem. (Default: "o")

markeredgecolor
             The edge color of the marker. (Default: "color" property)

markerfacecolor
             The color to use for "filling" the marker. (Default: "none" [unfilled])

markersize
             The size of the marker. (Default: 6)

baseline   The handle of the line object which implements the baseline. Use set
             with the returned handle to change graphic properties of the baseline.

basevalue   The y-value where the baseline is drawn. (Default: 0)

See also: [stem3], page 270, [bar], page 264, [hist], page 266, [plot], page 260, [stairs], page 268.

stem3 (*x, y, z*)   [Function File]
stem3 (..., *linespec*)   [Function File]
stem3 (..., "*filled*")   [Function File]
stem3 (..., *prop, val, ...*)   [Function File]
stem3 (*hax, ...*)   [Function File]
h = stem3 (...)   [Function File]

Plot a 3-D stem graph.

Stems are drawn from the height *z* to the location in the x-y plane determined by *x* and *y*. The default color is "b" (blue), the default line style is "-", and the default marker is "o".

The line style can be altered by the `linespec` argument in the same manner as the `plot` command. If the "filled" argument is present the markers at the top of the stems will be filled in.

Optional property/value pairs may be specified to control the appearance of the plot.

If the first argument *hax* is an axes handle, then plot into this axis, rather than the current axes returned by `gca`.

The optional return value *h* is a handle to the "stem series" hggroup containing the line and marker objects used for the plot. See [stem], page 269, for a description of the "stem series" object.

Example:

```
theta = 0:0.2:6;
stem3 (cos (theta), sin (theta), theta);
```

plots 31 stems with heights from 0 to 6 lying on a circle.

Implementation Note: Color definitions with RGB-triples are not valid.

See also: [stem], page 269, [bar], page 264, [hist], page 266, [plot], page 260.

scatter (*x*, *y*)	[Function File]
scatter (*x*, *y*, *s*)	[Function File]
scatter (*x*, *y*, *s*, *c*)	[Function File]
scatter (..., *style*)	[Function File]
scatter (..., "*filled*")	[Function File]
scatter (..., *prop*, *val*, ...)	[Function File]
scatter (*hax*, ...)	[Function File]
h = scatter (...)	[Function File]

Draw a 2-D scatter plot.

A marker is plotted at each point defined by the coordinates in the vectors *x* and *y*.

The size of the markers is determined by *s*, which can be a scalar or a vector of the same length as *x* and *y*. If *s* is not given, or is an empty matrix, then a default value of 8 points is used.

The color of the markers is determined by *c*, which can be a string defining a fixed color; a 3-element vector giving the red, green, and blue components of the color; a vector of the same length as *x* that gives a scaled index into the current colormap; or an Nx3 matrix defining the RGB color of each marker individually.

The marker to use can be changed with the *style* argument, that is a string defining a marker in the same manner as the `plot` command. If no marker is specified it defaults to "o" or circles. If the argument "`filled`" is given then the markers are filled.

Additional property/value pairs are passed directly to the underlying patch object.

If the first argument *hax* is an axes handle, then plot into this axis, rather than the current axes returned by `gca`.

The optional return value *h* is a graphics handle to the created patch object.

Example:

```
x = randn (100, 1);
y = randn (100, 1);
scatter (x, y, [], sqrt (x.^2 + y.^2));
```

See also: [scatter3], page 306, [patch], page 337, [plot], page 260.

plotmatrix (*x*, *y*)	[Function File]
plotmatrix (*x*)	[Function File]
plotmatrix (..., *style*)	[Function File]
plotmatrix (*hax*, ...)	[Function File]

`[h, ax, bigax, p, pax] = plotmatrix (...)`   [Function File]
  Scatter plot of the columns of one matrix against another.

  Given the arguments x and y, that have a matching number of rows, `plotmatrix` plots a set of axes corresponding to

        plot (x(:, i), y(:, j))

  Given a single argument x this is equivalent to

        plotmatrix (x, x)

  except that the diagonal of the set of axes will be replaced with the histogram `hist (x(:, i))`.

  The marker to use can be changed with the *style* argument, that is a string defining a marker in the same manner as the `plot` command.

  If the first argument *hax* is an axes handle, then plot into this axis, rather than the current axes returned by `gca`.

  The optional return value *h* provides handles to the individual graphics objects in the scatter plots, whereas *ax* returns the handles to the scatter plot axis objects. *bigax* is a hidden axis object that surrounds the other axes, such that the commands `xlabel`, `title`, etc., will be associated with this hidden axis. Finally, *p* returns the graphics objects associated with the histogram and *pax* the corresponding axes objects.

  Example:

        plotmatrix (randn (100, 3), "g+")

  See also: [scatter], page 271, [plot], page 260.

`pareto (y)`   [Function File]
`pareto (y, x)`   [Function File]
`pareto (hax, ...)`   [Function File]
`h = pareto (...)`   [Function File]
  Draw a Pareto chart.

  A Pareto chart is a bar graph that arranges information in such a way that priorities for process improvement can be established; It organizes and displays information to show the relative importance of data. The chart is similar to the histogram or bar chart, except that the bars are arranged in decreasing magnitude from left to right along the x-axis.

  The fundamental idea (Pareto principle) behind the use of Pareto diagrams is that the majority of an effect is due to a small subset of the causes. For quality improvement, the first few contributing causes (leftmost bars as presented on the diagram) to a problem usually account for the majority of the result. Thus, targeting these "major causes" for elimination results in the most cost-effective improvement scheme.

  Typically only the magnitude data y is present in which case x is taken to be the range $1 : \text{length } (y)$. If x is given it may be a string array, a cell array of strings, or a numerical vector.

  If the first argument *hax* is an axes handle, then plot into this axis, rather than the current axes returned by `gca`.

  The optional return value *h* is a 2-element vector with a graphics handle for the created bar plot and a second handle for the created line plot.

An example of the use of `pareto` is

```
Cheese = {"Cheddar", "Swiss", "Camembert", ...
          "Munster", "Stilton", "Blue"};
Sold = [105, 30, 70, 10, 15, 20];
pareto (Sold, Cheese);
```

See also: [bar], page 264, [barh], page 266, [hist], page 266, [pie], page 280, [plot], page 260.

`rose (th)`	[Function File]
`rose (th, nbins)`	[Function File]
`rose (th, bins)`	[Function File]
`rose (hax, ...)`	[Function File]
`h = rose (...)`	[Function File]
`[thout rout] = rose (...)`	[Function File]

Plot an angular histogram.

With one vector argument, *th*, plot the histogram with 20 angular bins. If *th* is a matrix then each column of *th* produces a separate histogram.

If *nbins* is given and is a scalar, then the histogram is produced with *nbin* bins. If *bins* is a vector, then the center of each bin is defined by the values of *bins* and the number of bins is given by the number of elements in *bins*.

If the first argument *hax* is an axes handle, then plot into this axis, rather than the current axes returned by `gca`.

The optional return value *h* is a vector of graphics handles to the line objects representing each histogram.

If two output arguments are requested then no plot is made and the polar vectors necessary to plot the histogram are returned instead.

```
[th, r] = rose ([2*randn(1e5,1), pi + 2*randn(1e5,1)]);
polar (th, r);
```

See also: [hist], page 266, [polar], page 280.

The `contour`, `contourf` and `contourc` functions produce two-dimensional contour plots from three-dimensional data.

`contour (z)`	[Function File]
`contour (z, vn)`	[Function File]
`contour (x, y, z)`	[Function File]
`contour (x, y, z, vn)`	[Function File]
`contour (..., style)`	[Function File]
`contour (hax, ...)`	[Function File]
`[c, h] = contour (...)`	[Function File]

Create a 2-D contour plot.

Plot level curves (contour lines) of the matrix *z*, using the contour matrix *c* computed by `contourc` from the same arguments; see the latter for their interpretation.

The appearance of contour lines can be defined with a line style *style* in the same manner as `plot`. Only line style and color are used; Any markers defined by *style* are ignored.

GNU Octave

If the first argument *hax* is an axes handle, then plot into this axis, rather than the current axes returned by gca.

The optional output *c* are the contour levels in contourc format.

The optional return value *h* is a graphics handle to the hggroup comprising the contour lines.

Example:
```
x = 0:2;
y = x;
z = x' * y;
contour (x, y, z, 2:3)
```

See also: [ezcontour], page 289, [contourc], page 274, [contourf], page 274, [contour3], page 275, [clabel], page 315, [meshc], page 293, [surfc], page 295, [caxis], page 287, [colormap], page 660, [plot], page 260.

contourf (*z*)	[Function File]
contourf (*z*, *vn*)	[Function File]
contourf (*x*, *y*, *z*)	[Function File]
contourf (*x*, *y*, *z*, *vn*)	[Function File]
contourf (..., *style*)	[Function File]
contourf (*hax*, ...)	[Function File]
[*c*, *h*] = contourf (...)	[Function File]

Create a 2-D contour plot with filled intervals.

Plot level curves (contour lines) of the matrix *z* and fill the region between lines with colors from the current colormap.

The level curves are taken from the contour matrix *c* computed by contourc for the same arguments; see the latter for their interpretation.

The appearance of contour lines can be defined with a line style *style* in the same manner as plot. Only line style and color are used; Any markers defined by *style* are ignored.

If the first argument *hax* is an axes handle, then plot into this axis, rather than the current axes returned by gca.

The optional output *c* are the contour levels in contourc format.

The optional return value *h* is a graphics handle to the hggroup comprising the contour lines.

The following example plots filled contours of the peaks function.
```
[x, y, z] = peaks (50);
contourf (x, y, z, -7:9)
```

See also: [ezcontourf], page 290, [contour], page 273, [contourc], page 274, [contour3], page 275, [clabel], page 315, [meshc], page 293, [surfc], page 295, [caxis], page 287, [colormap], page 660, [plot], page 260.

[*c*, *lev*] = contourc (*z*)	[Function File]
[*c*, *lev*] = contourc (*z*, *vn*)	[Function File]
[*c*, *lev*] = contourc (*x*, *y*, *z*)	[Function File]

[c, lev] = contourc (*x*, *y*, *z*, *vn*)   [Function File]
> Compute contour lines (isolines of constant Z value).
>
> The matrix *z* contains height values above the rectangular grid determined by *x* and *y*. If only a single input *z* is provided then *x* is taken to be 1:rows (z) and *y* is taken to be 1:columns (z).
>
> The optional input *vn* is either a scalar denoting the number of contour lines to compute or a vector containing the Z values where lines will be computed. When *vn* is a vector the number of contour lines is numel (*vn*). However, to compute a single contour line at a given value use *vn* = [val, val]. If *vn* is omitted it defaults to 10.
>
> The return value *c* is a 2x*n* matrix containing the contour lines in the following format
>
> ```
> c = [lev1, x1, x2, ..., levn, x1, x2, ...
>        len1, y1, y2, ..., lenn, y1, y2, ...]
> ```
>
> in which contour line *n* has a level (height) of *levn* and length of *lenn*.
>
> The optional return value *lev* is a vector with the Z values of of the contour levels.
>
> Example:
>
> ```
> x = 0:2;
> y = x;
> z = x' * y;
> contourc (x, y, z, 2:3)
>    ⇒    2.0000   2.0000   1.0000   3.0000   1.5000   2.0000
>         2.0000   1.0000   2.0000   2.0000   2.0000   1.5000
> ```
>
> **See also:** [contour], page 273, [contourf], page 274, [contour3], page 275, [clabel], page 315.

contour3 (*z*)   [Function File]
contour3 (*z*, *vn*)   [Function File]
contour3 (*x*, *y*, *z*)   [Function File]
contour3 (*x*, *y*, *z*, *vn*)   [Function File]
contour3 (..., *style*)   [Function File]
contour3 (*hax*, ...)   [Function File]
[c, h] = contour3 (...)   [Function File]
> Create a 3-D contour plot.
>
> contour3 plots level curves (contour lines) of the matrix *z* at a Z level corresponding to each contour. This is in contrast to contour which plots all of the contour lines at the same Z level and produces a 2-D plot.
>
> The level curves are taken from the contour matrix *c* computed by contourc for the same arguments; see the latter for their interpretation.
>
> The appearance of contour lines can be defined with a line style *style* in the same manner as plot. Only line style and color are used; Any markers defined by *style* are ignored.
>
> If the first argument *hax* is an axes handle, then plot into this axis, rather than the current axes returned by gca.
>
> The optional output *c* are the contour levels in contourc format.

The optional return value h is a graphics handle to the hggroup comprising the contour lines.

Example:

```
contour3 (peaks (19));
colormap cool;
hold on;
surf (peaks (19), "facecolor", "none", "edgecolor", "black");
```

See also: [contour], page 273, [contourc], page 274, [contourf], page 274, [clabel], page 315, [meshc], page 293, [surfc], page 295, [caxis], page 287, [colormap], page 660, [plot], page 260.

The `errorbar`, `semilogxerr`, `semilogyerr`, and `loglogerr` functions produce plots with error bar markers. For example,

```
x = 0:0.1:10;
y = sin (x);
yp =  0.1 .* randn (size (x));
ym = -0.1 .* randn (size (x));
errorbar (x, sin (x), ym, yp);
```

produces the figure shown in Figure 15.3.

Figure 15.3: Errorbar plot.

errorbar (*y, ey*)	[Function File]
errorbar (*y, ..., fmt*)	[Function File]
errorbar (*x, y, ey*)	[Function File]
errorbar (*x, y, err, fmt*)	[Function File]
errorbar (*x, y, lerr, uerr, fmt*)	[Function File]
errorbar (*x, y, ex, ey, fmt*)	[Function File]
errorbar (*x, y, lx, ux, ly, uy, fmt*)	[Function File]
errorbar (*x1, y1, ..., fmt, xn, yn, ...*)	[Function File]

errorbar (*hax*, ...)   [Function File]
h = errorbar (...)   [Function File]
> Create a 2-D plot with errorbars.

> Many different combinations of arguments are possible. The simplest form is

>> errorbar (*y*, *ey*)

> where the first argument is taken as the set of *y* coordinates, the second argument *ey* are the errors around the *y* values, and the *x* coordinates are taken to be the indices of the elements (`1:numel (y)`).

> The general form of the function is

>> errorbar (*x*, *y*, *err1*, ..., *fmt*, ...)

> After the *x* and *y* arguments there can be 1, 2, or 4 parameters specifying the error values depending on the nature of the error values and the plot format *fmt*.

err (scalar)
> When the error is a scalar all points share the same error value. The errorbars are symmetric and are drawn from *data-err* to *data+err*. The *fmt* argument determines whether *err* is in the x-direction, y-direction (default), or both.

err (vector or matrix)
> Each data point has a particular error value. The errorbars are symmetric and are drawn from *data*(n)-*err*(n) to *data*(n)+*err*(n).

lerr, *uerr* (scalar)
> The errors have a single low-side value and a single upper-side value. The errorbars are not symmetric and are drawn from *data-lerr* to *data+uerr*.

lerr, *uerr* (vector or matrix)
> Each data point has a low-side error and an upper-side error. The errorbars are not symmetric and are drawn from *data*(n)-*lerr*(n) to *data*(n)+*uerr*(n).

> Any number of data sets (*x1,y1*, *x2,y2*, ...) may appear as long as they are separated by a format string *fmt*.

> If *y* is a matrix, *x* and the error parameters must also be matrices having the same dimensions. The columns of *y* are plotted versus the corresponding columns of *x* and errorbars are taken from the corresponding columns of the error parameters.

> If *fmt* is missing, the yerrorbars (`"~"`) plot style is assumed.

> If the *fmt* argument is supplied then it is interpreted, as in normal plots, to specify the line style, marker, and color. In addition, *fmt* may include an errorbar style which **must precede** the ordinary format codes. The following errorbar styles are supported:

'~'   Set yerrorbars plot style (default).

'>'   Set xerrorbars plot style.

'~>'   Set xyerrorbars plot style.

'#~'   Set yboxes plot style.

'#'   Set xboxes plot style.

'#~>'   Set xyboxes plot style.

If the first argument *hax* is an axes handle, then plot into this axis, rather than the current axes returned by `gca`.

The optional return value *h* is a handle to the hggroup object representing the data plot and errorbars.

Note: For compatibility with MATLAB a line is drawn through all data points. However, most scientific errorbar plots are a scatter plot of points with errorbars. To accomplish this, add a marker style to the *fmt* argument such as ".". Alternatively, remove the line by modifying the returned graphic handle with `set (h, "linestyle", "none")`.

Examples:

```
errorbar (x, y, ex, ">.r")
```

produces an xerrorbar plot of *y* versus *x* with *x* errorbars drawn from *x-ex* to *x+ex*. The marker "." is used so no connecting line is drawn and the errorbars appear in red.

```
errorbar (x, y1, ey, "~",
          x, y2, ly, uy)
```

produces yerrorbar plots with *y1* and *y2* versus *x*. Errorbars for *y1* are drawn from *y1-ey* to *y1+ey*, errorbars for *y2* from *y2-ly* to *y2+uy*.

```
errorbar (x, y, lx, ux,
          ly, uy, "~>")
```

produces an xyerrorbar plot of *y* versus *x* in which *x* errorbars are drawn from *x-lx* to *x+ux* and *y* errorbars from *y-ly* to *y+uy*.

See also: [semilogxerr], page 278, [semilogyerr], page 279, [loglogerr], page 279, [plot], page 260.

`semilogxerr (y, ey)`	[Function File]
`semilogxerr (y, ..., fmt)`	[Function File]
`semilogxerr (x, y, ey)`	[Function File]
`semilogxerr (x, y, err, fmt)`	[Function File]
`semilogxerr (x, y, lerr, uerr, fmt)`	[Function File]
`semilogxerr (x, y, ex, ey, fmt)`	[Function File]
`semilogxerr (x, y, lx, ux, ly, uy, fmt)`	[Function File]
`semilogxerr (x1, y1, ..., fmt, xn, yn, ...)`	[Function File]
`semilogxerr (hax, ...)`	[Function File]
`h = semilogxerr (...)`	[Function File]

Produce 2-D plots using a logarithmic scale for the x-axis and errorbars at each data point.

Many different combinations of arguments are possible. The most common form is

```
semilogxerr (x, y, ey, fmt)
```

which produces a semi-logarithmic plot of *y* versus *x* with errors in the *y*-scale defined by *ey* and the plot format defined by *fmt*. See [errorbar], page 276, for available formats and additional information.

If the first argument *hax* is an axes handle, then plot into this axis, rather than the current axes returned by `gca`.

See also: [errorbar], page 276, [semilogyerr], page 279, [loglogerr], page 279.

`semilogyerr (y, ey)`	[Function File]
`semilogyerr (y, ..., fmt)`	[Function File]
`semilogyerr (x, y, ey)`	[Function File]
`semilogyerr (x, y, err, fmt)`	[Function File]
`semilogyerr (x, y, lerr, uerr, fmt)`	[Function File]
`semilogyerr (x, y, ex, ey, fmt)`	[Function File]
`semilogyerr (x, y, lx, ux, ly, uy, fmt)`	[Function File]
`semilogyerr (x1, y1, ..., fmt, xn, yn, ...)`	[Function File]
`semilogyerr (hax, ...)`	[Function File]
`h = semilogyerr (...)`	[Function File]

Produce 2-D plots using a logarithmic scale for the y-axis and errorbars at each data point.

Many different combinations of arguments are possible. The most common form is

```
semilogyerr (x, y, ey, fmt)
```

which produces a semi-logarithmic plot of *y* versus *x* with errors in the *y*-scale defined by *ey* and the plot format defined by *fmt*. See [errorbar], page 276, for available formats and additional information.

If the first argument *hax* is an axes handle, then plot into this axis, rather than the current axes returned by `gca`.

See also: [errorbar], page 276, [semilogxerr], page 278, [loglogerr], page 279.

`loglogerr (y, ey)`	[Function File]
`loglogerr (y, ..., fmt)`	[Function File]
`loglogerr (x, y, ey)`	[Function File]
`loglogerr (x, y, err, fmt)`	[Function File]
`loglogerr (x, y, lerr, uerr, fmt)`	[Function File]
`loglogerr (x, y, ex, ey, fmt)`	[Function File]
`loglogerr (x, y, lx, ux, ly, uy, fmt)`	[Function File]
`loglogerr (x1, y1, ..., fmt, xn, yn, ...)`	[Function File]
`loglogerr (hax, ...)`	[Function File]
`h = loglogerr (...)`	[Function File]

Produce 2-D plots on a double logarithm axis with errorbars.

Many different combinations of arguments are possible. The most common form is

```
loglogerr (x, y, ey, fmt)
```

which produces a double logarithm plot of *y* versus *x* with errors in the *y*-scale defined by *ey* and the plot format defined by *fmt*. See [errorbar], page 276, for available formats and additional information.

If the first argument *hax* is an axes handle, then plot into this axis, rather than the current axes returned by `gca`.

See also: [errorbar], page 276, [semilogxerr], page 278, [semilogyerr], page 279.

Finally, the `polar` function allows you to easily plot data in polar coordinates. However, the display coordinates remain rectangular and linear. For example,

```
polar (0:0.1:10*pi, 0:0.1:10*pi);
```

produces the spiral plot shown in Figure 15.4.

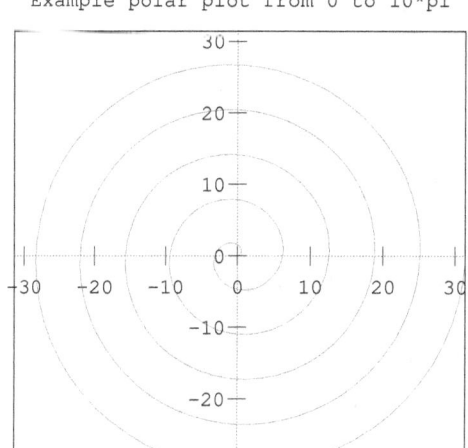

Figure 15.4: Polar plot.

polar (*theta, rho*)	[Function File]
polar (*theta, rho, fmt*)	[Function File]
polar (*cplx*)	[Function File]
polar (*cplx, fmt*)	[Function File]
polar (*hax*, ...)	[Function File]
h = polar (...)	[Function File]

Create a 2-D plot from polar coordinates *theta* and *rho*.

If a single complex input *cplx* is given then the real part is used for *theta* and the imaginary part is used for *rho*.

The optional argument *fmt* specifies the line format in the same way as `plot`.

If the first argument *hax* is an axes handle, then plot into this axis, rather than the current axes returned by `gca`.

The optional return value *h* is a graphics handle to the created plot.

See also: [rose], page 273, [compass], page 283, [plot], page 260.

pie (*x*)	[Function File]
pie (..., *explode*)	[Function File]
pie (..., *labels*)	[Function File]
pie (*hax*, ...);	[Function File]
h = pie (...);	[Function File]

Plot a 2-D pie chart.

When called with a single vector argument, produce a pie chart of the elements in x. The size of the ith slice is the percentage that the element xi represents of the total sum of x: `pct = x(i) / sum (x)`.

The optional input *explode* is a vector of the same length as x that, if non-zero, "explodes" the slice from the pie chart.

The optional input *labels* is a cell array of strings of the same length as x specifying the label for each slice.

If the first argument *hax* is an axes handle, then plot into this axis, rather than the current axes returned by `gca`.

The optional return value h is a list of handles to the patch and text objects generating the plot.

Note: If `sum (x)` ≤ 1 then the elements of x are interpreted as percentages directly and are not normalized by `sum (x)`. Furthermore, if the sum is less than 1 then there will be a missing slice in the pie plot to represent the missing, unspecified percentage.

See also: [pie3], page 281, [bar], page 264, [hist], page 266, [rose], page 273.

`pie3 (x)`	[Function File]
`pie3 (..., explode)`	[Function File]
`pie3 (..., labels)`	[Function File]
`pie3 (hax, ...);`	[Function File]
`h = pie3 (...);`	[Function File]

Plot a 3-D pie chart.

Called with a single vector argument, produces a 3-D pie chart of the elements in x. The size of the ith slice is the percentage that the element xi represents of the total sum of x: `pct = x(i) / sum (x)`.

The optional input *explode* is a vector of the same length as x that, if non-zero, "explodes" the slice from the pie chart.

The optional input *labels* is a cell array of strings of the same length as x specifying the label for each slice.

If the first argument *hax* is an axes handle, then plot into this axis, rather than the current axes returned by `gca`.

The optional return value h is a list of graphics handles to the patch, surface, and text objects generating the plot.

Note: If `sum (x)` ≤ 1 then the elements of x are interpreted as percentages directly and are not normalized by `sum (x)`. Furthermore, if the sum is less than 1 then there will be a missing slice in the pie plot to represent the missing, unspecified percentage.

See also: [pie], page 280, [bar], page 264, [hist], page 266, [rose], page 273.

`quiver (u, v)`	[Function File]
`quiver (x, y, u, v)`	[Function File]
`quiver (..., s)`	[Function File]
`quiver (..., style)`	[Function File]
`quiver (..., "filled")`	[Function File]
`quiver (hax, ...)`	[Function File]

h = quiver (...)   [Function File]
> Plot the (*u*, *v*) components of a vector field in an (*x*, *y*) meshgrid. If the grid is
> uniform, you can specify *x* and *y* as vectors.
>
> If *x* and *y* are undefined they are assumed to be `(1:m, 1:n)` where `[m, n]` = `size`
> `(u)`.
>
> The variable *s* is a scalar defining a scaling factor to use for the arrows of the field
> relative to the mesh spacing. A value of 0 disables all scaling. The default value is
> 0.9.
>
> The style to use for the plot can be defined with a line style *style* of the same format
> as the `plot` command. If a marker is specified then markers at the grid points of the
> vectors are drawn rather than arrows. If the argument `"filled"` is given then the
> markers are filled.
>
> If the first argument *hax* is an axes handle, then plot into this axis, rather than the
> current axes returned by `gca`.
>
> The optional return value *h* is a graphics handle to a quiver object. A quiver object
> regroups the components of the quiver plot (body, arrow, and marker), and allows
> them to be changed together.
>
> Example:
>
> ```
>     [x, y] = meshgrid (1:2:20);
>     h = quiver (x, y, sin (2*pi*x/10), sin (2*pi*y/10));
>     set (h, "maxheadsize", 0.33);
> ```
>
> **See also:** [quiver3], page 282, [compass], page 283, [feather], page 283, [plot],
> page 260.

quiver3 (*u*, *v*, *w*)   [Function File]
quiver3 (*x*, *y*, *z*, *u*, *v*, *w*)   [Function File]
quiver3 (..., *s*)   [Function File]
quiver3 (..., *style*)   [Function File]
quiver3 (..., "*filled*")   [Function File]
quiver3 (*hax*, ...)   [Function File]
h = quiver3 (...)   [Function File]
> Plot the (*u*, *v*, *w*) components of a vector field in an (*x*, *y*, *z*) meshgrid. If the grid is
> uniform, you can specify *x*, *y*, and *z* as vectors.
>
> If *x*, *y*, and *z* are undefined they are assumed to be `(1:m, 1:n, 1:p)` where `[m, n]`
> = `size (u)` and *p* = `max (size (w))`.
>
> The variable *s* is a scalar defining a scaling factor to use for the arrows of the field
> relative to the mesh spacing. A value of 0 disables all scaling. The default value is
> 0.9.
>
> The style to use for the plot can be defined with a line style *style* of the same format
> as the `plot` command. If a marker is specified then markers at the grid points of the
> vectors are drawn rather than arrows. If the argument `"filled"` is given then the
> markers are filled.
>
> If the first argument *hax* is an axes handle, then plot into this axis, rather than the
> current axes returned by `gca`.

The optional return value h is a graphics handle to a quiver object. A quiver object regroups the components of the quiver plot (body, arrow, and marker), and allows them to be changed together.

```
[x, y, z] = peaks (25);
surf (x, y, z);
hold on;
[u, v, w] = surfnorm (x, y, z / 10);
h = quiver3 (x, y, z, u, v, w);
set (h, "maxheadsize", 0.33);
```

See also: [quiver], page 281, [compass], page 283, [feather], page 283, [plot], page 260.

compass (*u*, *v*)	[Function File]
compass (*z*)	[Function File]
compass (..., *style*)	[Function File]
compass (*hax*, ...)	[Function File]
h = compass (...)	[Function File]

Plot the (*u*, *v*) components of a vector field emanating from the origin of a polar plot.

The arrow representing each vector has one end at the origin and the tip at [u(i), v(i)]. If a single complex argument z is given, then u = real (*z*) and v = imag (*z*).

The style to use for the plot can be defined with a line style *style* of the same format as the plot command.

If the first argument *hax* is an axes handle, then plot into this axis, rather than the current axes returned by gca.

The optional return value h is a vector of graphics handles to the line objects representing the drawn vectors.

```
a = toeplitz ([1;randn(9,1)], [1,randn(1,9)]);
compass (eig (a));
```

See also: [polar], page 280, [feather], page 283, [quiver], page 281, [rose], page 273, [plot], page 260.

feather (*u*, *v*)	[Function File]
feather (*z*)	[Function File]
feather (..., *style*)	[Function File]
feather (*hax*, ...)	[Function File]
h = feather (...)	[Function File]

Plot the (*u*, *v*) components of a vector field emanating from equidistant points on the x-axis.

If a single complex argument z is given, then u = real (*z*) and v = imag (*z*).

The style to use for the plot can be defined with a line style *style* of the same format as the plot command.

If the first argument *hax* is an axes handle, then plot into this axis, rather than the current axes returned by gca.

The optional return value h is a vector of graphics handles to the line objects representing the drawn vectors.

```
        phi = [0 : 15 : 360] * pi/180;
        feather (sin (phi), cos (phi));
```

See also: [plot], page 260, [quiver], page 281, [compass], page 283.

pcolor (*x, y, c*)	[Function File]
pcolor (*c*)	[Function File]
pcolor (*hax*, ...)	[Function File]
h = pcolor (...)	[Function File]

Produce a 2-D density plot.

A `pcolor` plot draws rectangles with colors from the matrix c over the two-dimensional region represented by the matrices x and y. x and y are the coordinates of the mesh's vertices and are typically the output of `meshgrid`. If x and y are vectors, then a typical vertex is $(x(j), y(i), c(i,j))$. Thus, columns of c correspond to different x values and rows of c correspond to different y values.

The values in c are scaled to span the range of the current colormap. Limits may be placed on the color axis by the command `caxis`, or by setting the `clim` property of the parent axis.

The face color of each cell of the mesh is determined by interpolating the values of c for each of the cell's vertices; Contrast this with `imagesc` which renders one cell for each element of c.

`shading` modifies an attribute determining the manner by which the face color of each cell is interpolated from the values of c, and the visibility of the cells' edges. By default the attribute is `"faceted"`, which renders a single color for each cell's face with the edge visible.

If the first argument *hax* is an axes handle, then plot into this axis, rather than the current axes returned by `gca`.

The optional return value h is a graphics handle to the created surface object.

See also: [caxis], page 287, [shading], page 306, [meshgrid], page 302, [contour], page 273, [imagesc], page 658.

area (*y*)	[Function File]
area (*x, y*)	[Function File]
area (..., *lvl*)	[Function File]
area (..., *prop, val*, ...)	[Function File]
area (*hax*, ...)	[Function File]
h = area (...)	[Function File]

Area plot of the columns of *y*.

This plot shows the contributions of each column value to the row sum. It is functionally similar to `plot (x, cumsum (y, 2))`, except that the area under the curve is shaded.

If the *x* argument is omitted it defaults to `1:rows (y)`. A value *lvl* can be defined that determines where the base level of the shading under the curve should be defined. The default level is 0.

Additional property/value pairs are passed directly to the underlying patch object.

If the first argument *hax* is an axes handle, then plot into this axis, rather than the current axes returned by `gca`.

The optional return value *h* is a graphics handle to the hggroup object comprising the area patch objects. The `"BaseValue"` property of the hggroup can be used to adjust the level where shading begins.

Example: Verify identity sin^2 + cos^2 = 1

```
t = linspace (0, 2*pi, 100)';
y = [sin(t).^2, cos(t).^2];
area (t, y);
legend ("sin^2", "cos^2", "location", "NorthEastOutside");
```

See also: [plot], page 260, [patch], page 337.

comet (*y*)	[Function File]
comet (*x*, *y*)	[Function File]
comet (*x*, *y*, *p*)	[Function File]
comet (*hax*, ...)	[Function File]

Produce a simple comet style animation along the trajectory provided by the input coordinate vectors (*x*, *y*). If *x* is not specified it defaults to the indices of *y*.

The speed of the comet may be controlled by *p*, which represents the time each point is displayed before moving to the next one. The default for *p* is 0.1 seconds.

If the first argument *hax* is an axes handle, then plot into this axis, rather than the current axes returned by `gca`.

See also: [comet3], page 285.

comet3 (*z*)	[Function File]
comet3 (*x*, *y*, *z*)	[Function File]
comet3 (*x*, *y*, *z*, *p*)	[Function File]
comet3 (*hax*, ...)	[Function File]

Produce a simple comet style animation along the trajectory provided by the input coordinate vectors (*x*, *y*, *z*). If only *z* is specified then *x*, *y* default to the indices of *z*.

The speed of the comet may be controlled by *p*, which represents the time each point is displayed before moving to the next one. The default for *p* is 0.1 seconds.

If the first argument *hax* is an axes handle, then plot into this axis, rather than the current axes returned by `gca`.

See also: [comet], page 285.

15.2.1.1 Axis Configuration

The axis function may be used to change the axis limits of an existing plot and various other axis properties, such as the aspect ratio and the appearance of tic marks.

axis ()	[Function File]
axis ([*x_lo x_hi*])	[Function File]
axis ([*x_lo x_hi y_lo y_hi*])	[Function File]
axis ([*x_lo x_hi y_lo y_hi z_lo z_hi*])	[Function File]
axis (*option*)	[Function File]

axis (..., *option*)   [Function File]
axis (*hax*, ...)   [Function File]
limits = axis ()   [Function File]

Set axis limits and appearance.

The argument *limits* should be a 2-, 4-, or 6-element vector. The first and second elements specify the lower and upper limits for the x-axis. The third and fourth specify the limits for the y-axis, and the fifth and sixth specify the limits for the z-axis.

Without any arguments, `axis` turns autoscaling on.

With one output argument, *limits* = axis returns the current axis limits.

The vector argument specifying limits is optional, and additional string arguments may be used to specify various axis properties. For example,

```
axis ([1, 2, 3, 4], "square");
```

forces a square aspect ratio, and

```
axis ("tic", "labely");
```

turns tic marks on for all axes and tic mark labels on for the y-axis only.

The following options control the aspect ratio of the axes.

"square"   Force a square aspect ratio.

"equal"   Force x distance to equal y-distance.

"normal"   Restore default aspect ratio.

The following options control the way axis limits are interpreted.

"auto"   Set the specified axes to have nice limits around the data or all if no axes are specified.

"manual"   Fix the current axes limits.

"tight"   Fix axes to the limits of the data.

"image"   Equivalent to "tight" and "equal".

The following options affect the appearance of tic marks.

"on"   Turn tic marks and labels on for all axes.

"off"   Turn tic marks off for all axes.

"tic[xyz]"
           Turn tic marks on for all axes, or turn them on for the specified axes and off for the remainder.

"label[xyz]"
           Turn tic labels on for all axes, or turn them on for the specified axes and off for the remainder.

"nolabel"
           Turn tic labels off for all axes.

Note, if there are no tic marks for an axis, there can be no labels.

The following options affect the direction of increasing values on the axes.

"ij"   Reverse y-axis, so lower values are nearer the top.

"xy"   Restore y-axis, so higher values are nearer the top.

If the first argument *hax* is an axes handle, then operate on this axes rather than the current axes returned by gca.

See also: [xlim], page 287, [ylim], page 287, [zlim], page 287, [daspect], page 307, [pbaspect], page 308, [box], page 316, [grid], page 316.

Similarly the axis limits of the colormap can be changed with the caxis function.

caxis ([*cmin cmax*])	[Function File]
caxis ("*auto*")	[Function File]
caxis ("*manual*")	[Function File]
caxis (*hax*, ...)	[Function File]
limits = caxis ()	[Function File]

Query or set color axis limits for plots.

The limits argument should be a 2-element vector specifying the lower and upper limits to assign to the first and last value in the colormap. Data values outside this range are clamped to the first and last colormap entries.

If the "auto" option is given then automatic colormap limits are applied. The automatic algorithm sets *cmin* to the minimum data value and *cmax* to the maximum data value. If "manual" is specified then the "climmode" property is set to "manual" and the numeric values in the "clim" property are used for limits.

If the first argument *hax* is an axes handle, then operate on this axis rather than the current axes returned by gca.

Called without arguments the current color axis limits are returned.

See also: [colormap], page 660.

The xlim, ylim, and zlim functions may be used to get or set individual axis limits. Each has the same form.

xlimits = xlim ()	[Function File]
xmode = xlim ("*mode*")	[Function File]
xlim ([*x_lo x_hi*])	[Function File]
xlim ("*auto*")	[Function File]
xlim ("*manual*")	[Function File]
xlim (*hax*, ...)	[Function File]

Query or set the limits of the x-axis for the current plot.

Called without arguments xlim returns the x-axis limits of the current plot. With the input query "mode", return the current x-limit calculation mode which is either "auto" or "manual".

If passed a 2-element vector [*x_lo x_hi*], the limits of the x-axis are set to these values and the mode is set to "manual".

The current plotting mode can be changed by using either "auto" or "manual" as the argument.

If the first argument *hax* is an axes handle, then operate on this axis rather than the current axes returned by `gca`.

See also: [ylim], page 287, [zlim], page 287, [axis], page 285, [set], page 342, [get], page 341, [gca], page 340.

15.2.1.2 Two-dimensional Function Plotting

Octave can plot a function from a function handle, inline function, or string defining the function without the user needing to explicitly create the data to be plotted. The function `fplot` also generates two-dimensional plots with linear axes using a function name and limits for the range of the x-coordinate instead of the x and y data. For example,

```
fplot (@sin, [-10, 10], 201);
```

produces a plot that is equivalent to the one above, but also includes a legend displaying the name of the plotted function.

`fplot (fn, limits)`	[Function File]
`fplot (..., tol)`	[Function File]
`fplot (..., n)`	[Function File]
`fplot (..., fmt)`	[Function File]
`[x, y] = fplot (...)`	[Function File]

Plot a function *fn* within the range defined by *limits*.

fn is a function handle, inline function, or string containing the name of the function to evaluate.

The limits of the plot are of the form `[xlo, xhi]` or `[xlo, xhi, ylo, yhi]`.

The next three arguments are all optional and any number of them may be given in any order.

tol is the relative tolerance to use for the plot and defaults to 2e-3 (.2%).

n is the minimum number of points to use. When *n* is specified, the maximum stepsize will be `xhi - xlo / n`. More than *n* points may still be used in order to meet the relative tolerance requirement.

The *fmt* argument specifies the linestyle to be used by the plot command.

If the first argument *hax* is an axes handle, then plot into this axis, rather than the current axes returned by `gca`.

With no output arguments the results are immediately plotted. With two output arguments the 2-D plot data is returned. The data can subsequently be plotted manually with `plot (x, y)`.

Example:

```
fplot (@cos, [0, 2*pi])
fplot ("[cos(x), sin(x)]", [0, 2*pi])
```

Note: `fplot` works best with continuous functions. Functions with discontinuities are unlikely to plot well. This restriction may be removed in the future.

See also: [ezplot], page 288, [plot], page 260.

Other functions that can create two-dimensional plots directly from a function include `ezplot`, `ezcontour`, `ezcontourf` and `ezpolar`.

ezplot (*f*)   [Function File]
ezplot (*f2v*)   [Function File]
ezplot (*fx*, *fy*)   [Function File]
ezplot (..., *dom*)   [Function File]
ezplot (..., *n*)   [Function File]
ezplot (*hax*, ...)   [Function File]
h = ezplot (...)   [Function File]

    Plot the 2-D curve defined by the function *f*.

    The function *f* may be a string, inline function, or function handle and can have either one or two variables. If *f* has one variable, then the function is plotted over the domain -2*pi < x < 2*pi with 500 points.

    If *f2v* is a function of two variables then the implicit function *f* (*x*,*y*) = 0 is calculated over the meshed domain -2*pi <= x | y <= 2*pi with 60 points in each dimension.

    For example:

        ezplot (@(*x*, *y*) x.^2 - y.^2 - 1)

    If two functions are passed as inputs then the parametric function

        x = *fx* (*t*)
        y = *fy* (*t*)

    is plotted over the domain -2*pi <= t <= 2*pi with 500 points.

    If *dom* is a two element vector, it represents the minimum and maximum values of both *x* and *y*, or *t* for a parametric plot. If *dom* is a four element vector, then the minimum and maximum values are [xmin xmax ymin ymax].

    n is a scalar defining the number of points to use in plotting the function.

    If the first argument *hax* is an axes handle, then plot into this axis, rather than the current axes returned by gca.

    The optional return value *h* is a vector of graphics handles to the created line objects.

    See also: [plot], page 260, [ezplot3], page 308, [ezpolar], page 290, [ezcontour], page 289, [ezcontourf], page 290, [ezmesh], page 309, [ezmeshc], page 309, [ezsurf], page 310, [ezsurfc], page 311.

ezcontour (*f*)   [Function File]
ezcontour (..., *dom*)   [Function File]
ezcontour (..., *n*)   [Function File]
ezcontour (*hax*, ...)   [Function File]
h = ezcontour (...)   [Function File]

    Plot the contour lines of a function.

    f is a string, inline function, or function handle with two arguments defining the function. By default the plot is over the meshed domain -2*pi <= x | y <= 2*pi with 60 points in each dimension.

    If *dom* is a two element vector, it represents the minimum and maximum values of both *x* and *y*. If *dom* is a four element vector, then the minimum and maximum values are [xmin xmax ymin ymax].

    n is a scalar defining the number of points to use in each dimension.

If the first argument *hax* is an axes handle, then plot into this axis, rather than the current axes returned by `gca`.

The optional return value *h* is a graphics handle to the created plot.

Example:

```
f = @(x,y) sqrt (abs (x .* y)) ./ (1 + x.^2 + y.^2);
ezcontour (f, [-3, 3]);
```

See also: [contour], page 273, [ezcontourf], page 290, [ezplot], page 288, [ezmeshc], page 309, [ezsurfc], page 311.

`ezcontourf (f)`	[Function File]
`ezcontourf (..., dom)`	[Function File]
`ezcontourf (..., n)`	[Function File]
`ezcontourf (hax, ...)`	[Function File]
`h = ezcontourf (...)`	[Function File]

Plot the filled contour lines of a function.

f is a string, inline function, or function handle with two arguments defining the function. By default the plot is over the meshed domain `-2*pi <= x | y <= 2*pi` with 60 points in each dimension.

If *dom* is a two element vector, it represents the minimum and maximum values of both *x* and *y*. If *dom* is a four element vector, then the minimum and maximum values are `[xmin xmax ymin ymax]`.

n is a scalar defining the number of points to use in each dimension.

If the first argument *hax* is an axes handle, then plot into this axis, rather than the current axes returned by `gca`.

The optional return value *h* is a graphics handle to the created plot.

Example:

```
f = @(x,y) sqrt (abs (x .* y)) ./ (1 + x.^2 + y.^2);
ezcontourf (f, [-3, 3]);
```

See also: [contourf], page 274, [ezcontour], page 289, [ezplot], page 288, [ezmeshc], page 309, [ezsurfc], page 311.

`ezpolar (f)`	[Function File]
`ezpolar (..., dom)`	[Function File]
`ezpolar (..., n)`	[Function File]
`ezpolar (hax, ...)`	[Function File]
`h = ezpolar (...)`	[Function File]

Plot a 2-D function in polar coordinates.

The function *f* is a string, inline function, or function handle with a single argument. The expected form of the function is `rho = f(theta)`. By default the plot is over the domain `0 <= theta <= 2*pi` with 500 points.

If *dom* is a two element vector, it represents the minimum and maximum values of *theta*.

n is a scalar defining the number of points to use in plotting the function.

If the first argument *hax* is an axes handle, then plot into this axis, rather than the current axes returned by `gca`.

The optional return value *h* is a graphics handle to the created plot.

Example:

```
ezpolar (@(t) sin (5/4 * t), [0, 8*pi]);
```

See also: [polar], page 280, [ezplot], page 288.

15.2.1.3 Two-dimensional Geometric Shapes

`rectangle ()`	[Function File]
`rectangle (..., "`*Position*`", `*pos*`)`	[Function File]
`rectangle (..., "`*Curvature*`", `*curv*`)`	[Function File]
`rectangle (..., "`*EdgeColor*`", `*ec*`)`	[Function File]
`rectangle (..., "`*FaceColor*`", `*fc*`)`	[Function File]
`rectangle (`*hax*`, ...)`	[Function File]
`h = rectangle (...)`	[Function File]

Draw a rectangular patch defined by *pos* and *curv*.

The variable *pos*`(1:2)` defines the lower left-hand corner of the patch and *pos*`(3:4)` defines its width and height. By default, the value of *pos* is `[0, 0, 1, 1]`.

The variable *curv* defines the curvature of the sides of the rectangle and may be a scalar or two-element vector with values between 0 and 1. A value of 0 represents no curvature of the side, whereas a value of 1 means that the side is entirely curved into the arc of a circle. If *curv* is a two-element vector, then the first element is the curvature along the x-axis of the patch and the second along y-axis.

If *curv* is a scalar, it represents the curvature of the shorter of the two sides of the rectangle and the curvature of the other side is defined by

```
min (pos(1:2)) / max (pos(1:2)) * curv
```

Additional property/value pairs are passed to the underlying patch command.

If the first argument *hax* is an axes handle, then plot into this axis, rather than the current axes returned by `gca`.

The optional return value *h* is a graphics handle to the created rectangle object.

See also: [patch], page 337, [line], page 337, [cylinder], page 312, [ellipsoid], page 312, [sphere], page 312.

15.2.2 Three-Dimensional Plots

The function `mesh` produces mesh surface plots. For example,

```
tx = ty = linspace (-8, 8, 41)';
[xx, yy] = meshgrid (tx, ty);
r = sqrt (xx .^ 2 + yy .^ 2) + eps;
tz = sin (r) ./ r;
mesh (tx, ty, tz);
```

produces the familiar "sombrero" plot shown in Figure 15.5. Note the use of the function `meshgrid` to create matrices of X and Y coordinates to use for plotting the Z data. The `ndgrid` function is similar to `meshgrid`, but works for N-dimensional matrices.

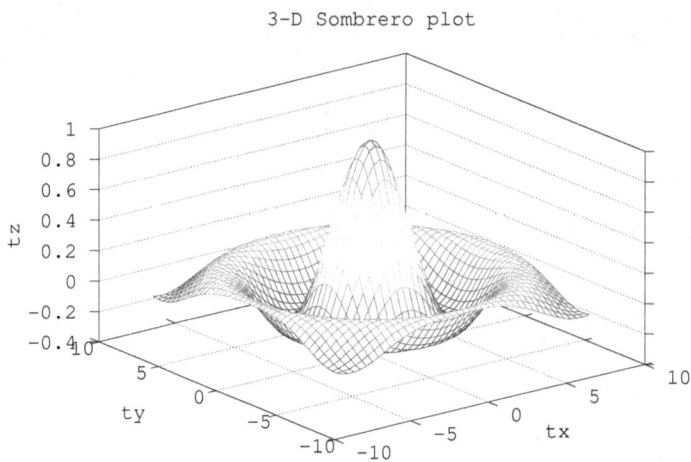

Figure 15.5: Mesh plot.

The `meshc` function is similar to `mesh`, but also produces a plot of contours for the surface.

The `plot3` function displays arbitrary three-dimensional data, without requiring it to form a surface. For example,

```
t = 0:0.1:10*pi;
r = linspace (0, 1, numel (t));
z = linspace (0, 1, numel (t));
plot3 (r.*sin(t), r.*cos(t), z);
```

displays the spiral in three dimensions shown in Figure 15.6.

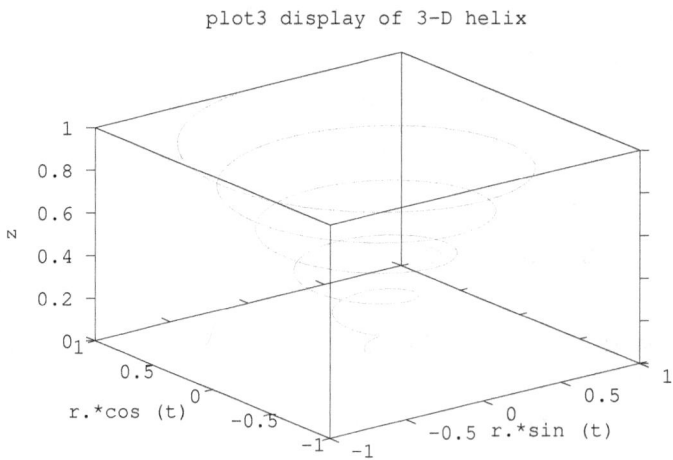

Figure 15.6: Three-dimensional spiral.

Finally, the `view` function changes the viewpoint for three-dimensional plots.

mesh (*x*, *y*, *z*)   [Function File]
mesh (*z*)   [Function File]
mesh (..., *c*)   [Function File]
mesh (..., *prop, val*, ...)   [Function File]
mesh (*hax*, ...)   [Function File]
h = mesh (...)   [Function File]

    Plot a 3-D wireframe mesh.

    The wireframe mesh is plotted using rectangles. The vertices of the rectangles [*x*, *y*] are typically the output of `meshgrid`. over a 2-D rectangular region in the x-y plane. *z* determines the height above the plane of each vertex. If only a single *z* matrix is given, then it is plotted over the meshgrid *x* = 1:`columns` (*z*), *y* = 1:`rows` (*z*). Thus, columns of *z* correspond to different *x* values and rows of *z* correspond to different *y* values.

    The color of the mesh is computed by linearly scaling the *z* values to fit the range of the current colormap. Use `caxis` and/or change the colormap to control the appearance.

    Optionally, the color of the mesh can be specified independently of *z* by supplying a color matrix, *c*.

    Any property/value pairs are passed directly to the underlying surface object.

    If the first argument *hax* is an axes handle, then plot into this axis, rather than the current axes returned by `gca`.

    The optional return value *h* is a graphics handle to the created surface object.

    See also: [ezmesh], page 309, [meshc], page 293, [meshz], page 294, [trimesh], page 628, [contour], page 273, [surf], page 295, [surface], page 339, [meshgrid], page 302, [hidden], page 294, [shading], page 306, [colormap], page 660, [caxis], page 287.

meshc (*x*, *y*, *z*)   [Function File]
meshc (*z*)   [Function File]
meshc (..., *c*)   [Function File]
meshc (..., *prop, val*, ...)   [Function File]
meshc (*hax*, ...)   [Function File]
h = meshc (...)   [Function File]

    Plot a 3-D wireframe mesh with underlying contour lines.

    The wireframe mesh is plotted using rectangles. The vertices of the rectangles [*x*, *y*] are typically the output of `meshgrid`. over a 2-D rectangular region in the x-y plane. *z* determines the height above the plane of each vertex. If only a single *z* matrix is given, then it is plotted over the meshgrid *x* = 1:`columns` (*z*), *y* = 1:`rows` (*z*). Thus, columns of *z* correspond to different *x* values and rows of *z* correspond to different *y* values.

    The color of the mesh is computed by linearly scaling the *z* values to fit the range of the current colormap. Use `caxis` and/or change the colormap to control the appearance.

    Optionally the color of the mesh can be specified independently of *z* by supplying a color matrix, *c*.

    Any property/value pairs are passed directly to the underlying surface object.

If the first argument *hax* is an axes handle, then plot into this axis, rather than the current axes returned by `gca`.

The optional return value *h* is a 2-element vector with a graphics handle to the created surface object and to the created contour plot.

See also: [ezmeshc], page 309, [mesh], page 292, [meshz], page 294, [contour], page 273, [surfc], page 295, [surface], page 339, [meshgrid], page 302, [hidden], page 294, [shading], page 306, [colormap], page 660, [caxis], page 287.

`meshz (x, y, z)`	[Function File]
`meshz (z)`	[Function File]
`meshz (..., c)`	[Function File]
`meshz (..., prop, val, ...)`	[Function File]
`meshz (hax, ...)`	[Function File]
`h = meshz (...)`	[Function File]

Plot a 3-D wireframe mesh with a surrounding curtain.

The wireframe mesh is plotted using rectangles. The vertices of the rectangles [*x*, *y*] are typically the output of `meshgrid`. over a 2-D rectangular region in the x-y plane. *z* determines the height above the plane of each vertex. If only a single *z* matrix is given, then it is plotted over the meshgrid `x = 1:columns (z)`, `y = 1:rows (z)`. Thus, columns of *z* correspond to different *x* values and rows of *z* correspond to different *y* values.

The color of the mesh is computed by linearly scaling the *z* values to fit the range of the current colormap. Use `caxis` and/or change the colormap to control the appearance.

Optionally the color of the mesh can be specified independently of *z* by supplying a color matrix, *c*.

Any property/value pairs are passed directly to the underlying surface object.

If the first argument *hax* is an axes handle, then plot into this axis, rather than the current axes returned by `gca`.

The optional return value *h* is a graphics handle to the created surface object.

See also: [mesh], page 292, [meshc], page 293, [contour], page 273, [surf], page 295, [surface], page 339, [waterfall], page 307, [meshgrid], page 302, [hidden], page 294, [shading], page 306, [colormap], page 660, [caxis], page 287.

`hidden`	[Command]
`hidden "on"`	[Command]
`hidden "off"`	[Command]
`mode = hidden (...)`	[Function File]

Control mesh hidden line removal.

When called with no argument the hidden line removal state is toggled. When called with one of the modes `"on"` or `"off"` the state is set accordingly.

The optional output argument *mode* is the current state.

Hidden Line Removal determines what graphic objects behind a mesh plot are visible. The default is for the mesh to be opaque and lines behind the mesh are not visible. If hidden line removal is turned off then objects behind the mesh can be seen through the faces (openings) of the mesh, although the mesh grid lines are still opaque.

See also: [mesh], page 292, [meshc], page 293, [meshz], page 294, [ezmesh], page 309, [ezmeshc], page 309, [trimesh], page 628, [waterfall], page 307.

surf (x, y, z)	[Function File]
surf (z)	[Function File]
surf ($\ldots$, c)	[Function File]
surf ($\ldots$, *prop*, *val*, $\ldots$)	[Function File]
surf (*hax*, $\ldots$)	[Function File]
h = surf ($\ldots$)	[Function File]

Plot a 3-D surface mesh.

The surface mesh is plotted using shaded rectangles. The vertices of the rectangles [x, y] are typically the output of meshgrid. over a 2-D rectangular region in the x-y plane. z determines the height above the plane of each vertex. If only a single z matrix is given, then it is plotted over the meshgrid x = 1:columns (z), y = 1:rows (z). Thus, columns of z correspond to different x values and rows of z correspond to different y values.

The color of the surface is computed by linearly scaling the z values to fit the range of the current colormap. Use caxis and/or change the colormap to control the appearance.

Optionally, the color of the surface can be specified independently of z by supplying a color matrix, c.

Any property/value pairs are passed directly to the underlying surface object.

If the first argument *hax* is an axes handle, then plot into this axis, rather than the current axes returned by gca.

The optional return value h is a graphics handle to the created surface object.

Note: The exact appearance of the surface can be controlled with the shading command or by using set to control surface object properties.

See also: [ezsurf], page 310, [surfc], page 295, [surfl], page 296, [surfnorm], page 297, [trisurf], page 628, [contour], page 273, [mesh], page 292, [surface], page 339, [meshgrid], page 302, [hidden], page 294, [shading], page 306, [colormap], page 660, [caxis], page 287.

surfc (x, y, z)	[Function File]
surfc (z)	[Function File]
surfc ($\ldots$, c)	[Function File]
surfc ($\ldots$, *prop*, *val*, $\ldots$)	[Function File]
surfc (*hax*, $\ldots$)	[Function File]
h = surfc ($\ldots$)	[Function File]

Plot a 3-D surface mesh with underlying contour lines.

The surface mesh is plotted using shaded rectangles. The vertices of the rectangles [x, y] are typically the output of meshgrid. over a 2-D rectangular region in the x-y plane. z determines the height above the plane of each vertex. If only a single z matrix is given, then it is plotted over the meshgrid x = 1:columns (z), y = 1:rows (z). Thus, columns of z correspond to different x values and rows of z correspond to different y values.

The color of the surface is computed by linearly scaling the *z* values to fit the range of the current colormap. Use `caxis` and/or change the colormap to control the appearance.

Optionally, the color of the surface can be specified independently of *z* by supplying a color matrix, *c*.

Any property/value pairs are passed directly to the underlying surface object.

If the first argument *hax* is an axes handle, then plot into this axis, rather than the current axes returned by `gca`.

The optional return value *h* is a graphics handle to the created surface object.

Note: The exact appearance of the surface can be controlled with the `shading` command or by using `set` to control surface object properties.

See also: [ezsurfc], page 311, [surf], page 295, [surfl], page 296, [surfnorm], page 297, [trisurf], page 628, [contour], page 273, [mesh], page 292, [surface], page 339, [meshgrid], page 302, [hidden], page 294, [shading], page 306, [colormap], page 660, [caxis], page 287.

`surfl (z)`	[Function File]
`surfl (x, y, z)`	[Function File]
`surfl (..., lsrc)`	[Function File]
`surfl (x, y, z, lsrc, P)`	[Function File]
`surfl (..., "cdata")`	[Function File]
`surfl (..., "light")`	[Function File]
`surfl (hax, ...)`	[Function File]
`h = surfl (...)`	[Function File]

Plot a 3-D surface using shading based on various lighting models.

The surface mesh is plotted using shaded rectangles. The vertices of the rectangles [*x*, *y*] are typically the output of `meshgrid`. over a 2-D rectangular region in the x-y plane. *z* determines the height above the plane of each vertex. If only a single *z* matrix is given, then it is plotted over the meshgrid `x = 1:columns (z)`, `y = 1:rows (z)`. Thus, columns of *z* correspond to different *x* values and rows of *z* correspond to different *y* values.

The default lighting mode `"cdata"`, changes the cdata property of the surface object to give the impression of a lighted surface. **Warning:** The alternative mode `"light"` mode which creates a light object to illuminate the surface is not implemented (yet).

The light source location can be specified using *lsrc*. It can be given as a 2-element vector [azimuth, elevation] in degrees, or as a 3-element vector [lx, ly, lz]. The default value is rotated 45 degrees counterclockwise to the current view.

The material properties of the surface can specified using a 4-element vector $P = [AM\ D\ SP\ exp]$ which defaults to $p = [0.55\ 0.6\ 0.4\ 10]$.

`"AM"` strength of ambient light
`"D"` strength of diffuse reflection
`"SP"` strength of specular reflection
`"EXP"` specular exponent

If the first argument *hax* is an axes handle, then plot into this axis, rather than the current axes returned by `gca`.

The optional return value *h* is a graphics handle to the created surface object.
Example:

```
colormap (bone (64));
surfl (peaks);
shading interp;
```

See also: [diffuse], page 302, [specular], page 302, [surf], page 295, [shading], page 306, [colormap], page 660, [caxis], page 287.

surfnorm (*x, y, z*)	[Function File]
surfnorm (*z*)	[Function File]
[*nx, ny, nz*] = surfnorm (...)	[Function File]
surfnorm (*h*, ...)	[Function File]

Find the vectors normal to a meshgridded surface. The meshed gridded surface is defined by *x*, *y*, and *z*. If *x* and *y* are not defined, then it is assumed that they are given by

```
[x, y] = meshgrid (1:rows (z),
                   1:columns (z));
```

If no return arguments are requested, a surface plot with the normal vectors to the surface is plotted. Otherwise the components of the normal vectors at the mesh gridded points are returned in *nx*, *ny*, and *nz*.

The normal vectors are calculated by taking the cross product of the diagonals of each of the quadrilaterals in the meshgrid to find the normal vectors of the centers of these quadrilaterals. The four nearest normal vectors to the meshgrid points are then averaged to obtain the normal to the surface at the meshgridded points.

An example of the use of **surfnorm** is

```
surfnorm (peaks (25));
```

See also: [surf], page 295, [quiver3], page 282.

[*fv*] = isosurface (*val, iso*)	[Function File]
[*fv*] = isosurface (*x, y, z, val, iso*)	[Function File]
[*fv*] = isosurface (..., "*noshare*", "*verbose*")	[Function File]
[*fvc*] = isosurface (..., *col*)	[Function File]
[*f, v*] = isosurface (*x, y, z, val, iso*)	[Function File]
[*f, v, c*] = isosurface (*x, y, z, val, iso, col*)	[Function File]
isosurface (*x, y, z, val, iso, col, opt*)	[Function File]

If called with one output argument and the first input argument *val* is a three-dimensional array that contains the data of an isosurface geometry and the second input argument *iso* keeps the isovalue as a scalar value then return a structure array *fv* that contains the fields *Faces* and *Vertices* at computed points [x, y, z] = meshgrid (1:1, 1:m, 1:n). The output argument *fv* can directly be taken as an input argument for the **patch** function.

If called with further input arguments *x*, *y* and *z* which are three–dimensional arrays with the same size than *val* then the volume data is taken at those given points.

The string input argument "**noshare**" is only for compatibility and has no effect. If given the string input argument "**verbose**" then print messages to the command line interface about the current progress.

If called with the input argument *col* which is a three-dimensional array of the same size than *val* then take those values for the interpolation of coloring the isosurface geometry. Add the field *FaceVertexCData* to the structure array *fv*.

If called with two or three output arguments then return the information about the faces *f*, vertices *v* and color data *c* as separate arrays instead of a single structure array.

If called with no output argument then directly process the isosurface geometry with the `patch` command.

For example,

```
[x, y, z] = meshgrid (1:5, 1:5, 1:5);
val = rand (5, 5, 5);
isosurface (x, y, z, val, .5);
```

will directly draw a random isosurface geometry in a graphics window. Another example for an isosurface geometry with different additional coloring

```
N = 15;    # Increase number of vertices in each direction
iso = .4;  # Change isovalue to .1 to display a sphere
lin = linspace (0, 2, N);
[x, y, z] = meshgrid (lin, lin, lin);
c = abs ((x-.5).^2 + (y-.5).^2 + (z-.5).^2);
figure (); # Open another figure window

subplot (2,2,1); view (-38, 20);
[f, v] = isosurface (x, y, z, c, iso);
p = patch ("Faces", f, "Vertices", v, "EdgeColor", "none");
set (gca, "PlotBoxAspectRatioMode", "manual", ...
          "PlotBoxAspectRatio", [1 1 1]);
# set (p, "FaceColor", "green", "FaceLighting", "phong");
# light ("Position", [1 1 5]); # Available with the JHandles package

subplot (2,2,2); view (-38, 20);
p = patch ("Faces", f, "Vertices", v, "EdgeColor", "blue");
set (gca, "PlotBoxAspectRatioMode", "manual", ...
          "PlotBoxAspectRatio", [1 1 1]);
# set (p, "FaceColor", "none", "FaceLighting", "phong");
# light ("Position", [1 1 5]);

subplot (2,2,3); view (-38, 20);
[f, v, c] = isosurface (x, y, z, c, iso, y);
p = patch ("Faces", f, "Vertices", v, "FaceVertexCData", c, ...
           "FaceColor", "interp", "EdgeColor", "none");
set (gca, "PlotBoxAspectRatioMode", "manual", ...
          "PlotBoxAspectRatio", [1 1 1]);
# set (p, "FaceLighting", "phong");
# light ("Position", [1 1 5]);

subplot (2,2,4); view (-38, 20);
p = patch ("Faces", f, "Vertices", v, "FaceVertexCData", c, ...
           "FaceColor", "interp", "EdgeColor", "blue");
set (gca, "PlotBoxAspectRatioMode", "manual", ...
          "PlotBoxAspectRatio", [1 1 1]);
# set (p, "FaceLighting", "phong");
# light ("Position", [1 1 5]);
```

See also: [isonormals], page 299, [isocolors], page 300.

`[n] = isonormals (val, v)`	[Function File]
`[n] = isonormals (val, p)`	[Function File]
`[n] = isonormals (x, y, z, val, v)`	[Function File]
`[n] = isonormals (x, y, z, val, p)`	[Function File]
`[n] = isonormals (..., "negate")`	[Function File]
`isonormals (..., p)`	[Function File]

If called with one output argument and the first input argument *val* is a three-dimensional array that contains the data for an isosurface geometry and the second input argument *v* keeps the vertices of an isosurface then return the normals *n* in form of a matrix with the same size than *v* at computed points `[x, y, z] = meshgrid (1:l, 1:m, 1:n)`. The output argument *n* can be taken to manually set *VertexNormals* of a patch.

If called with further input arguments *x*, *y* and *z* which are three–dimensional arrays with the same size than *val* then the volume data is taken at those given points. Instead of the vertices data *v* a patch handle *p* can be passed to this function.

If given the string input argument `"negate"` as last input argument then compute the reverse vector normals of an isosurface geometry.

If no output argument is given then directly redraw the patch that is given by the patch handle *p*.

For example:

```
function [] = isofinish (p)
  set (gca, "PlotBoxAspectRatioMode", "manual", ...
            "PlotBoxAspectRatio", [1 1 1]);
  set (p, "VertexNormals", -get (p,"VertexNormals")); # Revert normals
  set (p, "FaceColor", "interp");
  ## set (p, "FaceLighting", "phong");
  ## light ("Position", [1 1 5]); # Available with JHandles
endfunction

N = 15;   # Increase number of vertices in each direction
iso = .4; # Change isovalue to .1 to display a sphere
lin = linspace (0, 2, N);
[x, y, z] = meshgrid (lin, lin, lin);
c = abs ((x-.5).^2 + (y-.5).^2 + (z-.5).^2);
figure (); # Open another figure window

subplot (2,2,1); view (-38, 20);
[f, v, cdat] = isosurface (x, y, z, c, iso, y);
p = patch ("Faces", f, "Vertices", v, "FaceVertexCData", cdat, ...
           "FaceColor", "interp", "EdgeColor", "none");
isofinish (p); ## Call user function isofinish

subplot (2,2,2); view (-38, 20);
p = patch ("Faces", f, "Vertices", v, "FaceVertexCData", cdat, ...
           "FaceColor", "interp", "EdgeColor", "none");
isonormals (x, y, z, c, p); # Directly modify patch
isofinish (p);

subplot (2,2,3); view (-38, 20);
p = patch ("Faces", f, "Vertices", v, "FaceVertexCData", cdat, ...
           "FaceColor", "interp", "EdgeColor", "none");
n = isonormals (x, y, z, c, v); # Compute normals of isosurface
```

```
set (p, "VertexNormals", n);    # Manually set vertex normals
isofinish (p);

subplot (2,2,4); view (-38, 20);
p = patch ("Faces", f, "Vertices", v, "FaceVertexCData", cdat, ...
           "FaceColor", "interp", "EdgeColor", "none");
isonormals (x, y, z, c, v, "negate"); # Use reverse directly
isofinish (p);
```

See also: [isosurface], page 297, [isocolors], page 300.

[cd] = isocolors (c, v)	[Function File]
[cd] = isocolors (x, y, z, c, v)	[Function File]
[cd] = isocolors (x, y, z, r, g, b, v)	[Function File]
[cd] = isocolors (r, g, b, v)	[Function File]
[cd] = isocolors (..., p)	[Function File]
isocolors (...)	[Function File]

If called with one output argument and the first input argument c is a three-dimensional array that contains color values and the second input argument v keeps the vertices of a geometry then return a matrix cd with color data information for the geometry at computed points [x, y, z] = meshgrid (1:l, 1:m, 1:n). The output argument cd can be taken to manually set FaceVertexCData of a patch.

If called with further input arguments x, y and z which are three–dimensional arrays of the same size than c then the color data is taken at those given points. Instead of the color data c this function can also be called with RGB values r, g, b. If input argumnets x, y, z are not given then again meshgrid computed values are taken.

Optionally, the patch handle p can be given as the last input argument to all variations of function calls instead of the vertices data v. Finally, if no output argument is given then directly change the colors of a patch that is given by the patch handle p.

For example:

```
function [] = isofinish (p)
  set (gca, "PlotBoxAspectRatioMode", "manual", ...
            "PlotBoxAspectRatio", [1 1 1]);
  set (p, "FaceColor", "interp");
  ## set (p, "FaceLighting", "flat");
  ## light ("Position", [1 1 5]); ## Available with JHandles
endfunction

N = 15;    # Increase number of vertices in each direction
iso = .4;  # Change isovalue to .1 to display a sphere
lin = linspace (0, 2, N);
[x, y, z] = meshgrid (lin, lin, lin);
c = abs ((x-.5).^2 + (y-.5).^2 + (z-.5).^2);
figure (); # Open another figure window

subplot (2,2,1); view (-38, 20);
[f, v] = isosurface (x, y, z, c, iso);
p = patch ("Faces", f, "Vertices", v, "EdgeColor", "none");
```

```
cdat = rand (size (c));        # Compute random patch color data
isocolors (x, y, z, cdat, p); # Directly set colors of patch
isofinish (p);                 # Call user function isofinish

subplot (2,2,2); view (-38, 20);
p = patch ("Faces", f, "Vertices", v, "EdgeColor", "none");
[r, g, b] = meshgrid (lin, 2-lin, 2-lin);
cdat = isocolors (x, y, z, c, v); # Compute color data vertices
set (p, "FaceVertexCData", cdat); # Set color data manually
isofinish (p);

subplot (2,2,3); view (-38, 20);
p = patch ("Faces", f, "Vertices", v, "EdgeColor", "none");
cdat = isocolors (r, g, b, c, p); # Compute color data patch
set (p, "FaceVertexCData", cdat); # Set color data manually
isofinish (p);

subplot (2,2,4); view (-38, 20);
p = patch ("Faces", f, "Vertices", v, "EdgeColor", "none");
r = g = b = repmat ([1:N] / N, [N, 1, N]); # Black to white
cdat = isocolors (x, y, z, r, g, b, v);
set (p, "FaceVertexCData", cdat);
isofinish (p);
```

See also: [isosurface], page 297, [isonormals], page 299.

shrinkfaces (*p*, *sf*)	[Function File]
nfv = shrinkfaces (*p*, *sf*)	[Function File]
nfv = shrinkfaces (*fv*, *sf*)	[Function File]
nfv = shrinkfaces (*f*, *v*, *sf*)	[Function File]
[*nf*, *nv*] = shrinkfaces (...)	[Function File]

Reduce the faces area for a given patch, structure or explicit faces and points matrices by a scale factor *sf*. The structure *fv* must contain the fields "faces" and "vertices". If the factor *sf* is omitted then a default of 0.3 is used.

Given a patch handle as the first input argument and no output parameters, perform the shrinking of the patch faces in place and redraw the patch.

If called with one output argument, return a structure with fields "faces", "vertices", and "facevertexcdata" containing the data after shrinking which can then directly be used as an input argument for the patch function.

Performing the shrinking on faces which are not convex can lead to undesired results.

For example,

```
[phi r] = meshgrid (linspace (0, 1.5*pi, 16), linspace (1, 2, 4));
tri = delaunay (phi(:), r(:));
v = [r(:).*sin(phi(:)) r(:).*cos(phi(:))];
clf ()
p = patch ("Faces", tri, "Vertices", v, "FaceColor", "none");
fv = shrinkfaces (p);
patch (fv)
axis equal
grid on
```

draws a triangulated 3/4 circle and the corresponding shrunken version.

See also: [patch], page 337.

diffuse (*sx*, *sy*, *sz*, *lv*)   [Function File]
> Calculate diffuse reflection strength of a surface defined by the normal vector elements *sx*, *sy*, *sz*.
>
> The light source location vector *lv* can be given as 2-element vector [azimuth, elevation] in degrees or as 3-element vector [lx, ly, lz].
>
> **See also:** [specular], page 302, [surfl], page 296.

specular (*sx*, *sy*, *sz*, *lv*, *vv*)   [Function File]
specular (*sx*, *sy*, *sz*, *lv*, *vv*, *se*)   [Function File]
> Calculate specular reflection strength of a surface defined by the normal vector elements *sx*, *sy*, *sz* using Phong's approximation.
>
> The light source location and viewer location vectors can be specified using parameter *lv* and *vv* respectively. The location vectors can given as 2-element vectors [azimuth, elevation] in degrees or as 3-element vectors [x, y, z].
>
> An optional sixth argument describes the specular exponent (spread) *se*.
>
> **See also:** [diffuse], page 302, [surfl], page 296.

[*xx*, *yy*] = meshgrid (*x*, *y*)   [Function File]
[*xx*, *yy*, *zz*] = meshgrid (*x*, *y*, *z*)   [Function File]
[*xx*, *yy*] = meshgrid (*x*)   [Function File]
[*xx*, *yy*, *zz*] = meshgrid (*x*)   [Function File]
> Given vectors of *x* and *y* coordinates, return matrices *xx* and *yy* corresponding to a full 2-D grid.
>
> The rows of *xx* are copies of *x*, and the columns of *yy* are copies of *y*. If *y* is omitted, then it is assumed to be the same as *x*.
>
> If the optional *z* input is given, or *zz* is requested, then the output will be a full 3-D grid.
>
> `meshgrid` is most frequently used to produce input for a 2-D or 3-D function that will be plotted. The following example creates a surface plot of the "sombrero" function.
>
> ```
> f = @(x,y) sin (sqrt (x.^2 + y.^2)) ./ sqrt (x.^2 + y.^2);
> range = linspace (-8, 8, 41);
> [X, Y] = meshgrid (range, range);
> Z = f (X, Y);
> surf (X, Y, Z);
> ```

Programming Note: `meshgrid` is restricted to 2-D or 3-D grid generation. The `ndgrid` function will generate 1-D through N-D grids. However, the functions are not completely equivalent. If x is a vector of length M and y is a vector of length N, then `meshgrid` will produce an output grid which is NxM. `ndgrid` will produce an output which is MxN (transpose) for the same input. Some core functions expect `meshgrid` input and others expect `ndgrid` input. Check the documentation for the function in question to determine the proper input format.

See also: [ndgrid], page 303, [mesh], page 292, [contour], page 273, [surf], page 295.

`[y1, y2, ..., yn] = ndgrid (x1, x2, ..., xn)`	[Function File]
`[y1, y2, ..., yn] = ndgrid (x)`	[Function File]

Given n vectors $x1$, ..., xn, `ndgrid` returns n arrays of dimension n. The elements of the i-th output argument contains the elements of the vector xi repeated over all dimensions different from the i-th dimension. Calling ndgrid with only one input argument x is equivalent to calling ndgrid with all n input arguments equal to x:

$$[y1, y2, \ldots, yn] = \text{ndgrid} (x, \ldots, x)$$

Programming Note: `ndgrid` is very similar to the function `meshgrid` except that the first two dimensions are transposed in comparison to `meshgrid`. Some core functions expect `meshgrid` input and others expect `ndgrid` input. Check the documentation for the function in question to determine the proper input format.

See also: [meshgrid], page 302.

`plot3 (x, y, z)`	[Function File]
`plot3 (x, y, z, prop, value, ...)`	[Function File]
`plot3 (x, y, z, fmt)`	[Function File]
`plot3 (x, cplx)`	[Function File]
`plot3 (cplx)`	[Function File]
`plot3 (hax, ...)`	[Function File]
`h = plot3 (...)`	[Function File]

Produce 3-D plots.

Many different combinations of arguments are possible. The simplest form is

        plot3 (x, y, z)

in which the arguments are taken to be the vertices of the points to be plotted in three dimensions. If all arguments are vectors of the same length, then a single continuous line is drawn. If all arguments are matrices, then each column of is treated as a separate line. No attempt is made to transpose the arguments to make the number of rows match.

If only two arguments are given, as

        plot3 (x, cplx)

the real and imaginary parts of the second argument are used as the y and z coordinates, respectively.

If only one argument is given, as

        plot3 (cplx)

the real and imaginary parts of the argument are used as the y and z values, and they are plotted versus their index.

Arguments may also be given in groups of three as

```
plot3 (x1, y1, z1, x2, y2, z2, ...)
```

in which each set of three arguments is treated as a separate line or set of lines in three dimensions.

To plot multiple one- or two-argument groups, separate each group with an empty format string, as

```
plot3 (x1, c1, "", c2, "", ...)
```

Multiple property-value pairs may be specified which will affect the line objects drawn by plot3. If the *fmt* argument is supplied it will format the line objects in the same manner as plot.

If the first argument *hax* is an axes handle, then plot into this axis, rather than the current axes returned by gca.

The optional return value *h* is a graphics handle to the created plot.

Example:

```
z = [0:0.05:5];
plot3 (cos (2*pi*z), sin (2*pi*z), z, ";helix;");
plot3 (z, exp (2i*pi*z), ";complex sinusoid;");
```

See also: [ezplot3], page 308, [plot], page 260.

view (*azimuth*, *elevation*)	[Function File]
view ([*azimuth elevation*])	[Function File]
view ([*x y z*])	[Function File]
view (*2*)	[Function File]
view (*3*)	[Function File]
view (*hax*, ...)	[Function File]
[*azimuth*, *elevation*] = view ()	[Function File]

Query or set the viewpoint for the current axes.

The parameters *azimuth* and *elevation* can be given as two arguments or as 2-element vector. The viewpoint can also be specified with Cartesian coordinates x, y, and z.

The call view (2) sets the viewpoint to $azimuth = 0$ and $elevation = 90$, which is the default for 2-D graphs.

The call view (3) sets the viewpoint to $azimuth = -37.5$ and $elevation = 30$, which is the default for 3-D graphs.

If the first argument *hax* is an axes handle, then operate on this axis rather than the current axes returned by gca.

If no inputs are given, return the current *azimuth* and *elevation*.

slice (*x*, *y*, *z*, *v*, *sx*, *sy*, *sz*)	[Function File]
slice (*x*, *y*, *z*, *v*, *xi*, *yi*, *zi*)	[Function File]
slice (*v*, *sx*, *sy*, *sz*)	[Function File]
slice (*v*, *xi*, *yi*, *zi*)	[Function File]
slice (..., *method*)	[Function File]
slice (*hax*, ...)	[Function File]

h = slice (...)   [Function File]
> Plot slices of 3-D data/scalar fields.
>
> Each element of the 3-dimensional array *v* represents a scalar value at a location given by the parameters *x*, *y*, and *z*. The parameters *x*, *x*, and *z* are either 3-dimensional arrays of the same size as the array *v* in the "meshgrid" format or vectors. The parameters *xi*, etc. respect a similar format to *x*, etc., and they represent the points at which the array *vi* is interpolated using interp3. The vectors *sx*, *sy*, and *sz* contain points of orthogonal slices of the respective axes.
>
> If *x*, *y*, *z* are omitted, they are assumed to be x = 1:size (*v*, 2), y = 1:size (*v*, 1) and z = 1:size (*v*, 3).
>
> *method* is one of:
>
> "nearest"
> > Return the nearest neighbor.
>
> "linear"   Linear interpolation from nearest neighbors.
>
> "cubic"   Cubic interpolation from four nearest neighbors (not implemented yet).
>
> "spline"   Cubic spline interpolation—smooth first and second derivatives throughout the curve.
>
> The default method is "linear".
>
> If the first argument *hax* is an axes handle, then plot into this axis, rather than the current axes returned by gca.
>
> The optional return value *h* is a graphics handle to the created surface object.
>
> Examples:
>
> ```
>         [x, y, z] = meshgrid (linspace (-8, 8, 32));
>         v = sin (sqrt (x.^2 + y.^2 + z.^2)) ./ (sqrt (x.^2 + y.^2 + z.^2));
>         slice (x, y, z, v, [], 0, []);
>
>         [xi, yi] = meshgrid (linspace (-7, 7));
>         zi = xi + yi;
>         slice (x, y, z, v, xi, yi, zi);
> ```
>
> **See also:** [interp3], page 622, [surface], page 339, [pcolor], page 284.

ribbon (*y*)   [Function File]
ribbon (*x*, *y*)   [Function File]
ribbon (*x*, *y*, *width*)   [Function File]
ribbon (*hax*, ...)   [Function File]
h = ribbon (...)   [Function File]
> Plot a ribbon plot for the columns of *y* vs. *x*.
>
> The optional parameter *width* specifies the width of a single ribbon (default is 0.75). If *x* is omitted, a vector containing the row numbers is assumed (1:rows (Y)).
>
> If the first argument *hax* is an axes handle, then plot into this axis, rather than the current axes returned by gca.
>
> The optional return value *h* is a vector of graphics handles to the surface objects representing each ribbon.

See also: [surface], page 339, [waterfall], page 307.

shading (*type*)   [Function File]
shading (*hax*, *type*)   [Function File]
> Set the shading of patch or surface graphic objects.
>
> Valid arguments for *type* are
>
> "flat"   Single colored patches with invisible edges.
>
> "faceted"
> > Single colored patches with visible edges.
>
> "interp"   Color between patch vertices are interpolated and the patch edges are
> > invisible.
>
> If the first argument *hax* is an axes handle, then plot into this axis, rather than the
> current axes returned by gca.
>
> **See also:** [fill], page 338, [mesh], page 292, [patch], page 337, [pcolor], page 284, [surf],
> page 295, [surface], page 339, [hidden], page 294.

scatter3 (*x*, *y*, *z*)   [Function File]
scatter3 (*x*, *y*, *z*, *s*)   [Function File]
scatter3 (*x*, *y*, *z*, *s*, *c*)   [Function File]
scatter3 (..., *style*)   [Function File]
scatter3 (..., "*filled*")   [Function File]
scatter3 (..., *prop*, *val*)   [Function File]
scatter3 (*hax*, ...)   [Function File]
h = scatter3 (...)   [Function File]
> Draw a 3-D scatter plot.
>
> A marker is plotted at each point defined by the coordinates in the vectors *x*, *y*, and
> *z*.
>
> The size of the markers is determined by *s*, which can be a scalar or a vector of the
> same length as *x*, *y*, and *z*. If *s* is not given, or is an empty matrix, then a default
> value of 8 points is used.
>
> The color of the markers is determined by *c*, which can be a string defining a fixed
> color; a 3-element vector giving the red, green, and blue components of the color; a
> vector of the same length as *x* that gives a scaled index into the current colormap; or
> an Nx3 matrix defining the RGB color of each marker individually.
>
> The marker to use can be changed with the *style* argument, that is a string defining a
> marker in the same manner as the plot command. If no marker is specified it defaults
> to "o" or circles. If the argument "filled" is given then the markers are filled.
>
> Additional property/value pairs are passed directly to the underlying patch object.
>
> If the first argument *hax* is an axes handle, then plot into this axis, rather than the
> current axes returned by gca.
>
> The optional return value *h* is a graphics handle to the hggroup object representing
> the points.

```
[x, y, z] = peaks (20);
scatter3 (x(:), y(:), z(:), [], z(:));
```

See also: [scatter], page 271, [patch], page 337, [plot], page 260.

waterfall (*x*, *y*, *z*)	[Function File]
waterfall (*z*)	[Function File]
waterfall (..., *c*)	[Function File]
waterfall (..., *prop*, *val*, ...)	[Function File]
waterfall (*hax*, ...)	[Function File]
h = waterfall (...)	[Function File]

Plot a 3-D waterfall plot.

A waterfall plot is similar to a `meshz` plot except only mesh lines for the rows of *z* (x-values) are shown.

The wireframe mesh is plotted using rectangles. The vertices of the rectangles [*x*, *y*] are typically the output of `meshgrid`. over a 2-D rectangular region in the x-y plane. *z* determines the height above the plane of each vertex. If only a single *z* matrix is given, then it is plotted over the meshgrid `x = 1:columns (z)`, `y = 1:rows (z)`. Thus, columns of *z* correspond to different *x* values and rows of *z* correspond to different *y* values.

The color of the mesh is computed by linearly scaling the *z* values to fit the range of the current colormap. Use `caxis` and/or change the colormap to control the appearance.

Optionally the color of the mesh can be specified independently of *z* by supplying a color matrix, *c*.

Any property/value pairs are passed directly to the underlying surface object.

If the first argument *hax* is an axes handle, then plot into this axis, rather than the current axes returned by `gca`.

The optional return value *h* is a graphics handle to the created surface object.

See also: [meshz], page 294, [mesh], page 292, [meshc], page 293, [contour], page 273, [surf], page 295, [surface], page 339, [ribbon], page 305, [meshgrid], page 302, [hidden], page 294, [shading], page 306, [colormap], page 660, [caxis], page 287.

15.2.2.1 Aspect Ratio

For three-dimensional plots the aspect ratio can be set for data with `daspect` and for the plot box with `pbaspect`. See Section 15.2.1.1 [Axis Configuration], page 285, for controlling the x-, y-, and z-limits for plotting.

data_aspect_ratio = daspect ()	[Function File]
daspect (*data_aspect_ratio*)	[Function File]
daspect (*mode*)	[Function File]
data_aspect_ratio_mode = daspect ("*mode*")	[Function File]
daspect (*hax*, ...)	[Function File]

Query or set the data aspect ratio of the current axes.

The aspect ratio is a normalized 3-element vector representing the span of the x, y, and z-axis limits.

daspect (*mode*)

Set the data aspect ratio mode of the current axes. *mode* is either `"auto"` or `"manual"`.

daspect ("mode")

Return the data aspect ratio mode of the current axes.

daspect (*hax*, ...)

Operate on the axes in handle *hax* instead of the current axes.

See also: [axis], page 285, [pbaspect], page 308, [xlim], page 287, [ylim], page 287, [zlim], page 287.

plot_box_aspect_ratio = pbaspect ( )	[Function File]
pbaspect (*plot_box_aspect_ratio*)	[Function File]
pbaspect (*mode*)	[Function File]
plot_box_aspect_ratio_mode = pbaspect ("*mode*")	[Function File]
pbaspect (*hax*, ...)	[Function File]

Query or set the plot box aspect ratio of the current axes.

The aspect ratio is a normalized 3-element vector representing the rendered lengths of the x, y, and z axes.

pbaspect(*mode*)

Set the plot box aspect ratio mode of the current axes. *mode* is either `"auto"` or `"manual"`.

pbaspect ("mode")

Return the plot box aspect ratio mode of the current axes.

pbaspect (*hax*, ...)

Operate on the axes in handle *hax* instead of the current axes.

See also: [axis], page 285, [daspect], page 307, [xlim], page 287, [ylim], page 287, [zlim], page 287.

15.2.2.2 Three-dimensional Function Plotting

ezplot3 (*fx*, *fy*, *fz*)	[Function File]
ezplot3 (..., *dom*)	[Function File]
ezplot3 (..., *n*)	[Function File]
ezplot3 (*hax*, ...)	[Function File]
h = ezplot3 (...)	[Function File]

Plot a parametrically defined curve in three dimensions.

fx, *fy*, and *fz* are strings, inline functions, or function handles with one argument defining the function. By default the plot is over the domain `0 <= t <= 2*pi` with 500 points.

If *dom* is a two element vector, it represents the minimum and maximum values of *t*.

n is a scalar defining the number of points to use in plotting the function.

If the first argument *hax* is an axes handle, then plot into this axis, rather than the current axes returned by `gca`.

The optional return value *h* is a graphics handle to the created plot.

```
fx = @(t) cos (t);
fy = @(t) sin (t);
fz = @(t) t;
ezplot3 (fx, fy, fz, [0, 10*pi], 100);
```

See also: [plot3], page 303, [ezplot], page 288, [ezmesh], page 309, [ezsurf], page 310.

ezmesh (*f*)	[Function File]
ezmesh (*fx, fy, fz*)	[Function File]
ezmesh (..., *dom*)	[Function File]
ezmesh (..., *n*)	[Function File]
ezmesh (..., "*circ*")	[Function File]
ezmesh (*hax*, ...)	[Function File]
h = ezmesh (...)	[Function File]

Plot the mesh defined by a function.

f is a string, inline function, or function handle with two arguments defining the function. By default the plot is over the meshed domain `-2*pi <= x | y <= 2*pi` with 60 points in each dimension.

If three functions are passed, then plot the parametrically defined function [*fx (s, t), fy (s, t), fz (s, t)*].

If *dom* is a two element vector, it represents the minimum and maximum values of both *x* and *y*. If *dom* is a four element vector, then the minimum and maximum values are [xmin xmax ymin ymax].

n is a scalar defining the number of points to use in each dimension.

If the argument "circ" is given, then the function is plotted over a disk centered on the middle of the domain *dom*.

If the first argument *hax* is an axes handle, then plot into this axis, rather than the current axes returned by gca.

The optional return value *h* is a graphics handle to the created surface object.

Example 1: 2-argument function

```
f = @(x,y) sqrt (abs (x .* y)) ./ (1 + x.^2 + y.^2);
ezmesh (f, [-3, 3]);
```

Example 2: parametrically defined function

```
fx = @(s,t) cos (s) .* cos (t);
fy = @(s,t) sin (s) .* cos (t);
fz = @(s,t) sin (t);
ezmesh (fx, fy, fz, [-pi, pi, -pi/2, pi/2], 20);
```

See also: [mesh], page 292, [ezmeshc], page 309, [ezplot], page 288, [ezsurf], page 310, [ezsurfc], page 311, [hidden], page 294.

ezmeshc (*f*)	[Function File]
ezmeshc (*fx, fy, fz*)	[Function File]
ezmeshc (..., *dom*)	[Function File]
ezmeshc (..., *n*)	[Function File]
ezmeshc (..., "*circ*")	[Function File]

```
ezmeshc (hax, ...)                                                      [Function File]
h = ezmeshc (...)                                                       [Function File]
```
    Plot the mesh and contour lines defined by a function.

    f is a string, inline function, or function handle with two arguments defining the function. By default the plot is over the meshed domain `-2*pi <= x | y <= 2*pi` with 60 points in each dimension.

    If three functions are passed, then plot the parametrically defined function `[fx (s, t), fy (s, t), fz (s, t)]`.

    If *dom* is a two element vector, it represents the minimum and maximum values of both *x* and *y*. If *dom* is a four element vector, then the minimum and maximum values are `[xmin xmax ymin ymax]`.

    n is a scalar defining the number of points to use in each dimension.

    If the argument `"circ"` is given, then the function is plotted over a disk centered on the middle of the domain *dom*.

    If the first argument *hax* is an axes handle, then plot into this axis, rather than the current axes returned by `gca`.

    The optional return value *h* is a 2-element vector with a graphics handle for the created mesh plot and a second handle for the created contour plot.

    Example: 2-argument function

```
f = @(x,y) sqrt (abs (x .* y)) ./ (1 + x.^2 + y.^2);
ezmeshc (f, [-3, 3]);
```

    See also: [meshc], page 293, [ezmesh], page 309, [ezplot], page 288, [ezsurf], page 310, [ezsurfc], page 311, [hidden], page 294.

```
ezsurf (f)                                                              [Function File]
ezsurf (fx, fy, fz)                                                     [Function File]
ezsurf (..., dom)                                                       [Function File]
ezsurf (..., n)                                                         [Function File]
ezsurf (..., "circ")                                                    [Function File]
ezsurf (hax, ...)                                                       [Function File]
h = ezsurf (...)                                                        [Function File]
```
    Plot the surface defined by a function.

    f is a string, inline function, or function handle with two arguments defining the function. By default the plot is over the meshed domain `-2*pi <= x | y <= 2*pi` with 60 points in each dimension.

    If three functions are passed, then plot the parametrically defined function `[fx (s, t), fy (s, t), fz (s, t)]`.

    If *dom* is a two element vector, it represents the minimum and maximum values of both *x* and *y*. If *dom* is a four element vector, then the minimum and maximum values are `[xmin xmax ymin ymax]`.

    n is a scalar defining the number of points to use in each dimension.

    If the argument `"circ"` is given, then the function is plotted over a disk centered on the middle of the domain *dom*.

If the first argument *hax* is an axes handle, then plot into this axis, rather than the current axes returned by `gca`.

The optional return value *h* is a graphics handle to the created surface object.

Example 1: 2-argument function

```
f = @(x,y) sqrt (abs (x .* y)) ./ (1 + x.^2 + y.^2);
ezsurf (f, [-3, 3]);
```

Example 2: parametrically defined function

```
fx = @(s,t) cos (s) .* cos (t);
fy = @(s,t) sin (s) .* cos (t);
fz = @(s,t) sin (t);
ezsurf (fx, fy, fz, [-pi, pi, -pi/2, pi/2], 20);
```

See also: [surf], page 295, [ezsurfc], page 311, [ezplot], page 288, [ezmesh], page 309, [ezmeshc], page 309, [shading], page 306.

`ezsurfc (f)`	[Function File]
`ezsurfc (fx, fy, fz)`	[Function File]
`ezsurfc (..., dom)`	[Function File]
`ezsurfc (..., n)`	[Function File]
`ezsurfc (..., "circ")`	[Function File]
`ezsurfc (hax, ...)`	[Function File]
`h = ezsurfc (...)`	[Function File]

Plot the surface and contour lines defined by a function.

f is a string, inline function, or function handle with two arguments defining the function. By default the plot is over the meshed domain `-2*pi <= x | y <= 2*pi` with 60 points in each dimension.

If three functions are passed, then plot the parametrically defined function [*fx* (*s*, *t*), *fy* (*s*, *t*), *fz* (*s*, *t*)].

If *dom* is a two element vector, it represents the minimum and maximum values of both *x* and *y*. If *dom* is a four element vector, then the minimum and maximum values are [`xmin xmax ymin ymax`].

n is a scalar defining the number of points to use in each dimension.

If the argument `"circ"` is given, then the function is plotted over a disk centered on the middle of the domain *dom*.

If the first argument *hax* is an axes handle, then plot into this axis, rather than the current axes returned by `gca`.

The optional return value *h* is a 2-element vector with a graphics handle for the created surface plot and a second handle for the created contour plot.

Example:

```
f = @(x,y) sqrt (abs (x .* y)) ./ (1 + x.^2 + y.^2);
ezsurfc (f, [-3, 3]);
```

See also: [surfc], page 295, [ezsurf], page 310, [ezplot], page 288, [ezmesh], page 309, [ezmeshc], page 309, [shading], page 306.

15.2.2.3 Three-dimensional Geometric Shapes

`cylinder`	[Command]
`cylinder (r)`	[Function File]
`cylinder (r, n)`	[Function File]
`cylinder (hax, ...)`	[Function File]
`[x, y, z] = cylinder (...)`	[Function File]

Plot a 3-D unit cylinder.

The optional input r is a vector specifying the radius along the unit z-axis. The default is [1 1] indicating radius 1 at Z == 0 and at Z == 1.

The optional input n determines the number of faces around the the circumference of the cylinder. The default value is 20.

If the first argument *hax* is an axes handle, then plot into this axis, rather than the current axes returned by `gca`.

If outputs are requested `cylinder` returns three matrices in `meshgrid` format, such that `surf (x, y, z)` generates a unit cylinder.

Example:

```
[x, y, z] = cylinder (10:-1:0, 50);
surf (x, y, z);
title ("a cone");
```

See also: [ellipsoid], page 312, [rectangle], page 291, [sphere], page 312.

`sphere ()`	[Function File]
`sphere (n)`	[Function File]
`sphere (hax, ...)`	[Function File]
`[x, y, z] = sphere (...)`	[Function File]

Plot a 3-D unit sphere.

The optional input n determines the number of faces around the the circumference of the sphere. The default value is 20.

If the first argument *hax* is an axes handle, then plot into this axis, rather than the current axes returned by `gca`.

If outputs are requested `sphere` returns three matrices in `meshgrid` format such that `surf (x, y, z)` generates a unit sphere.

Example:

```
[x, y, z] = sphere (40);
surf (3*x, 3*y, 3*z);
axis equal;
title ("sphere of radius 3");
```

See also: [cylinder], page 312, [ellipsoid], page 312, [rectangle], page 291.

`ellipsoid (xc, yc, zc, xr, yr, zr, n)`	[Function File]
`ellipsoid (..., n)`	[Function File]
`ellipsoid (hax, ...)`	[Function File]
`[x, y, z] = ellipsoid (...)`	[Function File]

Plot a 3-D ellipsoid.

The inputs *xc*, *yc*, *zc* specify the center of the ellipsoid. The inputs *xr*, *yr*, *zr* specify the semi-major axis lengths.

The optional input *n* determines the number of faces around the the circumference of the cylinder. The default value is 20.

If the first argument *hax* is an axes handle, then plot into this axis, rather than the current axes returned by `gca`.

If outputs are requested `ellipsoid` returns three matrices in `meshgrid` format, such that `surf (x, y, z)` generates the ellipsoid.

See also: [cylinder], page 312, [rectangle], page 291, [sphere], page 312.

15.2.3 Plot Annotations

You can add titles, axis labels, legends, and arbitrary text to an existing plot. For example:

```
x = -10:0.1:10;
plot (x, sin (x));
title ("sin(x) for x = -10:0.1:10");
xlabel ("x");
ylabel ("sin (x)");
text (pi, 0.7, "arbitrary text");
legend ("sin (x)");
```

The functions `grid` and `box` may also be used to add grid and border lines to the plot. By default, the grid is off and the border lines are on.

`title (`*string*`)`	[Function File]
`title (`*string*`, `*prop*`, `*val*`, ...)`	[Function File]
`title (`*hax*`, ...)`	[Function File]
h `= title (...)`	[Function File]

Specify the string used as a title for the current axis.

An optional list of *property*/*value* pairs can be used to change the appearance of the created title text object.

If the first argument *hax* is an axes handle, then plot into this axis, rather than the current axes returned by `gca`.

The optional return value *h* is a graphics handle to the created text object.

See also: [xlabel], page 315, [ylabel], page 315, [zlabel], page 315, [text], page 315.

`legend (`*str1*`, `*str2*`, ...)`	[Function File]
`legend (`*matstr*`)`	[Function File]
`legend (`*cellstr*`)`	[Function File]
`legend (..., "`*location*`", `*pos*`)`	[Function File]
`legend (..., "`*orientation*`", `*orient*`)`	[Function File]
`legend (`*hax*`, ...)`	[Function File]
`legend (`*hobjs*`, ...)`	[Function File]
`legend (`*hax*`, `*hobjs*`, ...)`	[Function File]
`legend ("`*option*`")`	[Function File]
`[`*hleg*`, `*hleg_obj*`, `*hplot*`, `*labels*`] = legend (...)`	[Function File]

Display a legend for the current axes using the specified strings as labels.

Legend entries may be specified as individual character string arguments, a character array, or a cell array of character strings.

If the first argument *hax* is an axes handle, then plot into this axis, rather than the current axes returned by `gca`. If the handles, *hobjs*, are not specified then the legend's strings will be associated with the axes' descendants. `legend` works on line graphs, bar graphs, etc. A plot must exist before legend is called.

The optional parameter *pos* specifies the location of the legend as follows:

pos	location of the legend
north	center top
south	center bottom
east	right center
west	left center
northeast	right top (default)
northwest	left top
southeast	right bottom
southwest	left bottom
outside	can be appended to any location string

The optional parameter *orient* determines if the key elements are placed vertically or horizontally. The allowed values are `"vertical"` (default) or `"horizontal"`.

The following customizations are available using *option*:

`"show"`   Show legend on the plot

`"hide"`   Hide legend on the plot

`"toggle"`   Toggles between `"hide"` and `"show"`

`"boxon"`   Show a box around legend (default)

`"boxoff"`   Hide the box around legend

`"right"`   Place label text to the right of the keys (default)

`"left"`   Place label text to the left of the keys

`"off"`   Delete the legend object

The optional output values are

hleg   The graphics handle of the legend object.

hleg_obj   Graphics handles to the text and line objects which make up the legend.

hplot   Graphics handles to the plot objects which were used in making the legend.

labels   A cell array of strings of the labels in the legend.

The legend label text is either provided in the call to `legend` or is taken from the DisplayName property of graphics objects. If no labels or DisplayNames are available, then the label text is simply `"data1"`, `"data2"`, ..., `"dataN"`.

Implementation Note: A legend is implemented as an additional axes object of the current figure with the `"tag"` set to `"legend"`. Properties of the legend object may be manipulated directly by using `set`.

```
text (x, y, string)                                                      [Function File]
text (x, y, z, string)                                                   [Function File]
text (..., prop, val, ...)                                               [Function File]
h = text (...)                                                           [Function File]
```
> Create a text object with text *string* at position *x*, *y*, (*z*) on the current axes.
>
> Multiple locations can be specified if *x*, *y*, (*z*) are vectors. Multiple strings can be specified with a character matrix or a cell array of strings.
>
> Optional property/value pairs may be used to control the appearance of the text.
>
> The optional return value *h* is a vector of graphics handles to the created text objects.
>
> **See also:** [gtext], page 332, [title], page 313, [xlabel], page 315, [ylabel], page 315, [zlabel], page 315.

See Section 15.3.3.5 [Text Properties], page 353 for the properties that you can set.

```
xlabel (string)                                                          [Function File]
xlabel (string, property, val, ...)                                      [Function File]
xlabel (hax, ...)                                                        [Function File]
h = xlabel (...)                                                         [Function File]
```
> Specify the string used to label the x-axis of the current axis.
>
> An optional list of *property*/*value* pairs can be used to change the properties of the created text label.
>
> If the first argument *hax* is an axes handle, then operate on this axis rather than the current axes returned by `gca`.
>
> The optional return value *h* is a graphics handle to the created text object.
>
> **See also:** [ylabel], page 315, [zlabel], page 315, [datetick], page 707, [title], page 313, [text], page 315.

```
clabel (c, h)                                                            [Function File]
clabel (c, h, v)                                                         [Function File]
clabel (c, h, "manual")                                                  [Function File]
clabel (c)                                                               [Function File]
clabel (..., prop, val, ...)                                             [Function File]
h = clabel (...)                                                         [Function File]
```
> Add labels to the contours of a contour plot.
>
> The contour levels are specified by the contour matrix *c* which is returned by `contour`, `contourc`, `contourf`, and `contour3`. Contour labels are rotated to match the local line orientation and centered on the line. The position of labels along the contour line is chosen randomly.
>
> If the argument *h* is a handle to a contour group object, then label this plot rather than the one in the current axes returned by `gca`.
>
> By default, all contours are labeled. However, the contours to label can be specified by the vector *v*. If the `"manual"` argument is given then the contours to label can be selected with the mouse.

Additional property/value pairs that are valid properties of text objects can be given and are passed to the underlying text objects. Moreover, the contour group property "LabelSpacing" is available which determines the spacing between labels on a contour to be specified. The default is 144 points, or 2 inches.

The optional return value *h* is a vector of graphics handles to the text objects representing each label. The "userdata" property of the text objects contains the numerical value of the contour label.

An example of the use of clabel is

```
[c, h] = contour (peaks (), -4 : 6);
clabel (c, h, -4:2:6, "fontsize", 12);
```

See also: [contour], page 273, [contourf], page 274, [contour3], page 275, [meshc], page 293, [surfc], page 295, [text], page 315.

box *on*   [Command]
box *off*   [Command]
box   [Command]
box (*hax*, . . .)   [Function File]
    Control display of the axis border.

    The argument may be either "on" or "off". If it is omitted, the current box state is toggled.

    If the first argument *hax* is an axes handle, then operate on this axis rather than the current axes returned by gca.

    See also: [axis], page 285, [grid], page 316.

grid   [Command]
grid *on*   [Command]
grid *off*   [Command]
grid *minor*   [Command]
grid *minor on*   [Command]
grid *minor off*   [Command]
grid (*hax*, . . .)   [Function File]
    Control the display of plot grid lines.

    The function state input may be either "on" or "off". If it is omitted, the current grid state is toggled.

    When the first argument is "minor" all subsequent commands modify the minor grid rather than the major grid.

    If the first argument *hax* is an axes handle, then operate on this axis rather than the current axes returned by gca.

    To control the grid lines for an individual axis use the **set** function. For example:

```
set (gca, "ygrid", "on");
```

    See also: [axis], page 285, [box], page 316.

colorbar   [Command]
colorbar (*loc*)   [Function File]

```
colorbar (delete_option)                                   [Function File]
colorbar (hcb, ...)                                        [Function File]
colorbar (hax, ...)                                        [Function File]
colorbar (..., "peer", hax, ...)                           [Function File]
colorbar (..., "location", loc, ...)                       [Function File]
colorbar (..., prop, val, ...)                             [Function File]
h = colorbar (...)                                         [Function File]
```
Add a colorbar to the current axes.

A colorbar displays the current colormap along with numerical rulings so that the color scale can be interpreted.

The optional input *loc* determines the location of the colorbar. Valid values for *loc* are

"EastOutside"
Place the colorbar outside the plot to the right. This is the default.

"East"   Place the colorbar inside the plot to the right.

"WestOutside"
Place the colorbar outside the plot to the left.

"West"   Place the colorbar inside the plot to the left.

"NorthOutside"
Place the colorbar above the plot.

"North"   Place the colorbar at the top of the plot.

"SouthOutside"
Place the colorbar under the plot.

"South"   Place the colorbar at the bottom of the plot.

To remove a colorbar from a plot use any one of the following keywords for the *delete_option*: "delete", "hide", "off".

If the argument "peer" is given, then the following argument is treated as the axes handle in which to add the colorbar. Alternatively, If the first argument *hax* is an axes handle, then the colorbar is added to this axis, rather than the current axes returned by gca.

If the first argument *hcb* is a handle to a colorbar object, then operate on this colorbar directly.

Additional property/value pairs are passed directly to the underlying axes object.

The optional return value *h* is a graphics handle to the created colorbar object.

Implementation Note: A colorbar is created as an additional axes to the current figure with the "tag" property set to "colorbar". The created axes object has the extra property "location" which controls the positioning of the colorbar.

See also: [colormap], page 660.

15.2.4 Multiple Plots on One Page

Octave can display more than one plot in a single figure. The simplest way to do this is to use the `subplot` function to divide the plot area into a series of subplot windows that are indexed by an integer. For example,

```
subplot (2, 1, 1)
fplot (@sin, [-10, 10]);
subplot (2, 1, 2)
fplot (@cos, [-10, 10]);
```

creates a figure with two separate axes, one displaying a sine wave and the other a cosine wave. The first call to subplot divides the figure into two plotting areas (two rows and one column) and makes the first plot area active. The grid of plot areas created by `subplot` is numbered in column-major order (top to bottom, left to right).

subplot (*rows*, *cols*, *index*)	[Function File]
subplot (*rcn*)	[Function File]
subplot (*hax*)	[Function File]
subplot (..., "*align*")	[Function File]
subplot (..., "*replace*")	[Function File]
subplot (..., "*position*", *pos*)	[Function File]
subplot (..., *prop*, *val*, ...)	[Function File]
hax = subplot (...)	[Function File]

Set up a plot grid with *rows* by *cols* subwindows and set the current axes for plotting (`gca`) to the location given by *index*.

If only one numeric argument is supplied, then it must be a three digit value specifying the number of rows in digit 1, the number of columns in digit 2, and the plot index in digit 3.

The plot index runs row-wise; First, all columns in a row are numbered and then the next row is filled.

For example, a plot with 2x3 grid will have plot indices running as follows:

1	2	3
4	5	6

index may also be a vector. In this case, the new axis will enclose the grid locations specified. The first demo illustrates this:

```
demo ("subplot", 1)
```

The index of the subplot to make active may also be specified by its axes handle, *hax*, returned from a previous `subplot` command.

If the option `"align"` is given then the plot boxes of the subwindows will align, but this may leave no room for axis tick marks or labels.

If the option `"replace"` is given then the subplot axis will be reset, rather than just switching the current axis for plotting to the requested subplot.

The "position" property can be used to exactly position the subplot axes within the current figure. The option *pos* is a 4-element vector [x, y, width, height] that determines the location and size of the axes. The values in *pos* are normalized in the range [0,1].

Any property/value pairs are passed directly to the underlying axes object.

If the output *hax* is requested, subplot returns the axis handle for the subplot. This is useful for modifying the properties of a subplot using set.

See also: [axes], page 336, [plot], page 260, [gca], page 340, [set], page 342.

15.2.5 Multiple Plot Windows

You can open multiple plot windows using the figure function. For example,

```
figure (1);
fplot (@sin, [-10, 10]);
figure (2);
fplot (@cos, [-10, 10]);
```

creates two figures, with the first displaying a sine wave and the second a cosine wave. Figure numbers must be positive integers.

figure	[Command]
figure *n*	[Command]
figure (*n*)	[Function File]
figure (..., "*property*", *value*, ...)	[Function File]
h = figure (...)	[Function File]

    Create a new figure window for plotting.

    If no arguments are specified, a new figure with the next available number is created.

    If called with an integer *n*, and no such numbered figure exists, then a new figure with the specified number is created. If the figure already exists then it is made visible and becomes the current figure for plotting.

    Multiple property-value pairs may be specified for the figure object, but they must appear in pairs.

    The optional return value *h* is a graphics handle to the created figure object.

    See also: [axes], page 336, [gcf], page 340, [clf], page 322, [close], page 323.

15.2.6 Manipulation of Plot Windows

By default, Octave refreshes the plot window when a prompt is printed, or when waiting for input. The drawnow function is used to cause a plot window to be updated.

drawnow ()	[Built-in Function]
drawnow ("*expose*")	[Built-in Function]
drawnow (*term*, *file*, *mono*, *debug_file*)	[Built-in Function]

    Update figure windows and their children. The event queue is flushed and any callbacks generated are executed. With the optional argument "expose", only graphic objects are updated and no other events or callbacks are processed. The third calling form of drawnow is for debugging and is undocumented.

Only figures that are modified will be updated. The `refresh` function can also be used to force an update of the current figure, even if it is not modified.

`refresh ()`   [Function File]
`refresh (h)`   [Function File]
    Refresh a figure, forcing it to be redrawn.

    When called without an argument the current figure is redrawn. Otherwise, the figure with graphic handle *h* is redrawn.

    See also: [drawnow], page 319.

Normally, high-level plot functions like `plot` or `mesh` call `newplot` to initialize the state of the current axes so that the next plot is drawn in a blank window with default property settings. To have two plots superimposed over one another, use the `hold` function. For example,

```
hold on;
x = -10:0.1:10;
plot (x, sin (x));
plot (x, cos (x));
hold off;
```

displays sine and cosine waves on the same axes. If the hold state is off, consecutive plotting commands like this will only display the last plot.

`newplot ()`   [Function File]
`newplot (hfig)`   [Function File]
`newplot (hax)`   [Function File]
`hax = newplot (...)`   [Function File]
    Prepare graphics engine to produce a new plot.

    This function is called at the beginning of all high-level plotting functions. It is not normally required in user programs. `newplot` queries the `"NextPlot"` field of the current figure and axis to determine what to do.

Figure NextPlot	Action
`"new"`	Create a new figure and make it the current figure.
`"add"` (default)	Add new graphic objects to the current figure.
`"replacechildren"`	Delete child objects whose HandleVisibility is set to `"on"`. Set NextPlot property to `"add"`. This typically clears a figure, but leaves in place hidden objects such as menubars. This is equivalent to `clf`.
`"replace"`	Delete all child objects of the figure and reset all figure properties to their defaults. However, the following four properties are not reset: Position, Units, PaperPosition, PaperUnits. This is equivalent to `clf reset`.

Axis NextPlot	**Action**
`"add"`	Add new graphic objects to the current axes. This is equivalent to `hold on`.
`"replacechildren"`	Delete child objects whose HandleVisibility is set to `"on"`, but leave axis properties unmodified. This typically clears a plot, but preserves special settings such as log scaling for axes. This is equivalent to `cla`.
`"replace"` (default)	Delete all child objects of the axis and reset all axis properties to their defaults. However, the following properties are not reset: Position, Units. This is equivalent to `cla reset`.

If the optional input *hfig* or *hax* is given then prepare the specified figure or axes rather than the current figure and axes.

The optional return value *hax* is a graphics handle to the created axes object (not figure).

Caution: Calling `newplot` may change the current figure and current axis.

`hold`	[Command]
`hold` *on*	[Command]
`hold` *off*	[Command]
`hold` *all*	[Command]
`hold (hax, ...)`	[Function File]

Toggle or set the `"hold"` state of the plotting engine which determines whether new graphic objects are added to the plot or replace the existing objects.

`hold on`	Retain plot data and settings so that subsequent plot commands are displayed on a single graph.
`hold all`	Retain plot line color, line style, data, and settings so that subsequent plot commands are displayed on a single graph with the next line color and style.
`hold off`	Restore default graphics settings which clear the graph and reset axis properties before each new plot command. (default).
`hold`	Toggle the current hold state.

When given the additional argument *hax*, the hold state is modified for this axis rather than the current axes returned by `gca`.

To query the current hold state use the `ishold` function.

See also: [ishold], page 321, [cla], page 322, [clf], page 322, [newplot], page 320.

`ishold`	[Command]
`ishold (hax)`	[Function File]
`ishold (hfig)`	[Function File]

Return true if the next plot will be added to the current plot, or false if the plot device will be cleared before drawing the next plot.

If the first argument is an axes handle *hax* or figure handle *hfig* then operate on this plot rather than the current one.

See also: [hold], page 321, [newplot], page 320.

To clear the current figure, call the `clf` function. To clear the current axis, call the `cla` function. To bring the current figure to the top of the window stack, call the `shg` function. To delete a graphics object, call `delete` on its index. To close the figure window, call the `close` function.

`clf`	[Command]
`clf` *reset*	[Command]
`clf (hfig)`	[Function File]
`clf (hfig, "reset")`	[Function File]
`h = clf (...)`	[Function File]

Clear the current figure window.

`clf` operates by deleting child graphics objects with visible handles (HandleVisibility = "on").

If the optional argument `"reset"` is specified, delete all child objects including those with hidden handles and reset all figure properties to their defaults. However, the following properties are not reset: Position, Units, PaperPosition, PaperUnits.

If the first argument *hfig* is a figure handle, then operate on this figure rather than the current figure returned by `gcf`.

The optional return value *h* is the graphics handle of the figure window that was cleared.

See also: [cla], page 322, [close], page 323, [delete], page 323, [reset], page 361.

`cla`	[Command]
`cla` *reset*	[Command]
`cla (hax)`	[Function File]
`cla (hax, "reset")`	[Function File]

Clear the current axes.

`cla` operates by deleting child graphic objects with visible handles (HandleVisibility = "on").

If the optional argument `"reset"` is specified, delete all child objects including those with hidden handles and reset all axis properties to their defaults. However, the following properties are not reset: Position, Units.

If the first argument *hax* is an axes handle, then operate on this axis rather than the current axes returned by `gca`.

See also: [clf], page 322, [delete], page 323, [reset], page 361.

`shg`	[Command]

Show the graph window.

Currently, this is the same as executing `drawnow`.

See also: [drawnow], page 319, [figure], page 319.

delete (*file*)   [Function File]
delete (*handle*)   [Function File]
>   Delete the named file or graphics handle.
>
>   Deleting graphics objects is the proper way to remove features from a plot without
>   clearing the entire figure.
>
>   **See also:** [clf], page 322, [cla], page 322, [unlink], page 708.

close   [Command]
close (*h*)   [Command]
close *all*   [Command]
close *all hidden*   [Command]
>   Close figure window(s).
>
>   When called with no arguments, close the current figure. This is equivalent to `close`
>   (`gcf`). If the input *h* is a graphic handle, or vector of graphics handles, then close
>   each figure in *h*.
>
>   If the argument `"all"` is given then all figures with visible handles (HandleVisibility
>   = `"on"`) are closed.
>
>   If the argument `"all hidden"` is given then all figures, including hidden ones, are
>   closed.
>
>   Implementation Note: `close` operates by calling the function specified by the
>   `"closerequestfcn"` property for each figure. By default, the function `closereq` is
>   used. It is possible that the function invoked will delay or abort removing the figure.
>   To remove a figure without executing any callback functions use `delete`. When
>   writing a callback function to close a window do not use `close` to avoid recursion.
>
>   **See also:** [closereq], page 323, [delete], page 323.

closereq ()   [Function File]
>   Close the current figure and delete all graphics objects associated with it.
>
>   By default, the `"closerequestfcn"` property of a new plot figure points to this
>   function.
>
>   **See also:** [close], page 323, [delete], page 323.

15.2.7 Use of the `interpreter` Property

All text objects—such as titles, labels, legends, and text—include the property
`"interpreter"`, this property determines the manner in which special control sequences
in the text are rendered. If the interpreter is set to `"none"`, then no rendering occurs.
Currently the `"latex"` interpreter is not implemented and is equivalent to `"none"`.

The `"tex"` option implements a subset of TeX functionality when rendering text. This
allows the insertion of special glyphs such as Greek characters or mathematical symbols.
The special characters are inserted with a code following a backslash (\) character, as in
the table Table 15.1.

In addition, the formatting of the text can be changed within the string by using the
codes

\bf   Bold font

\it	Italic font
\sl	Oblique Font
\rm	Normal font

These may be used in conjunction with the { and } characters to limit the change in the font to part of the string. For example,

```
xlabel ('{\bf H} = a {\bf V}')
```

where the character 'a' will not appear in a bold font. Note that to avoid having Octave interpret the backslash characters in the strings, the strings should be in single quotes.

It is also possible to change the fontname and size within the text

\fontname{*fontname*}	Specify the font to use
\fontsize{*size*}	Specify the size of the font to use

Finally, superscripting and subscripting can be controlled with the '^' and '_' characters. If the '^' or '_' is followed by a { character, then all of the block surrounded by the { } pair is super- or sub-scripted. Without the { } pair, only the character immediately following the '^' or '_' is super- or sub-scripted.

Greek Lowercase Letters

Code	Sym	Code	Sym	Code	Sym
\alpha	α	\beta	β	\gamma	γ
\delta	δ	\epsilon	ϵ	\zeta	ζ
\eta	η	\theta	θ	\vartheta	ϑ
\iota	ι	\kappa	κ	\lambda	λ
\mu	μ	\nu	ν	\xi	ξ
\o	o	\pi	π	\varpi	ϖ
\rho	ρ	\sigma	σ	\varsigma	ς
\tau	τ	\upsilon	υ	\phi	ϕ
\chi	χ	\psi	ψ	\omega	ω

Greek Uppercase Letters

Code	Sym	Code	Sym	Code	Sym
\Gamma	Γ	\Delta	Δ	\Theta	Θ
\Lambda	Λ	\Xi	Ξ	\Pi	Π
\Sigma	Σ	\Upsilon	Υ	\Phi	Φ
\Psi	Ψ	\Omega	Ω		

Misc Symbols Type Ord

Code	Sym	Code	Sym	Code	Sym
\aleph	$\aleph$	\wp	$\wp$	\Re	$\Re$
\Im	$\Im$	\partial	∂	\infty	∞
\prime	$\prime$	\nabla	∇	\surd	$\surd$
\angle	$\angle$	\forall	$\forall$	\exists	$\exists$
\neg	$\neg$	\clubsuit	$\clubsuit$	\diamondsuit	$\diamondsuit$
\heartsuit	$\heartsuit$	\spadesuit	$\spadesuit$		

"Large" Operators

Code	Sym	Code	Sym	Code	Sym
\int	$\int$				

Binary operators

Code	Sym	Code	Sym	Code	Sym
\pm	$\pm$	\cdot	$\cdot$	\times	$\times$
\ast	$\ast$	\circ	$\circ$	\bullet	$\bullet$
\div	$\div$	\cap	$\cap$	\cup	$\cup$
\vee	$\vee$	\wedge	$\wedge$	\oplus	$\oplus$
\otimes	$\otimes$	\oslash	$\oslash$		

Table 15.1: Available special characters in TeX mode

Relations

Code	Sym	Code	Sym	Code	Sym
\leq	≤	\subset	⊂	\subseteq	⊆
\in	∈	\geq	≥	\supset	⊃
\supscteq	⊇	\ni	∋	\mid	|
\equiv	≡	\sim	~	\approx	≈
\cong	≅	\propto	∝	\perp	⊥

Arrows

Code	Sym	Code	Sym	Code	Sym
\leftarrow	←	\Leftarrow	⇐	\rightarrow	→
\Rightarrow	⇒	\leftrightarrow	↔	\uparrow	↑
\downarrow	↓				

Openings and Closings

Code	Sym	Code	Sym	Code	Sym
\lfloor	⌊	\langle	⟨	\lceil	⌈
\rfloor	⌋	\rangle	⟩	\rceil	⌉

Alternate Names

Code	Sym	Code	Sym	Code	Sym
\neq	≠				

Other (not in Appendix F Tables)

Code	Sym	Code	Sym	Code	Sym
\ldots	. . .	\0	⊘	\copyright	©
\deg	°				

Table 15.1: Available special characters in TeX mode (cont.)

A complete example showing the capabilities of the extended text is

```
x = 0:0.01:3;
plot (x, erf (x));
hold on;
plot (x,x,"r");
axis ([0, 3, 0, 1]);
text (0.65, 0.6175, strcat ('\leftarrow x = {2/\surd\pi}',
' {\fontsize{16}\int_{\fontsize{8}0}^{\fontsize{8}x}}',
' e^{-t^2} dt} = 0.6175'))
```

The result of which can be seen in Figure 15.7

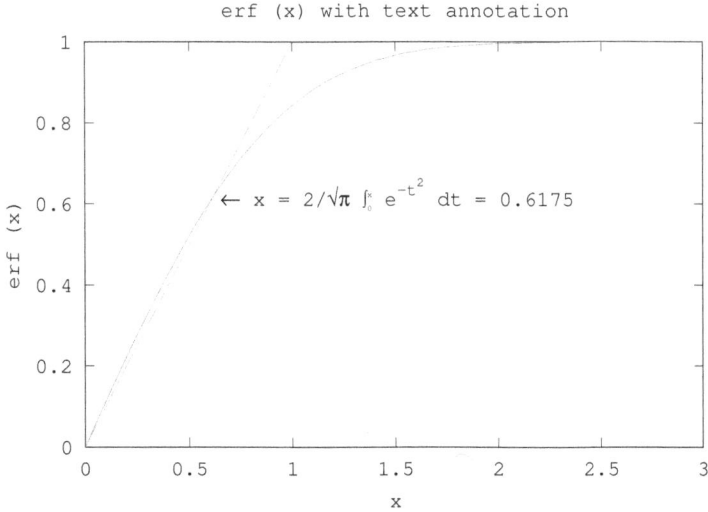

Figure 15.7: Example of inclusion of text with the TEX interpreter

15.2.8 Printing and Saving Plots

The `print` command allows you to send plots to you printer and to save plots in a variety of formats. For example,

        print -dpsc

prints the current figure to a color PostScript printer. And,

        print -deps foo.eps

saves the current figure to an encapsulated PostScript file called 'foo.eps'.

`print ()`	[Function File]
`print (options)`	[Function File]
`print (filename, options)`	[Function File]
`print (h, filename, options)`	[Function File]

Print a plot, or save it to a file.

Both output formatted for printing (PDF and PostScript), and many bitmapped and vector image formats are supported.

filename defines the name of the output file. If the file name has no suffix, one is inferred from the specified device and appended to the file name. If no filename is specified, the output is sent to the printer.

h specifies the handle of the figure to print. If no handle is specified the current figure is used.

For output to a printer, PostScript file, or PDF file, the paper size is specified by the figure's **papersize** property. The location and size of the image on the page are specified by the figure's **paperposition** property. The orientation of the page is specified by the figure's **paperorientation** property.

The width and height of images are specified by the figure's **paperpositon(3:4)** property values.

The `print` command supports many *options*:

-f*h*
: Specify the handle, *h*, of the figure to be printed. The default is the current figure.

-P*printer*
: Set the *printer* name to which the plot is sent if no *filename* is specified.

-G*ghostscript_command*
: Specify the command for calling Ghostscript. For Unix and Windows the defaults are `"gs"` and `"gswin32c"`, respectively.

-color
-mono
: Color or monochrome output.

-solid
-dashed
: Force all lines to be solid or dashed, respectively.

-portrait
-landscape
: Specify the orientation of the plot for printed output. For non-printed output the aspect ratio of the output corresponds to the plot area defined by the `"paperposition"` property in the orientation specified. This option is equivalent to changing the figure's `"paperorientation"` property.

-TextAlphaBits=*n*
-GraphicsAlphaBits=*n*
: Octave is able to produce output for various printers, bitmaps, and vector formats by using Ghostscript. For bitmap and printer output anti-aliasing is applied using Ghostscript's TextAlphaBits and GraphicsAlphaBits options. The default number of bits for each is 4. Allowed values for *N* are 1, 2, or 4.

-d*device*
: The available output format is specified by the option *device*, and is one of:

  ps
  ps2
  psc
  psc2
  : PostScript (level 1 and 2, mono and color). The FLTK graphics toolkit generates PostScript level 3.0.

  eps
  eps2
  epsc
  epsc2
  : Encapsulated PostScript (level 1 and 2, mono and color). The FLTK graphic toolkit generates PostScript level 3.0.

`tex`	
`epslatex`	
`epslatexstandalone`	
`pstex`	
`pslatex`	
`pdflatex`	Generate a LaTeX (or TeX) file for labels and eps/ps/pdf for graphics. The file produced by `epslatexstandalone` can be processed directly by LaTeX. The other formats are intended to be included in a LaTeX (or TeX) document. The `tex` device is the same as the `epslatex` device. The `pdflatex` device is only available for the FLTK graphics toolkit.
`tikz`	Generate a LaTeX file using PGF/TikZ. For the FLTK toolkit the result is PGF.
`ill`	
`aifm`	Adobe Illustrator (Obsolete for Gnuplot versions > 4.2)
`cdr`	
`corel`	CorelDraw
`dxf`	AutoCAD
`emf`	
`meta`	Microsoft Enhanced Metafile
`fig`	XFig. For the Gnuplot graphics toolkit, the additional options '`-textspecial`' or '`-textnormal`' can be used to control whether the special flag should be set for the text in the figure. (default is '`-textnormal`')
`hpgl`	HP plotter language
`mf`	Metafont
`png`	Portable network graphics
`jpg`	
`jpeg`	JPEG image
`gif`	GIF image (only available for the Gnuplot graphics toolkit)
`pbm`	PBMplus
`svg`	Scalable vector graphics
`pdf`	Portable document format

If the device is omitted, it is inferred from the file extension, or if there is no filename it is sent to the printer as PostScript.

-dghostscript_device

Additional devices are supported by Ghostscript. Some examples are;

`ljet2p`	HP LaserJet IIP
`ljet3`	HP LaserJet III

deskjet	HP DeskJet and DeskJet Plus
cdj550	HP DeskJet 550C
paintjet	HP PointJet
pcx24b	24-bit color PCX file format
ppm	Portable Pixel Map file format
pdfwrite	Produces pdf output from eps

For a complete list, type `system ("gs -h")` to see what formats and devices are available.

When Ghostscript output is sent to a printer the size is determined by the figure's `"papersize"` property. When the output is sent to a file the size is determined by the plot box defined by the figure's `"paperposition"` property.

`-append`   Append PostScript or PDF output to a pre-existing file of the same type.

`-rNUM`   Resolution of bitmaps in pixels per inch. For both metafiles and SVG the default is the screen resolution; for other formats it is 150 dpi. To specify screen resolution, use `"-r0"`.

`-loose`
`-tight`   Force a tight or loose bounding box for eps files. The default is loose.

`-preview`   Add a preview to eps files. Supported formats are:

`-interchange`
Provide an interchange preview.

`-metalfile`
Provide a metafile preview.

`-pict`   Provide pict preview.

`-tiff`   Provide a tiff preview.

`-Sxsize,ysize`
Plot size in pixels for EMF, GIF, JPEG, PBM, PNG, and SVG. For PS, EPS, PDF, and other vector formats the plot size is in points. This option is equivalent to changing the size of the plot box associated with the `"paperposition"` property. When using the command form of the print function you must quote the xsize,ysize option. For example, by writing `"-S640,480"`.

`-Ffontname`
`-Ffontname:size`
`-F:size`   Use fontname and/or fontsize for all text. fontname is ignored for some devices: dxf, fig, hpgl, etc.

The filename and options can be given in any order.

Example: Print to a file using the svg device.

```
figure (1);
clf ();
surf (peaks);
print -dsvg figure1.svg
```

Example: Print to an HP DeskJet 550C.

```
clf ();
surf (peaks);
print -dcdj550
```

See also: [saveas], page 331, [orient], page 331, [figure], page 319.

saveas (*h*, *filename*)   [Function File]
saveas (*h*, *filename*, *fmt*)   [Function File]

Save graphic object *h* to the file *filename* in graphic format *fmt*.

fmt should be one of the following formats:

ps	PostScript
eps	Encapsulated PostScript
jpg	JPEG Image
png	PNG Image
emf	Enhanced Meta File
pdf	Portable Document Format

All device formats specified in `print` may also be used. If *fmt* is omitted it is extracted from the extension of *filename*. The default format is `"pdf"`.

```
clf ();
surf (peaks);
saveas (1, "figure1.png");
```

See also: [print], page 327, [orient], page 331.

orient (*orientation*)   [Function File]
orient (*hfig*, *orientation*)   [Function File]
orientation = orient ()   [Function File]
orientation = orient (*hfig*)   [Function File]

Query or set the print orientation for figure *hfig*.

Valid values for *orientation* are `"portrait"`, `"landscape"`, and `"tall"`.

The `"landscape"` option changes the orientation so the plot width is larger than the plot height. The `"paperposition"` is also modified so that the plot fills the page, while leaving a 0.25 inch border.

The `"tall"` option sets the orientation to `"portrait"` and fills the page with the plot, while leaving a 0.25 inch border.

The `"portrait"` option (default) changes the orientation so the plot height is larger than the plot width. It also restores the default `"paperposition"` property.

When called with no arguments, return the current print orientation.

If the argument *hfig* is omitted, then operate on the current figure returned by `gcf`.

See also: [print], page 327, [saveas], page 331.

15.2.9 Interacting with Plots

The user can select points on a plot with the `ginput` function or selection the position at which to place text on the plot with the `gtext` function using the mouse. Menus may also be created and populated with specific user commands via the `uimenu` function.

`[x, y, buttons] = ginput (n)`	[Function File]
`[x, y, buttons] = ginput ()`	[Function File]

Return the position and type of mouse button clicks and/or key strokes in the current figure window.

If *n* is defined, then capture *n* events before returning. When *n* is not defined `ginput` will loop until the return key RET is pressed.

The return values *x*, *y* are the coordinates where the mouse was clicked in the units of the current axes. The return value *button* is 1, 2, or 3 for the left, middle, or right button. If a key is pressed the ASCII value is returned in *button*.

See also: [gtext], page 332, [waitforbuttonpress], page 332.

`waitforbuttonpress ()`	[Function File]
`b = waitforbuttonpress ()`	[Function File]

Wait for mouse click or key press over the current figure window.

The return value of *b* is 0 if a mouse button was pressed or 1 if a key was pressed.

See also: [waitfor], page 694, [ginput], page 332, [kbhit], page 229.

`gtext (s)`	[Function File]
`gtext ({s1, s2, ...})`	[Function File]
`gtext ({s1; s2; ...})`	[Function File]
`gtext (..., prop, val, ...)`	[Function File]
`h = gtext (...)`	[Function File]

Place text on the current figure using the mouse.

The text is defined by the string *s*. If *s* is a cell string organized as a row vector then each string of the cell array is written to a separate line. If *s* is organized as a column vector then one string element of the cell array is placed for every mouse click.

Optional property/value pairs are passed directly to the underlying text objects.

The optional return value *h* is a graphics handle to the created text object(s).

See also: [ginput], page 332, [text], page 315.

`uimenu (property, value, ...)`	[Function File]
`uimenu (h, property, value, ...)`	[Function File]

Create a uimenu object and return a handle to it. If *h* is omitted then a top-level menu for the current figure is created. If *h* is given then a submenu relative to *h* is created.

uimenu objects have the following specific properties:

`"accelerator"`

> A string containing the key combination together with CTRL to execute this menu entry (e.g., `"x"` for CTRL+x).

"callback"

> Is the function called when this menu entry is executed. It can be either a function string (e.g., "myfun"), a function handle (e.g., @myfun) or a cell array containing the function handle and arguments for the callback function (e.g., {@myfun, arg1, arg2}).

"checked"

> Can be set "on" or "off". Sets a mark at this menu entry.

"enable"   Can be set "on" or "off". If disabled the menu entry cannot be selected and it is grayed out.

"foregroundcolor"

> A color value setting the text color for this menu entry.

"label"   A string containing the label for this menu entry. A "&"-symbol can be used to mark the "accelerator" character (e.g., "E&xit")

"position"

> An scalar value containing the relative menu position. The entry with the lowest value is at the first position starting from left or top.

"separator"

> Can be set "on" or "off". If enabled it draws a separator line above the current position. It is ignored for top level entries.

Examples:

```
f = uimenu ("label", "&File", "accelerator", "f");
e = uimenu ("label", "&Edit", "accelerator", "e");
uimenu (f, "label", "Close", "accelerator", "q", ...
          "callback", "close (gcf)");
uimenu (e, "label", "Toggle &Grid", "accelerator", "g", ...
          "callback", "grid (gca)");
```

See also: [figure], page 319.

15.2.10 Test Plotting Functions

The functions sombrero and peaks provide a way to check that plotting is working. Typing either sombrero or peaks at the Octave prompt should display a three-dimensional plot.

sombrero ()	[Function File]
sombrero (*n*)	[Function File]
z = sombrero (...)	[Function File]
[*x, y, z*] = sombrero (...)	[Function File]

> Plot the familiar 3-D sombrero function.
>
> The function plotted is
>
> $$z = \frac{\sin(\sqrt{(x^2 + y^2)})}{\sqrt{(x^2 + y^2)}}$$

Called without a return argument, sombrero plots the surface of the above function over the meshgrid [-8,8] using surf.

If n is a scalar the plot is made with n grid lines. The default value for n is 41.

When called with output arguments, return the data for the function evaluated over the meshgrid. This can subsequently be plotted with `surf (x, y, z)`.

See also: [peaks], page 334, [meshgrid], page 302, [mesh], page 292, [surf], page 295.

`peaks ()`	[Function File]
`peaks (n)`	[Function File]
`peaks (x, y)`	[Function File]
`z = peaks (...)`	[Function File]
`[x, y, z] = peaks (...)`	[Function File]

Plot a function with lots of local maxima and minima.

The function has the form

$$f(x,y) = 3(1-x)^2 e^{\left(-x^2-(y+1)^2\right)} - 10\left(\frac{x}{5} - x^3 - y^5\right)\right) - \frac{1}{3}e^{\left(-(x+1)^2-y^2\right)}$$

Called without a return argument, `peaks` plots the surface of the above function using `surf`.

If n is a scalar, `peaks` plots the value of the above function on an n-by-n mesh over the range [-3,3]. The default value for n is 49.

If n is a vector, then it represents the grid values over which to calculate the function. If x and y are specified then the function value is calculated over the specified grid of vertices.

When called with output arguments, return the data for the function evaluated over the meshgrid. This can subsequently be plotted with `surf (x, y, z)`.

See also: [sombrero], page 333, [meshgrid], page 302, [mesh], page 292, [surf], page 295.

15.3 Graphics Data Structures

15.3.1 Introduction to Graphics Structures

The graphics functions use pointers, which are of class graphics_handle, in order to address the data structures which control graphical displays. A graphics handle may point any one of a number of different object types. The objects are the graphics data structures. The types of objects are: `figure`, `axes`, `line`, `text`, `patch`, `surface`, `text` and `image`.

Each of these objects has a function by the same name. and, each of these functions returns a graphics handle pointing to an object of corresponding type. In addition there are several functions which operate on properties of the graphics objects and which return handles: the functions `plot` and `plot3` return a handle pointing to an object of type line, the function `subplot` returns a handle pointing to an object of type axes, the function `fill` returns a handle pointing to an object of type patch, the functions `area`, `bar`, `barh`, `contour`, `contourf`, `contour3`, `surf`, `mesh`, `surfc`, `meshc`, `errorbar`, `quiver`, `quiver3`, `scatter`, `scatter3`, `stair`, `stem`, `stem3` each return a handle as documented in [Data Sources], page 367.

The graphics objects are arranged in a hierarchy:

1. The root is at 0. i.e., `get (0)` returns the properties of the root object.

2. Below the root are `figure` objects.

3. Below the `figure` objects are `axes`.

4. Below the `axes` objects are `line`, `text`, `patch`, `surface`, and `image` objects.

Graphics handles may be distinguished from function handles (see Section 11.11.1 [Function Handles], page 190) by means of the function `ishandle`. `ishandle` returns true if its argument is a handle of a graphics object. In addition, the figure object may be tested using `isfigure`. `isfigure` returns true only if its argument is a handle of a figure. The `whos` function can be used to show the object type of each currently defined graphics handle. (Note: this is not true today, but it is, I hope, considered an error in whos. It may be better to have whos just show graphics_handle as the class, and provide a new function which, given a graphics handle, returns its object type. This could generalize the ishandle() functions and, in fact, replace them.)

The `get` and `set` commands are used to obtain and set the values of properties of graphics objects. In addition, the `get` command may be used to obtain property names.

For example, the property `"type"` of the graphics object pointed to by the graphics handle h may be displayed by:

```
get (h, "type")
```

The properties and their current values are returned by `get (h)` where h is a handle of a graphics object. If only the names of the allowed properties are wanted they may be displayed by: `get (h, "")`.

Thus, for example:

```
h = figure ();
get (h, "type")
ans = figure
get (h, "");
error: get: ambiguous figure property name ; possible matches:
```

__enhanced__	hittest	resize
__graphics_toolkit__	integerhandle	resizefcn
__guidata__	interruptible	selected
__modified__	inverthardcopy	selectionhighlight
__myhandle__	keypressfcn	selectiontype
__plot_stream__	keyreleasefcn	tag
alphamap	menubar	toolbar
beingdeleted	mincolormap	type
busyaction	name	uicontextmenu
buttondownfcn	nextplot	units
children	numbertitle	userdata
clipping	outerposition	visible
closerequestfcn	paperorientation	windowbuttondownfcn
color	paperposition	windowbuttonmotionfcn
colormap	paperpositionmode	windowbuttonupfcn
createfcn	papersize	windowkeypressfcn
currentaxes	papertype	windowkeyreleasefcn
currentcharacter	paperunits	windowscrollwheelfcn
currentobject	parent	windowstyle
currentpoint	pointer	wvisual
deletefcn	pointershapedata	wvisualmode

```
dockcontrols          pointershapehotspot    xdisplay
doublebuffer          position               xvisual
filename              renderer               xvisualmode
handlevisibility      renderermode
```

The root figure has index 0. Its properties may be displayed by: `get (0, "")`.

The uses of `get` and `set` are further explained in [get], page 341, [set], page 342.

res = isprop (h, "prop")   [Function File]

> Return true if *prop* is a property of the object with handle *h*.
>
> *h* may also be an array of handles in which case *res* will be a logical array indicating whether each handle has the property *prop*.
>
> **See also:** [get], page 341, [set], page 342, [ismethod], page 675, [isobject], page 674.

15.3.2 Graphics Objects

The hierarchy of graphics objects was explained above. See Section 15.3.1 [Introduction to Graphics Structures], page 334. Here the specific objects are described, and the properties contained in these objects are discussed. Keep in mind that graphics objects are always referenced by *handle*.

root figure   the top level of the hierarchy and the parent of all figure objects. The handle index of the root figure is 0.

figure   A figure window.

axes   A set of axes. This object is a child of a figure object and may be a parent of line, text, image, patch, or surface objects.

line   A line in two or three dimensions.

text   Text annotations.

image   A bitmap image.

patch   A filled polygon, currently limited to two dimensions.

surface   A three-dimensional surface.

15.3.2.1 Creating Graphics Objects

You can create axes, line, patch, and surface objects directly using the `axes`, `line`, `patch`, `fill`, and `surface` functions. These objects become children of the current axes object.

axes ()   [Function File]
axes (property, value, ...)   [Function File]
axes (hax)   [Function File]
h = axes (...)   [Function File]

> Create an axes object and return a handle to it, or set the current axes to *hax*.
>
> Called without any arguments, or with *property*/*value* pairs, construct a new axes. For accepted properties and corresponding values, see [set], page 342.
>
> Called with a single axes handle argument *hax*, the function makes *hax* the current axis. It also restacks the axes in the corresponding figure so that *hax* is the first

entry in the list of children. This causes *hax* to be displayed on top of any other axes objects (Z-order stacking).

See also: gca, set, get.

`line ()`	[Function File]
`line (x, y)`	[Function File]
`line (x, y, property, value, ...)`	[Function File]
`line (x, y, z)`	[Function File]
`line (x, y, z, property, value, ...)`	[Function File]
`line (property, value, ...)`	[Function File]
`line (hax, ...)`	[Function File]
`h = line (...)`	[Function File]

Create line object from *x* and *y* (and possibly *z*) and insert in the current axes.

Multiple property-value pairs may be specified for the line object, but they must appear in pairs.

If the first argument *hax* is an axes handle, then plot into this axis, rather than the current axes returned by `gca`.

The optional return value *h* is a graphics handle (or vector of handles) to the line objects created.

See also: [image], page 657, [patch], page 337, [rectangle], page 291, [surface], page 339, [text], page 315.

`patch ()`	[Function File]
`patch (x, y, c)`	[Function File]
`patch (x, y, z, c)`	[Function File]
`patch (fv)`	[Function File]
`patch ("Faces", faces, "Vertices", verts, ...)`	[Function File]
`patch (..., prop, val, ...)`	[Function File]
`patch (hax, ...)`	[Function File]
`h = patch (...)`	[Function File]

Create patch object in the current axes with vertices at locations (*x*, *y*) and of color *c*.

If the vertices are matrices of size MxN then each polygon patch has M vertices and a total of N polygons will be created. If some polygons do not have M vertices use NaN to represent "no vertex". If the *z* input is present then 3-D patches will be created.

The color argument *c* can take many forms. To create polygons which all share a single color use a string value (e.g., `"r"` for red), a scalar value which is scaled by `caxis` and indexed into the current colormap, or a 3-element RGB vector with the precise TrueColor.

If *c* is a vector of length N then the ith polygon will have a color determined by scaling entry $c(i)$ according to `caxis` and then indexing into the current colormap. More complicated coloring situations require directly manipulating patch property/value pairs.

Instead of specifying polygons by matrices *x* and *y*, it is possible to present a unique list of vertices and then a list of polygon faces created from those vertices. In this case the "Vertices" matrix will be an Nx2 (2-D patch) or Nx3 (3-D path). The MxN "Faces" matrix describes M polygons having N vertices—each row describes a single polygon and each column entry is an index into the "Vertices" matrix to identify a vertex. The patch object can be created by directly passing the property/value pairs "Vertices"/*verts*, "Faces"/*faces* as inputs.

A third input form is to create a structure *fv* with the fields "vertices", "faces", and optionally "facevertexcdata".

If the first argument *hax* is an axes handle, then plot into this axis, rather than the current axes returned by gca.

The optional return value *h* is a graphics handle to the created patch object.

Implementation Note: Patches are highly configurable objects. To truly customize them requires setting patch properties directly. Useful patch properties are: "cdata", "edgecolor", "facecolor", "faces", "facevertexcdata".

See also: [fill], page 338, [get], page 341, [set], page 342.

fill (*x*, *y*, *c*)	[Function File]
fill (*x1*, *y1*, *c1*, *x2*, *y2*, *c2*)	[Function File]
fill (..., *prop*, *val*)	[Function File]
fill (*hax*, ...)	[Function File]
h = fill (...)	[Function File]

Create one or more filled 2-D polygons.

The inputs *x* and *y* are the coordinates of the polygon vertices. If the inputs are matrices then the rows represent different vertices and each column produces a different polygon. fill will close any open polygons before plotting.

The input *c* determines the color of the polygon. The simplest form is a single color specification such as a plot format or an RGB-triple. In this case the polygon(s) will have one unique color. If *c* is a vector or matrix then the color data is first scaled using caxis and then indexed into the current colormap. A row vector will color each polygon (a column from matrices *x* and *y*) with a single computed color. A matrix *c* of the same size as *x* and *y* will compute the color of each vertex and then interpolate the face color between the vertices.

Multiple property/value pairs for the underlying patch object may be specified, but they must appear in pairs.

If the first argument *hax* is an axes handle, then plot into this axis, rather than the current axes returned by gca.

The optional return value *h* is a vector of graphics handles to the created patch objects.

Example: red square

```
vertices = [0 0
            1 0
            1 1
            0 1];
fill (vertices(:,1), vertices(:,2), "r");
axis ([-0.5 1.5, -0.5 1.5])
axis equal
```

See also: [patch], page 337, [caxis], page 287, [colormap], page 660.

surface (*x*, *y*, *z*, *c*)   [Function File]
surface (*x*, *y*, *z*)   [Function File]
surface (*z*, *c*)   [Function File]
surface (*z*)   [Function File]
surface (..., *prop*, *val*, ...)   [Function File]
surface (*hax*, ...)   [Function File]
h = surface (...)   [Function File]

Create a surface graphic object given matrices *x* and *y* from `meshgrid` and a matrix of values *z* corresponding to the *x* and *y* coordinates of the surface.

If *x* and *y* are vectors, then a typical vertex is (x(j), y(i), z(i,j)). Thus, columns of *z* correspond to different *x* values and rows of *z* correspond to different *y* values. If only a single input *z* is given then *x* is taken to be `1:rows (z)` and *y* is `1:columns (z)`.

Any property/value input pairs are assigned to the surface object.

If the first argument *hax* is an axes handle, then plot into this axis, rather than the current axes returned by `gca`.

The optional return value *h* is a graphics handle to the created surface object.

See also: [surf], page 295, [mesh], page 292, [patch], page 337, [line], page 337.

15.3.2.2 Handle Functions

To determine whether a variable is a graphics object index, or an index to an axes or figure, use the functions `ishandle`, `isaxes`, and `isfigure`.

ishandle (*h*)   [Built-in Function]

Return true if *h* is a graphics handle and false otherwise.

h may also be a matrix of handles in which case a logical array is returned that is true where the elements of *h* are graphics handles and false where they are not.

See also: [isaxes], page 339, [isfigure], page 340.

ishghandle (*h*)   [Function File]

Return true if *h* is a graphics handle and false otherwise.

This function is equivalent to `ishandle` and is provided for compatibility with MAT-LAB.

See also: [ishandle], page 339.

isaxes (*h*)   [Function File]

Return true if *h* is an axes graphics handle and false otherwise.

If *h* is a matrix then return a logical array which is true where the elements of *h* are axes graphics handles and false where they are not.

See also: [isaxes], page 339, [ishandle], page 339.

isfigure (*h*)   [Function File]
>   Return true if *h* is a figure graphics handle and false otherwise.
>
>   If *h* is a matrix then return a logical array which is true where the elements of *h* are figure graphics handles and false where they are not.
>
>   **See also:** [isaxes], page 339, [ishandle], page 339.

The function `gcf` returns an index to the current figure object, or creates one if none exists. Similarly, `gca` returns the current axes object, or creates one (and its parent figure object) if none exists.

h = **gcf** ()   [Function File]
>   Return a handle to the current figure.
>
>   The current figure is the default target for graphics output. If multiple figures exist, `gcf` returns the last created figure or the last figure that was clicked on with the mouse.
>
>   If a current figure does not exist, create one and return its handle. The handle may then be used to examine or set properties of the figure. For example,
>
> ```
> fplot (@sin, [-10, 10]);
> fig = gcf ();
> set (fig, "numbertitle", "off", "name", "sin plot")
> ```
>
>   plots a sine wave, finds the handle of the current figure, and then renames the figure window to describe the contents.
>
>   Note: To find the current figure without creating a new one if it does not exist, query the `"CurrentFigure"` property of the root graphics object.
>
> ```
> get (0, "currentfigure");
> ```
>
>   **See also:** [gca], page 340, [gco], page 341, [gcbf], page 363, [gcbo], page 363, [get], page 341, [set], page 342.

h = **gca** ()   [Function File]
>   Return a handle to the current axis object.
>
>   The current axis is the default target for graphics output. In the case of a figure with multiple axes, `gca` returns the last created axes or the last axes that was clicked on with the mouse.
>
>   If no current axes object exists, create one and return its handle. The handle may then be used to examine or set properties of the axes. For example,
>
> ```
> ax = gca ();
> set (ax, "position", [0.5, 0.5, 0.5, 0.5]);
> ```
>
>   creates an empty axes object and then changes its location and size in the figure window.
>
>   Note: To find the current axis without creating a new axes object if it does not exist, query the `"CurrentAxes"` property of a figure.
>
> ```
> get (gcf, "currentaxes");
> ```
>
>   **See also:** [gcf], page 340, [gco], page 341, [gcbf], page 363, [gcbo], page 363, [get], page 341, [set], page 342.

h = gco ()   [Function File]
h = gco (*fig*)   [Function File]

> Return a handle to the current object of the current figure, or a handle to the current object of the figure with handle *fig*.

> The current object of a figure is the object that was last clicked on. It is stored in the "CurrentObject" property of the target figure.

> If the last mouse click did not occur on any child object of the figure, then the current object is the figure itself.

> If no mouse click occurred in the target figure, this function returns an empty matrix.

> Programming Note: The value returned by this function is not necessarily the same as the one returned by gcbo during callback execution. An executing callback can be interrupted by another callback and the current object may be changed.

> **See also:** [gcbo], page 363, [gca], page 340, [gcf], page 340, [gcbf], page 363, [get], page 341, [set], page 342.

The get and set functions may be used to examine and set properties for graphics objects. For example,

```
get (0)
   ⇒ ans =
      {
        type = root
        currentfigure = [](0x0)
        children = [](0x0)
        visible = on
        ...
      }
```

returns a structure containing all the properties of the root figure. As with all functions in Octave, the structure is returned by value, so modifying it will not modify the internal root figure plot object. To do that, you must use the set function. Also, note that in this case, the currentfigure property is empty, which indicates that there is no current figure window.

The get function may also be used to find the value of a single property. For example,

```
get (gca (), "xlim")
   ⇒ [ 0 1 ]
```

returns the range of the x-axis for the current axes object in the current figure.

To set graphics object properties, use the set function. For example,

```
set (gca (), "xlim", [-10, 10]);
```

sets the range of the x-axis for the current axes object in the current figure to '[-10, 10]'. Additionally, calling set with a graphics object index as the only argument returns a structure containing the default values for all the properties for the given object type. For example,

```
set (gca ())
```

returns a structure containing the default property values for axes objects.

`val = get (h)`   [Built-in Function]
`val = get (h, p)`   [Built-in Function]
> Return the value of the named property p from the graphics handle h. If p is omitted,
> return the complete property list for h. If h is a vector, return a cell array including
> the property values or lists respectively.
>
> **See also:** [set], page 342.

`set (h, property, value, ...)`   [Built-in Function]
`set (h, properties, values)`   [Built-in Function]
`set (h, pv)`   [Built-in Function]
> Set named property values for the graphics handle (or vector of graphics handles) h.
> There are three ways how to give the property names and values:
>
> - as a comma separated list of *property*, *value* pairs
>
>   Here, each *property* is a string containing the property name, each *value* is a
>   value of the appropriate type for the property.
>
> - as a cell array of strings *properties* containing property names and a cell array
>   *values* containing property values.
>
>   In this case, the number of columns of *values* must match the number of elements
>   in *properties*. The first column of *values* contains values for the first entry in
>   *properties*, etc. The number of rows of *values* must be 1 or match the number
>   of elements of h. In the first case, each handle in h will be assigned the same
>   values. In the latter case, the first handle in h will be assigned the values from
>   the first row of *values* and so on.
>
> - as a structure array *pv*
>
>   Here, the field names of *pv* represent the property names, and the field values
>   give the property values. In contrast to the previous case, all elements of *pv* will
>   be set in all handles in h independent of the dimensions of *pv*.
>
> **See also:** [get], page 341.

`parent = ancestor (h, type)`   [Function File]
`parent = ancestor (h, type, "toplevel")`   [Function File]
> Return the first ancestor of handle object h whose type matches *type*, where *type* is
> a character string. If *type* is a cell array of strings, return the first parent whose type
> matches any of the given type strings.
>
> If the handle object h itself is of type *type*, return h.
>
> If `"toplevel"` is given as a third argument, return the highest parent in the object
> hierarchy that matches the condition, instead of the first (nearest) one.
>
> **See also:** [findobj], page 359, [findall], page 359, [allchild], page 342.

`h = allchild (handles)`   [Function File]
> Find all children, including hidden children, of a graphics object.
>
> This function is similar to `get (h, "children")`, but also returns hidden objects
> (HandleVisibility = `"off"`). If *handles* is a scalar, h will be a vector. Otherwise, h
> will be a cell matrix of the same size as *handles* and each cell will contain a vector of
> handles.

See also: [findall], page 359, [findobj], page 359, [get], page 341, [set], page 342.

`findfigs ()`   [Function File]

    Find all visible figures that are currently off the screen and move them onto the screen.

    See also: [allchild], page 342, [figure], page 319, [get], page 341, [set], page 342.

Figures can be printed or saved in many graphics formats with `print` and `saveas`. Occasionally, however, it may be useful to save the original Octave handle graphic directly so that further modifications can be made such as modifying a title or legend.

This can be accomplished with the following functions by

```
fig_struct = hdl2struct (gcf);
save myplot.fig -struct fig_struct;
...
fig_struct = load ("myplot.fig");
struct2hdl (fig_struct);
```

`s = hdl2struct (h)`   [Function File]

    Return a structure, *s*, whose fields describe the properties of the object, and its children, associated with the handle, *h*.

    The fields of the structure *s* are `"type"`, `"handle"`, `"properties"`, `"children"`, and `"special"`.

    See also: [struct2hdl], page 343, [findobj], page 359.

`h = struct2hdl (s)`   [Function File]
`h = struct2hdl (s, p)`   [Function File]
`h = struct2hdl (s, p, hilev)`   [Function File]

    Construct a graphics handle object *h* from the structure *s*.

    The structure must contain the fields `"handle"`, `"type"`, `"children"`, `"properties"`, and `"special"`. If the handle of an existing figure or axes is specified, *p*, the new object will be created as a child of that object. If no parent handle is provided then a new figure and the necessary children will be constructed using the default values from the root figure.

    A third boolean argument *hilev* can be passed to specify whether the function should preserve listeners/callbacks, e.g., for legends or hggroups. The default is false.

    See also: [hdl2struct], page 343, [findobj], page 359.

`hnew = copyobj (horig)`   [Function File]
`hnew = copyobj (horig, hparent)`   [Function File]

    Construct a copy of the graphic object associated with handle *horig* and return a handle *hnew* to the new object.

    If a parent handle *hparent* (root, figure, axes, or hggroup) is specified, the copied object will be created as a child of *hparent*.

    See also: [struct2hdl], page 343, [hdl2struct], page 343, [findobj], page 359.

15.3.3 Graphics Object Properties

In this Section the object properties are discussed in detail, starting with the root figure properties and continuing through the graphics object hierarchy.

15.3.3.1 Root Figure Properties

The root figure properties are:

__modified__
              — Values: "on", "off"

__myhandle__
beingdeleted
              — Values: "on", "off"

busyaction
buttondownfcn
callbackobject
children

clipping   — Values: "on", "off"

createfcn
currentfigure
deletefcn
handlevisibility
              — Values: "on", "off"

hittest   — Values: "on", "off"

interruptible
              — Values: "on", "off"

parent

screendepth
screenpixelsperinch
screensize
selected

selectionhighlight
screendepth
screenpixelsperinch
showhiddenhandles
              — Values: "on", "off"

tag

type

uicontextmenu
units

userdata

visible

15.3.3.2 Figure Properties

The `figure` properties are:

`__graphics_toolkit__`
> — The graphics toolkit currently in use.

`__enhanced__`
`__modified__`
`__myhandle__`
`__plot_stream__`
`alphamap`

`beingdeleted`
> — Values: `"on"`, `"off"`

`busyaction`
`buttondownfcn`
`children`   Handle to children.

`clipping`   — Values: `"on"`, `"off"`

`closerequestfcn`
> — Handle of function to call on close.

`color`

`colormap`   An N-by-3 matrix containing the color map for the current axes.

`paperorientation`
`createfcn`
`currentaxes`
> Handle to graphics object of current axes.

`currentcharacter`
`currentobject`
`currentpoint`
> Holds the coordinates of the point over which the mouse pointer was when the mouse button was pressed. If a mouse callback function is defined, `"currentpoint"` holds the coordinates of the point over which the mouse pointer is when the function gets called.

`deletefcn`
`dockcontrols`
> — Values: `"on"`, `"off"`

`doublebuffer`
> — Values: `"on"`, `"off"`

`filename`

`handlevisibility`
> — Values: `"on"`, `"off"`

`hittest`

integerhandle
interruptible
          — Values: "on", "off"

inverthardcopy
keypressfcn
          see "keypressfcn"

keyreleasefcn
          With "keypressfcn", the keyboard callback functions. These callback func-
          tions get called when a key is pressed/released respectively. The functions are
          called with two input arguments. The first argument holds the handle of the
          calling figure. The second argument holds the event structure which has the
          following members:

          Character
                    The ASCII value of the key

          Key   lowercase value of the key

          Modifier A cell array containing strings representing the modifiers pressed
                    with the key. Possible values are "shift", "alt", and "control".

menubar

mincolormap
name

nextplot   May be one of

          "new"

          "add"

          "replace"
          "replacechildren"

numbertitle
paperorientation
          Indicates the orientation for printing. Either "landscape" or "portrait".

paperposition
paperpositionmode
papersize
papertype
paperunits
pointer

pointershapecdata
pointershapehotspot
position

renderer

renderermode
resize

```
resizefcn
selected

selectionhighlight
         — Values: "on", "off"

selectiontype
tag

toolbar

type

units

userdata

visible    Either "on" or "off" to toggle display of the figure.

windowbuttondownfcn
         See "windowbuttonupfcn"

windowbuttonmotionfcn
         See "windowbuttonupfcn"

windowbuttonupfcn
```

With "windowbuttondownfcn" and "windowbuttonmotionfcn", the mouse callback functions. These callback functions get called when the mouse button is pressed, dragged, and released respectively. When these callback functions are called, the "currentpoint" property holds the current coordinates of the cursor.

```
windowscrollwheelfcn
windowstyle
wvisual

wvisualmode
xdisplay

xvisual

xvisualmode
```

15.3.3.3 Axes Properties

The axes properties are:

```
__modified__
__myhandle__
activepositionproperty
alim

alimmode

ambientlightcolor
beingdeleted
box        Box surrounding axes. — Values: "on", "off"
```

busyaction
buttondownfcn
cameraposition
camerapositionmode
cameratarget
cameratargetmode
cameraupvector
cameraupvectormode
cameraviewangle
cameraviewanglemode
children

clim   Two-element vector defining the limits for the c axis of an image. See `pcolor`
            property. Setting this property also forces the corresponding mode property to
            be set to "manual".

climmode   Either "manual" or "auto".

clipping

color

colororder
createfcn
currentpoint
            Holds the coordinates of the point over which the mouse pointer was when
            the mouse button was pressed. If a mouse callback function is defined,
            "currentpoint" holds the coordinates of the point over which the mouse
            pointer is when the function gets called.

dataaspectratio
            A two-element vector specifying the relative height and width of the data dis-
            played in the axes. Setting `dataaspectratio` to '1, 2]' causes the length of
            one unit as displayed on the y-axis to be the same as the length of 2 units
            on the x-axis. Setting `dataaspectratio` also forces the `dataaspectratiomode`
            property to be set to "manual".

dataaspectratiomode
            Either "manual" or "auto".

deletefcn
drawmode

fontangle
fontname

fontsize

fontunits
fontweight
gridlinestyle
handlevisibility
hittest

```
interpreter
interruptible
layer

linestyleorder
linewidth
minorgridlinestyle
nextplot    May be one of

            "add"

            "replace"
            "replacechildren"

outerposition
```

A vector specifying the position of the plot, including titles, axes and legend. The four elements of the vector are the coordinates of the lower left corner and width and height of the plot, in units normalized to the width and height of the plot window. For example, `[0.2, 0.3, 0.4, 0.5]` sets the lower left corner of the axes at $(0.2, 0.3)$ and the width and height to be 0.4 and 0.5 respectively. See also the `position` property.

```
parent

plotboxaspectratio
plotboxaspectratiomode
position
```

A vector specifying the position of the plot, excluding titles, axes and legend. The four elements of the vector are the coordinates of the lower left corner and width and height of the plot, in units normalized to the width and height of the plot window. For example, `[0.2, 0.3, 0.4, 0.5]` sets the lower left corner of the axes at $(0.2, 0.3)$ and the width and height to be 0.4 and 0.5 respectively. See also the `outerposition` property.

```
projection
selected

selectionhighlight
tag

tickdir

tickdirmode
ticklength
tightinset
title       Index of text object for the axes title.

type

uicontextmenu
units

userdata

view        A three element vector specifying the view point for three-dimensional plots.

visible     Either "on" or "off" to toggle display of the axes.
```

```
x_normrendertransform
x_projectiontransform
x_rendertransform
x_viewporttransform
x_viewtransform
xaxislocation
```
Either "top" or "bottom".

`xcolor`

`xdir`   Either "forward" or "reverse".

`xgrid`   Either "on" or "off" to toggle display of grid lines.

`xlabel`   Indices to text objects for the axes labels.

`xlim`   Two-element vector defining the limits for the x-axis. Setting this property also forces the corresponding mode property to be set to "manual".

`xlimmode`   Either "manual" or "auto".

`xminorgrid`
Either "on" or "off" to toggle display of minor grid lines.

`xminortick`
`xscale`   Either "linear" or "log".

`xtick`   Set position of tick marks. Setting this property also forces the corresponding mode property to be set to "manual".

`xticklabel`
Setting this property also forces the corresponding mode property to be set to "manual".

`xticklabelmode`
Either "manual" or "auto".

`xtickmode`
Either "manual" or "auto".

`yaxislocation`
Either "left" or "right"

`ycolor`

`ydir`   Either "forward" or "reverse".

`ygrid`   Either "on" or "off" to toggle display of grid lines.

`ylabel`   Indices to text objects for the axes labels.

`ylim`   Two-element vectors defining the limits for the x, y, and z axes and the Setting one of these properties also forces the corresponding mode property to be set to "manual".

`ylimmode`   Either "manual" or "auto".

`yminorgrid`
Either "on" or "off" to toggle display of minor grid lines.

yminortick

yscale   Either "linear" or "log".

ytick   Set position of tick marks. Setting this property also forces the corresponding
             mode property to be set to "manual".

yticklabel
             Setting this property also forces the corresponding mode property to be set to
             "manual".

yticklabelmode
             Either "manual" or "auto".

ytickmode
             Either "manual" or "auto".

zcolor

zdir   Either "forward" or "reverse".

zgrid   Either "on" or "off" to toggle display of grid lines.

zlabel   Indices to text objects for the axes labels.

zlim   Two-element vector defining the limits for z-axis. Setting this property also
             forces the corresponding mode property to be set to "manual".

zlimmode   Either "manual" or "auto".

zminorgrid
             Either "on" or "off" to toggle display of minor grid lines.

zminortick

zscale   Either "linear" or "log".

ztick   Set position of tick marks. Setting this property also forces the corresponding
             mode property to be set to "manual".

zticklabel
             Setting this property also forces the corresponding mode property to be set to
             "manual".

zticklabelmode
             Either "manual" or "auto".

ztickmode
             Either "manual" or "auto".

15.3.3.4 Line Properties

The line properties are:

__modified__
__myhandle__
beingdeleted
busyaction
buttondownfcn
children

clipping

color   The RGB color of the line, or a color name. See Section 15.4.1 [Colors], page 361.

createfcn
deletefcn
displayname
            The text of the legend entry corresponding to this line.

erasemode
handlevisibility
hittest

interpreter
interruptible
ldata   The lower errorbar in the y direction to be plotted.

linestyle
linewidth
            See Section 15.4.2 [Line Styles], page 361.

linewidth
marker

markeredgecolor
markerfacecolor
markersize
            See Section 15.4.3 [Marker Styles], page 361.

parent

selected

selectionhighlight
tag

type

udata   The upper errorbar in the y direction to be plotted.

uicontextmenu
userdata

visible

xdata   The data to be plotted.

xdatasource
xldata   The lower errorbar to be plotted.

xlim

xliminclude
xudata   The upper errorbar to be plotted.

ydata   The data to be plotted.

ydatasource
ylim

yliminclude
zdata   The data to be plotted.

zdatasource
zlim

zliminclude

15.3.3.5 Text Properties

The text properties are:

__modified__
__myhandle__
backgroundcolor
beingdeleted
busyaction
buttondownfcn
children

clipping

color   The color of the text. See Section 15.4.1 [Colors], page 361.

createfcn
deletefcn
displayname
             The text of the legend entry corresponding to this line.

edgecolor
editing

erasemode
fontangle
             Flag whether the font is italic or normal. Valid values are "normal", "italic",
             and "oblique".

fontname   The font used for the text.

fontsize   The size of the font, in points to use.

fontunits
fontweight
             Flag whether the font is bold, etc. Valid values are "normal", "bold", "demi",
             or "light".

handlevisibility
hittest

horizontalalignment
             May be "left", "center", or "right".

interpreter
             Determines how the text is rendered. Valid values are "none", "tex", or
             "latex".

```
interruptible
linestyle
linewidth
margin
```

parent

position   The coordinates of the text object.

rotation   The angle of rotation for the displayed text, measured in degrees.

selected

```
selectionhighlight
string       The character string contained by the text object.
```

tag

type

```
uicontextmenu
units        May be "normalized" or "graph".
```

userdata

```
verticalalignment
visible
```

xlim

```
xliminclude
ylim
```

```
yliminclude
zlim
```

zliminclude

15.3.3.6 Image Properties

The image properties are:

```
__modified__
__myhandle__
beingdeleted
busyaction
buttondownfcn
```

cdata   The data for the image. Each pixel of the image corresponds to an element
             of cdata. The value of an element of cdata specifies the row-index into the
             colormap of the axes object containing the image. The color value found in the
             color map for the given index determines the color of the pixel.

```
cdatamapping
children
```

clim

```
climinclude
clipping
```

```
createfcn
deletefcn
handlevisibility
hittest

interruptible
parent

selected

selectionhighlight
tag

type

uicontextmenu
userdata

visible
```

xdata   Two-element vector specifying the range of the x-coordinates for the image.

```
xlim
```

```
xliminclude
```
ydata   Two-element vector specifying the range of the y-coordinates for the image.

```
ylim
```

```
yliminclude
```

15.3.3.7 Patch Properties

The `patch` properties are:

```
__modified__
__myhandle__
alim

aliminclude
alphadatamapping
ambientstrength
backfacelighting
beingdeleted
busyaction
buttondownfcn
```
cdata   Data defining the patch object.

```
cdatamapping
children

clim

climinclude
clipping
```

```
createfcn
deletefcn
diffusestrength
displayname
```
        The text of the legend entry corresponding to this line.

```
edgealpha
edgecolor
```
        The color of the line defining the patch. See Section 15.4.1 [Colors], page 361.

```
edgelighting
erasemode
facealpha
```
        A number in the range [0, 1] indicating the transparency of the patch.

```
facecolor
```
        The fill color of the patch. See Section 15.4.1 [Colors], page 361.

```
facelighting
faces

facevertexalphadata
facevertexcdata
handlevisibility
hittest

interpreter
interruptible
linestyle
```
        See Section 15.4.2 [Line Styles], page 361.

```
linewidth
```
        See Section 15.4.2 [Line Styles], page 361.

```
marker
```
   See Section 15.4.3 [Marker Styles], page 361.

```
markeredgecolor
```
        See Section 15.4.3 [Marker Styles], page 361.

```
markerfacecolor
```
        See Section 15.4.3 [Marker Styles], page 361.

```
markersize
```
        See Section 15.4.3 [Marker Styles], page 361.

```
normalmode
parent

selected

selectionhighlight
specularcolorreflectance
specularexponent
specularstrength
tag
```

type

uicontextmenu
userdata

vertexnormals
vertices

visible

xdata   Data defining the patch object.

xlim

xliminclude
ydata   Data defining the patch object.

ylim

yliminclude
zdata   Data defining the patch object.

zlim

zliminclude

15.3.3.8 Surface Properties

The surface properties are:

__modified__
__myhandle__
alim

aliminclude
alphadata
alphadatamapping
ambientstrength
backfacelighting
beingdeleted
busyaction
buttondownfcn
cdata

cdatamapping
cdatasource
children

clim

climinclude
clipping

createfcn
deletefcn
diffusestrength
displayname
        The text of the legend entry corresponding to this surface.

edgealpha
edgecolor
edgelighting
erasemode
facealpha
facecolor
facelighting
handlevisibility
hittest

interpreter
interruptible
linestyle
linewidth
marker

markeredgecolor
markerfacecolor
markersize
meshstyle
normalmode
parent

selected

selectionhighlight
specularcolorreflectance
specularexponent
specularstrength
tag

type

uicontextmenu
userdata

vertexnormals
visible

xdata   The data determining the surface. The xdata and ydata elements are vectors
            and zdata must be a matrix.

xdatasource
xlim

xliminclude
ydata   The data determining the surface. The xdata and ydata elements are vectors
            and zdata must be a matrix.

ydatasource
ylim

yliminclude
zdata   The data determining the surface. The xdata and ydata elements are vectors
            and zdata must be a matrix.

```
zdatasource
zlim

zliminclude
```

15.3.4 Searching Properties

h = findobj ()	[Function File]
h = findobj (*prop_name*, *prop_value*, ...)	[Function File]
h = findobj (*prop_name*, *prop_value*, "-*logical_op*",	[Function File]
prop_name, *prop_value*)	
h = findobj ("-*property*", *prop_name*)	[Function File]
h = findobj ("-*regexp*", *prop_name*, *pattern*)	[Function File]
h = findobj (*hlist*, ...)	[Function File]
h = findobj (*hlist*, "flat", ...)	[Function File]
h = findobj (*hlist*, "-*depth*", *d*, ...)	[Function File]

Find graphics object with specified property values.

The simplest form is

```
findobj (prop_name, prop_value)
```

which returns the handles of all objects which have a property named *prop_name* that has the value *prop_value*. If multiple property/value pairs are specified then only objects meeting all of the conditions are returned.

The search can be limited to a particular set of objects and their descendants, by passing a handle or set of handles *hlist* as the first argument.

The depth of the object hierarchy to search can be limited with the "-depth" argument. An example of searching only three generations of children is:

```
findobj (hlist, "-depth", 3, prop_name, prop_value)
```

Specifying a depth *d* of 0, limits the search to the set of objects passed in *hlist*. A depth *d* of 0 is equivalent to the "flat" argument.

A specified logical operator may be applied to the pairs of *prop_name* and *prop_value*. The supported logical operators are: "-and", "-or", "-xor", "-not".

Objects may also be matched by comparing a regular expression to the property values, where property values that match regexp (*prop_value*, *pattern*) are returned.

Finally, objects may be matched by property name only by using the "-property" option.

Implementation Note: The search only includes objects with visible handles (HandleVisibility = "on"). See [findall], page 359, to search for all objects including hidden ones.

See also: [findall], page 359, [allchild], page 342, [get], page 341, [set], page 342.

h = findall ()	[Function File]
h = findall (*prop_name*, *prop_value*, ...)	[Function File]
h = findall (*prop_name*, *prop_value*, "-*logical_op*",	[Function File]
prop_name, *prop_value*)	
h = findall ("-*property*", *prop_name*)	[Function File]
h = findall ("-*regexp*", *prop_name*, *pattern*)	[Function File]

h = findall (*hlist*, ...)   [Function File]

h = findall (*hlist*, "*flat*", ...)   [Function File]

h = findall (*hlist*, "-*depth*", *d*, ...)   [Function File]

Find graphics object, including hidden ones, with specified property values.

The return value h is a list of handles to the found graphic objects.

findall performs the same search as findobj, but it includes hidden objects (HandleVisibility = "off"). For full documentation, see [findobj], page 359.

See also: [findobj], page 359, [allchild], page 342, [get], page 341, [set], page 342.

15.3.5 Managing Default Properties

Object properties have two classes of default values, *factory defaults* (the initial values) and *user-defined defaults*, which may override the factory defaults.

Although default values may be set for any object, they are set in parent objects and apply to child objects, of the specified object type. For example, setting the default color property of line objects to "green", for the root object, will result in all line objects inheriting the color "green" as the default value.

```
set (0, "defaultlinecolor", "green");
```

sets the default line color for all objects. The rule for constructing the property name to set a default value is

```
default + object-type + property-name
```

This rule can lead to some strange looking names, for example defaultlinelinewidth" specifies the default linewidth property for line objects.

The example above used the root figure object, 0, so the default property value will apply to all line objects. However, default values are hierarchical, so defaults set in a figure objects override those set in the root figure object. Likewise, defaults set in axes objects override those set in figure or root figure objects. For example,

```
subplot (2, 1, 1);
set (0, "defaultlinecolor", "red");
set (1, "defaultlinecolor", "green");
set (gca (), "defaultlinecolor", "blue");
line (1:10, rand (1, 10));
subplot (2, 1, 2);
line (1:10, rand (1, 10));
figure (2)
line (1:10, rand (1, 10));
```

produces two figures. The line in first subplot window of the first figure is blue because it inherits its color from its parent axes object. The line in the second subplot window of the first figure is green because it inherits its color from its parent figure object. The line in the second figure window is red because it inherits its color from the global root figure parent object.

To remove a user-defined default setting, set the default property to the value "remove". For example,

```
set (gca (), "defaultlinecolor", "remove");
```

removes the user-defined default line color setting from the current axes object. To quickly remove all user-defined defaults use the **reset** function.

reset (*h, property*)   [Built-in Function]
>   Remove any defaults set for the handle *h*. The default figure properties of "position", "units", "windowstyle" and "paperunits" and the default axes properties of "position" and "units" are not reset.
>
>   **See also:** [cla], page 322, [clf], page 322.

Getting the "default" property of an object returns a list of user-defined defaults set for the object. For example,

>   get (gca (), "default");

returns a list of user-defined default values for the current axes object.

Factory default values are stored in the root figure object. The command

>   get (0, "factory");

returns a list of factory defaults.

15.4 Advanced Plotting

15.4.1 Colors

Colors may be specified as RGB triplets with values ranging from zero to one, or by name. Recognized color names include "blue", "black", "cyan", "green", "magenta", "red", "white", and "yellow".

15.4.2 Line Styles

Line styles are specified by the following properties:

linestyle
>   May be one of
>
> | "-" | Solid line. [default] |
> |--------|-----------------------|
> | "--" | Dashed line. |
> | ":" | Dotted line. |
> | "-." | A dash-dot line. |
> | "none" | No line. Points will still be marked using the current Marker Style. |

linewidth
>   A number specifying the width of the line. The default is 1. A value of 2 is twice as wide as the default, etc.

15.4.3 Marker Styles

Marker styles are specified by the following properties:

marker   A character indicating a plot marker to be place at each data point, or "none", meaning no markers should be displayed.

markeredgecolor

> The color of the edge around the marker, or "auto", meaning that the edge color is the same as the face color. See Section 15.4.1 [Colors], page 361.

markerfacecolor

> The color of the marker, or "none" to indicate that the marker should not be filled. See Section 15.4.1 [Colors], page 361.

markersize

> A number specifying the size of the marker. The default is 1. A value of 2 is twice as large as the default, etc.

The `colstyle` function will parse a `plot`-style specification and will return the color, line, and marker values that would result.

[*style, color, marker, msg*] = colstyle (*linespec*)   [Function File]

> Parse *linespec* and return the line style, color, and markers given. In the case of an error, the string *msg* will return the text of the error.

15.4.4 Callbacks

Callback functions can be associated with graphics objects and triggered after certain events occur. The basic structure of all callback function is

```
function mycallback (src, data)
...
endfunction
```

where `src` gives a handle to the source of the callback, and `code` gives some event specific data. This can then be associated with an object either at the objects creation or later with the `set` function. For example,

```
plot (x, "DeleteFcn", @(s, e) disp ("Window Deleted"))
```

where at the moment that the plot is deleted, the message "Window Deleted" will be displayed.

Additional user arguments can be passed to callback functions, and will be passed after the 2 default arguments. For example:

```
plot (x, "DeleteFcn", {@mycallback, "1"})
...
function mycallback (src, data, a1)
  fprintf ("Closing plot %d\n", a1);
endfunction
```

The basic callback functions that are available for all graphics objects are

- CreateFcn This is the callback that is called at the moment of the objects creation. It is not called if the object is altered in any way, and so it only makes sense to define this callback in the function call that defines the object. Callbacks that are added to `CreateFcn` later with the `set` function will never be executed.

- DeleteFcn This is the callback that is called at the moment an object is deleted.

- ButtonDownFcn This is the callback that is called if a mouse button is pressed while the pointer is over this object. Note, that the gnuplot interface does not respect this callback.

The object and figure that the event occurred in that resulted in the callback being called can be found with the `gcbo` and `gcbf` functions.

h = gcbo ()   [Function File]
[*h*, *fig*] = gcbo ()   [Function File]
> Return a handle to the object whose callback is currently executing.
>
> If no callback is executing, this function returns the empty matrix. This handle is obtained from the root object property `"CallbackObject"`.
>
> When called with a second output argument, return the handle of the figure containing the object whose callback is currently executing. If no callback is executing the second output is also set to the empty matrix.
>
> **See also:** [gcbf], page 363, [gco], page 341, [gca], page 340, [gcf], page 340, [get], page 341, [set], page 342.

fig = gcbf ()   [Function File]
> Return a handle to the figure containing the object whose callback is currently executing.
>
> If no callback is executing, this function returns the empty matrix. The handle returned by this function is the same as the second output argument of `gcbo`.
>
> **See also:** [gcbo], page 363, [gcf], page 340, [gco], page 341, [gca], page 340, [get], page 341, [set], page 342.

Callbacks can equally be added to properties with the `addlistener` function described below.

15.4.5 Application-defined Data

Octave has a provision for attaching application-defined data to a graphics handle. The data can be anything which is meaningful to the application, and will be completely ignored by Octave.

setappdata (*h*, *name*, *value*)   [Function File]
> Set the named application data to *value* for the object(s) with handle(s) *h*. If the application data with the specified name does not exist, it is created.
>
> **See also:** [getappdata], page 363, [guidata], page 693, [get], page 341, [set], page 342, [getpref], page 695, [setpref], page 695.

value = getappdata (*h*, *name*)   [Function File]
appdata = getappdata (*h*)   [Function File]
> Return the *value* for named application data for the object(s) with handle(s) *h*.
>
> `getappdata(h)` returns a structure, *appdata*, whose fields correspond to the appdata properties.
>
> **See also:** [setappdata], page 363, [guidata], page 693, [get], page 341, [set], page 342, [getpref], page 695, [setpref], page 695.

rmappdata (*h*, *name*)   [Function File]
> Delete the named application data for the object(s) with handle(s) *h*.

V = isappdata (*h*, *name*)   [Function File]
> Return true if the named application data, *name*, exists for the object with handle *h*.

> **See also:** [getappdata], page 363, [setappdata], page 363, [rmappdata], page 363.

15.4.6 Object Groups

A number of Octave high level plot functions return groups of other graphics objects or they return graphics objects that have their properties linked in such a way that changes to one of the properties results in changes in the others. A graphic object that groups other objects is an `hggroup`

hggroup ()   [Function File]
hggroup (*hax*)   [Function File]
hggroup (..., *property*, *value*, ...)   [Function File]
h = hggroup (...)   [Function File]
> Create handle graphics group object with axes parent *hax*.

> If no parent is specified, the group is created in the current axes.

> Multiple property/value pairs may be specified for the hggroup, but they must appear in pairs.

> The optional return value *h* is a graphics handle to the created hggroup object.

> Programming Note: An hggroup is a way to group base graphics objects such as line objects or patch objects into a single unit which can react appropriately. For example, the individual lines of a contour plot are collected into a single hggroup so that they can be made visible/invisible with a single command, set (`hg_handle`, `"visible"`, `"off"`).

> **See also:** [addproperty], page 364, [addlistener], page 365.

For example a simple use of a `hggroup` might be
```
x = 0:0.1:10;
hg = hggroup ();
plot (x, sin (x), "color", [1, 0, 0], "parent", hg);
hold on
plot (x, cos (x), "color", [0, 1, 0], "parent", hg);
set (hg, "visible", "off");
```
which groups the two plots into a single object and controls their visibility directly. The default properties of an `hggroup` are the same as the set of common properties for the other graphics objects. Additional properties can be added with the `addproperty` function.

addproperty (*name*, *h*, *type*)   [Built-in Function]
addproperty (*name*, *h*, *type*, *arg*, ...)   [Built-in Function]
> Create a new property named *name* in graphics object *h*. *type* determines the type of the property to create. *args* usually contains the default value of the property, but additional arguments might be given, depending on the type of the property.

> The supported property types are:

> string   A string property. *arg* contains the default string value.

any	An un-typed property. This kind of property can hold any octave value. *args* contains the default value.

radio   A string property with a limited set of accepted values. The first argument must be a string with all accepted values separated by a vertical bar ('|'). The default value can be marked by enclosing it with a '{' '}' pair. The default value may also be given as an optional second string argument.

boolean   A boolean property. This property type is equivalent to a radio property with "on|off" as accepted values. *arg* contains the default property value.

double   A scalar double property. *arg* contains the default value.

handle   A handle property. This kind of property holds the handle of a graphics object. *arg* contains the default handle value. When no default value is given, the property is initialized to the empty matrix.

data   A data (matrix) property. *arg* contains the default data value. When no default value is given, the data is initialized to the empty matrix.

color   A color property. *arg* contains the default color value. When no default color is given, the property is set to black. An optional second string argument may be given to specify an additional set of accepted string values (like a radio property).

type may also be the concatenation of a core object type and a valid property name for that object type. The property created then has the same characteristics as the referenced property (type, possible values, hidden state...). This allows to clone an existing property into the graphics object *h*.

Examples:

```
addproperty ("my_property", gcf, "string", "a string value");
addproperty ("my_radio", gcf, "radio", "val_1|val_2|{val_3}");
addproperty ("my_style", gcf, "linelinestyle", "--");
```

See also: [addlistener], page 365, [hggroup], page 364.

Once a property in added to an `hggroup`, it is not linked to any other property of either the children of the group, or any other graphics object. Add so to control the way in which this newly added property is used, the **addlistener** function is used to define a callback function that is executed when the property is altered.

addlistener (*h*, *prop*, *fcn*)   [Built-in Function]

Register *fcn* as listener for the property *prop* of the graphics object *h*. Property listeners are executed (in order of registration) when the property is set. The new value is already available when the listeners are executed.

prop must be a string naming a valid property in *h*.

fcn can be a function handle, a string or a cell array whose first element is a function handle. If *fcn* is a function handle, the corresponding function should accept at least 2 arguments, that will be set to the object handle and the empty matrix respectively. If *fcn* is a string, it must be any valid octave expression. If *fcn* is a cell array, the first element must be a function handle with the same signature as described above. The next elements of the cell array are passed as additional arguments to the function.

Example:

```
function my_listener (h, dummy, p1)
  fprintf ("my_listener called with p1=%s\n", p1);
endfunction

addlistener (gcf, "position", {@my_listener, "my string"})
```

See also: [addproperty], page 364, [hggroup], page 364.

dellistener (*h, prop, fcn*)   [Built-in Function]

Remove the registration of *fcn* as a listener for the property *prop* of the graphics object *h*. The function *fcn* must be the same variable (not just the same value), as was passed to the original call to `addlistener`.

If *fcn* is not defined then all listener functions of *prop* are removed.

Example:

```
function my_listener (h, dummy, p1)
  fprintf ("my_listener called with p1=%s\n", p1);
endfunction

c = {@my_listener, "my string"};
addlistener (gcf, "position", c);
dellistener (gcf, "position", c);
```

An example of the use of these two functions might be

```
x = 0:0.1:10;
hg = hggroup ();
h = plot (x, sin (x), "color", [1, 0, 0], "parent", hg);
addproperty ("linestyle", hg, "linelinestyle", get (h, "linestyle"));
addlistener (hg, "linestyle", @update_props);
hold on
plot (x, cos (x), "color", [0, 1, 0], "parent", hg);

function update_props (h, d)
  set (get (h, "children"), "linestyle", get (h, "linestyle"));
endfunction
```

that adds a `linestyle` property to the `hggroup` and propagating any changes its value to the children of the group. The `linkprop` function can be used to simplify the above to be

```
x = 0:0.1:10;
hg = hggroup ();
h1 = plot (x, sin (x), "color", [1, 0, 0], "parent", hg);
addproperty ("linestyle", hg, "linelinestyle", get (h, "linestyle"));
hold on
h2 = plot (x, cos (x), "color", [0, 1, 0], "parent", hg);
hlink = linkprop ([hg, h1, h2], "color");
```

hlink = linkprop (*h*, *prop*)   [Function File]
hlink = linkprop (*h*, {*prop1*, *prop2*, ...})   [Function File]

>   Link graphics object properties, such that a change in one is propagated to the others.
>
>   *prop* can be a string for a single property, or a cell array of strings for multiple properties. *h* is an array of graphics handles which will have their properties linked.
>
>   An example of the use of `linkprop` is
>
>   ```
>   x = 0:0.1:10;
>   subplot (1,2,1);
>   h1 = plot (x, sin (x));
>   subplot (1,2,2);
>   h2 = plot (x, cos (x));
>   hlink = linkprop ([h1, h2], {"color","linestyle"});
>   set (h1, "color", "green");
>   set (h2, "linestyle", "--");
>   ```

These capabilities are used in a number of basic graphics objects. The `hggroup` objects created by the functions of Octave contain one or more graphics object and are used to:

- group together multiple graphics objects,
- create linked properties between different graphics objects, and
- to hide the nominal user data, from the actual data of the objects.

For example the **stem** function creates a stem series where each `hggroup` of the stem series contains two line objects representing the body and head of the stem. The `ydata` property of the `hggroup` of the stem series represents the head of the stem, whereas the body of the stem is between the baseline and this value. For example

```
h = stem (1:4)
get (h, "xdata")
⇒ [ 1   2   3   4]'
get (get (h, "children")(1), "xdata")
⇒ [ 1   1 NaN   2   2 NaN   3   3 NaN   4   4 NaN]'
```

shows the difference between the `xdata` of the `hggroup` of a stem series object and the underlying line.

The basic properties of such group objects is that they consist of one or more linked `hggroup`, and that changes in certain properties of these groups are propagated to other members of the group. Whereas, certain properties of the members of the group only apply to the current member.

In addition the members of the group can also be linked to other graphics objects through callback functions. For example the baseline of the **bar** or **stem** functions is a line object, whose length and position are automatically adjusted, based on changes to the corresponding hggroup elements.

15.4.6.1 Data Sources in Object Groups

All of the group objects contain data source parameters. There are string parameters that contain an expression that is evaluated to update the relevant data property of the group when the `refreshdata` function is called.

refreshdata ()   [Function File]
refreshdata (*h*)   [Function File]
refreshdata (*h*, *workspace*)   [Function File]
> Evaluate any 'datasource' properties of the current figure and update the plot if the corresponding data has changed.
>
> If the first argument *h* is a list of graphic handles, then operate on these objects rather than the current figure returned by gcf.
>
> The optional second argument *workspace* can take the following values:
>
> "base"   Evaluate the datasource properties in the base workspace. (default).
>
> "caller"   Evaluate the datasource properties in the workspace of the function that called refreshdata.
>
> An example of the use of refreshdata is:
>
> ```
> x = 0:0.1:10;
> y = sin (x);
> plot (x, y, "ydatasource", "y");
> for i = 1 : 100
>   pause (0.1);
>   y = sin (x + 0.1*i);
>   refreshdata ();
> endfor
> ```

15.4.6.2 Area Series

Area series objects are created by the **area** function. Each of the **hggroup** elements contains a single patch object. The properties of the area series are

basevalue
> The value where the base of the area plot is drawn.

linewidth
linestyle
> The line width and style of the edge of the patch objects making up the areas. See Section 15.4.2 [Line Styles], page 361.

edgecolor
facecolor
> The line and fill color of the patch objects making up the areas. See Section 15.4.1 [Colors], page 361.

xdata
ydata   The x and y coordinates of the original columns of the data passed to **area** prior to the cumulative summation used in the **area** function.

xdatasource
ydatasource
> Data source variables.

15.4.6.3 Bar Series

Bar series objects are created by the `bar` or `barh` functions. Each `hggroup` element contains a single patch object. The properties of the bar series are

```
showbaseline
baseline
basevalue
```
> The property `showbaseline` flags whether the baseline of the bar series is displayed (default is `"on"`). The handle of the graphics object representing the baseline is given by the `baseline` property and the y-value of the baseline by the `basevalue` property.
>
> Changes to any of these properties are propagated to the other members of the bar series and to the baseline itself. Equally, changes in the properties of the base line itself are propagated to the members of the corresponding bar series.

```
barwidth
barlayout
horizontal
```
> The property `barwidth` is the width of the bar corresponding to the *width* variable passed to `bar` or *barh*. Whether the bar series is `"grouped"` or `"stacked"` is determined by the `barlayout` property and whether the bars are horizontal or vertical by the `horizontal` property.
>
> Changes to any of these property are propagated to the other members of the bar series.

```
linewidth
linestyle
```
> The line width and style of the edge of the patch objects making up the bars. See Section 15.4.2 [Line Styles], page 361.

```
edgecolor
facecolor
```
> The line and fill color of the patch objects making up the bars. See Section 15.4.1 [Colors], page 361.

`xdata`   The nominal x positions of the bars. Changes in this property and propagated to the other members of the bar series.

`ydata`   The y value of the bars in the `hggroup`.

```
xdatasource
ydatasource
```
> Data source variables.

15.4.6.4 Contour Groups

Contour group objects are created by the `contour`, `contourf` and `contour3` functions. The are equally one of the handles returned by the `surfc` and `meshc` functions. The properties of the contour group are contourmatrix

> A read only property that contains the data return by `contourc` used to create the contours of the plot.

fill

> A radio property that can have the values `"on"` or `"off"` that flags whether the contours to plot are to be filled.

zlevelmode
zlevel

> The radio property `zlevelmode` can have the values `"none"`, `"auto"`, or `"manual"`. When its value is `"none"` there is no z component to the plotted contours. When its value is `"auto"` the z value of the plotted contours is at the same value as the contour itself. If the value is `"manual"`, then the z value at which to plot the contour is determined by the `zlevel` property.

levellistmode
levellist
levelstepmode
levelstep

> If `levellistmode` is `"manual"`, then the levels at which to plot the contours is determined by `levellist`. If `levellistmode` is set to `"auto"`, then the distance between contours is determined by `levelstep`. If both `levellistmode` and `levelstepmode` are set to `"auto"`, then there are assumed to be 10 equal spaced contours.

textlistmode
textlist
textstepmode
textstep

> If `textlistmode` is `"manual"`, then the labeled contours is determined by `textlist`. If `textlistmode` is set to `"auto"`, then the distance between labeled contours is determined by `textstep`. If both `textlistmode` and `textstepmode` are set to `"auto"`, then there are assumed to be 10 equal spaced labeled contours.

showtext   Flag whether the contour labels are shown or not.

labelspacing

> The distance between labels on a single contour in points.

linewidth
linestyle
linecolor

> The properties of the contour lines. The properties `linewidth` and `linestyle` are similar to the corresponding properties for lines. The property `linecolor` is a color property (see Section 15.4.1 [Colors], page 361), that can also have the values of `"none"` or `"auto"`. If `linecolor` is `"none"`, then no contour line is drawn. If `linecolor` is `"auto"` then the line color is determined by the colormap.

xdata
ydata
zdata   The original x, y, and z data of the contour lines.

```
xdatasource
ydatasource
zdatasource
```
       Data source variables.

15.4.6.5 Error Bar Series

Error bar series are created by the `errorbar` function. Each `hggroup` element contains two line objects representing the data and the errorbars separately. The properties of the error bar series are

`color`   The RGB color or color name of the line objects of the error bars. See Section 15.4.1 [Colors], page 361.

```
linewidth
linestyle
```
       The line width and style of the line objects of the error bars. See Section 15.4.2 [Line Styles], page 361.

```
marker
markeredgecolor
markerfacecolor
markersize
```
       The line and fill color of the markers on the error bars. See Section 15.4.1 [Colors], page 361.

```
xdata
ydata
ldata
udata
xldata
xudata
```
       The original x, y, l, u, xl, xu data of the error bars.

```
xdatasource
ydatasource
ldatasource
udatasource
xldatasource
xudatasource
```
       Data source variables.

15.4.6.6 Line Series

Line series objects are created by the `plot` and `plot3` functions and are of the type `line`. The properties of the line series with the ability to add data sources.

`color`   The RGB color or color name of the line objects. See Section 15.4.1 [Colors], page 361.

```
linewidth
linestyle
```
       The line width and style of the line objects. See Section 15.4.2 [Line Styles], page 361.

```
marker
markeredgecolor
markerfacecolor
markersize
```
> The line and fill color of the markers. See Section 15.4.1 [Colors], page 361.

```
xdata
ydata
zdata
```
> The original x, y and z data.

```
xdatasource
ydatasource
zdatasource
```
> Data source variables.

15.4.6.7 Quiver Group

Quiver series objects are created by the `quiver` or `quiver3` functions. Each `hggroup` element of the series contains three line objects as children representing the body and head of the arrow, together with a marker as the point of origin of the arrows. The properties of the quiver series are

```
autoscale
autoscalefactor
```
> Flag whether the length of the arrows is scaled or defined directly from the u, v and w data. If the arrow length is flagged as being scaled by the `autoscale` property, then the length of the autoscaled arrow is controlled by the `autoscalefactor`.

```
maxheadsize
```
> This property controls the size of the head of the arrows in the quiver series. The default value is 0.2.

```
showarrowhead
```
> Flag whether the arrow heads are displayed in the quiver plot.

```
color
```
> The RGB color or color name of the line objects of the quiver. See Section 15.4.1 [Colors], page 361.

```
linewidth
linestyle
```
> The line width and style of the line objects of the quiver. See Section 15.4.2 [Line Styles], page 361.

```
marker
markerfacecolor
markersize
```
> The line and fill color of the marker objects at the original of the arrows. See Section 15.4.1 [Colors], page 361.

```
xdata
ydata
zdata
```
> The origins of the values of the vector field.

```
udata
vdata
wdata        The values of the vector field to plot.

xdatasource
ydatasource
zdatasource
udatasource
vdatasource
wdatasource
```
             Data source variables.

15.4.6.8 Scatter Group

Scatter series objects are created by the scatter or scatter3 functions. A single hggroup element contains as many children as there are points in the scatter plot, with each child representing one of the points. The properties of the stem series are

```
linewidth
```
             The line width of the line objects of the points. See Section 15.4.2 [Line Styles], page 361.

```
marker
markeredgecolor
markerfacecolor
```
             The line and fill color of the markers of the points. See Section 15.4.1 [Colors], page 361.

```
xdata
ydata
zdata        The original x, y and z data of the stems.
```

cdata   The color data for the points of the plot. Each point can have a separate color, or a unique color can be specified.

sizedata   The size data for the points of the plot. Each point can its own size or a unique size can be specified.

```
xdatasource
ydatasource
zdatasource
cdatasource
sizedatasource
```
             Data source variables.

15.4.6.9 Stair Group

Stair series objects are created by the stair function. Each hggroup element of the series contains a single line object as a child representing the stair. The properties of the stair series are

color   The RGB color or color name of the line objects of the stairs. See Section 15.4.1 [Colors], page 361.

linewidth
linestyle
> The line width and style of the line objects of the stairs. See Section 15.4.2
> [Line Styles], page 361.

marker
markeredgecolor
markerfacecolor
markersize
> The line and fill color of the markers on the stairs. See Section 15.4.1 [Colors],
> page 361.

xdata
ydata   The original x and y data of the stairs.

xdatasource
ydatasource
> Data source variables.

15.4.6.10 Stem Series

Stem series objects are created by the `stem` or `stem3` functions. Each `hggroup` element contains a single line object as a child representing the stems. The properties of the stem series are

showbaseline
baseline
basevalue
> The property `showbaseline` flags whether the baseline of the stem series is
> displayed (default is `"on"`). The handle of the graphics object representing
> the baseline is given by the `baseline` property and the y-value (or z-value for
> `stem3`) of the baseline by the `basevalue` property.
>
> Changes to any of these property are propagated to the other members of the
> stem series and to the baseline itself. Equally changes in the properties of the
> base line itself are propagated to the members of the corresponding stem series.

color   The RGB color or color name of the line objects of the stems. See Section 15.4.1
> [Colors], page 361.

linewidth
linestyle
> The line width and style of the line objects of the stems. See Section 15.4.2
> [Line Styles], page 361.

marker
markeredgecolor
markerfacecolor
markersize
> The line and fill color of the markers on the stems. See Section 15.4.1 [Colors],
> page 361.

```
xdata
ydata
zdata          The original x, y and z data of the stems.

xdatasource
ydatasource
zdatasource
               Data source variables.
```

15.4.6.11 Surface Group

Surface group objects are created by the `surf` or `mesh` functions, but are equally one of the handles returned by the `surfc` or `meshc` functions. The surface group is of the type `surface`.

The properties of the surface group are

```
edgecolor
facecolor
```
        The RGB color or color name of the edges or faces of the surface. See Section 15.4.1 [Colors], page 361.

```
linewidth
linestyle
```
        The line width and style of the lines on the surface. See Section 15.4.2 [Line Styles], page 361.

```
marker
markeredgecolor
markerfacecolor
markersize
```
        The line and fill color of the markers on the surface. See Section 15.4.1 [Colors], page 361.

```
xdata
ydata
zdata
cdata          The original x, y, z and c data.

xdatasource
ydatasource
zdatasource
cdatasource
               Data source variables.
```

15.4.7 Graphics Toolkits

`name = graphics_toolkit ()`	[Function File]
`name = graphics_toolkit (hlist)`	[Function File]
`graphics_toolkit (name)`	[Function File]
`graphics_toolkit (hlist, name)`	[Function File]

        Query or set the default graphics toolkit which is assigned to new figures.

With no inputs, return the current default graphics toolkit. If the input is a list of figure graphic handles, *hlist*, then return the name of the graphics toolkit in use for each figure.

When called with a single input *name* set the default graphics toolkit to *name*. If the toolkit is not already loaded, it is initialized by calling the function `__init_name__`. If the first input is a list of figure handles, *hlist*, then the graphics toolkit is set to *name* for these figures only.

See also: [available_graphics_toolkits], page 376.

available_graphics_toolkits ()   [Built-in Function]
    Return a cell array of registered graphics toolkits.

    See also: [graphics_toolkit], page 375, [register_graphics_toolkit], page 376.

loaded_graphics_toolkits ()   [Built-in Function]
    Return a cell array of the currently loaded graphics toolkits.

    See also: [available_graphics_toolkits], page 376.

register_graphics_toolkit (*toolkit*)   [Built-in Function]
    List *toolkit* as an available graphics toolkit.

    See also: [available_graphics_toolkits], page 376.

15.4.7.1 Customizing Toolkit Behavior

The specific behavior of the backend toolkit may be modified using the following utility functions. Note: Not all functions apply to every graphics toolkit.

[*prog*, *args*] = gnuplot_binary ()   [Loadable Function]
[*old_prog*, *old_args*] = gnuplot_binary (*new_prog*,   [Loadable Function]
        ***arg1*, ...)**
    Query or set the name of the program invoked by the plot command when the graphics toolkit is set to "gnuplot". Additional arguments to pass to the external plotting program may also be given. The default value is "gnuplot" with no additional arguments. See Appendix G [Installation], page 833.

    See also: [graphics_toolkit], page 375.

mode = gui_mode ()   [Built-in Function]
gui_mode (*mode*)   [Built-in Function]
    Query or set the GUI mode for the current graphics toolkit. The *mode* argument can be one of the following strings:

    "2d"   Allows panning and zooming of current axes.

    "3d"   Allows rotating and zooming of current axes.

    "none"   Mouse inputs have no effect.

    This function is currently implemented only for the FLTK graphics toolkit.

    See also: [mouse_wheel_zoom], page 377.

val = mouse_wheel_zoom ()   [Loadable Function]
old_val = mouse_wheel_zoom (*new_val*)   [Loadable Function]
mouse_wheel_zoom (*new_val*, "*local*")   [Loadable Function]

> Query or set the mouse wheel zoom factor.
>
> The zoom factor is a number in the range (0,1) which is the percentage of the current axis limits that will be used when zooming. For example, if the current x-axis limits are [0, 50] and mouse_wheel_zoom is 0.4 (40%), then a zoom operation will change the limits by 20.
>
> When called from inside a function with the "local" option, the variable is changed locally for the function and any subroutines it calls. The original variable value is restored when exiting the function.
>
> This function is currently implemented only for the FLTK graphics toolkit.
>
> **See also:** [gui_mode], page 376.

16 Matrix Manipulation

There are a number of functions available for checking to see if the elements of a matrix meet some condition, and for rearranging the elements of a matrix. For example, Octave can easily tell you if all the elements of a matrix are finite, or are less than some specified value. Octave can also rotate the elements, extract the upper- or lower-triangular parts, or sort the columns of a matrix.

16.1 Finding Elements and Checking Conditions

The functions **any** and **all** are useful for determining whether any or all of the elements of a matrix satisfy some condition. The **find** function is also useful in determining which elements of a matrix meet a specified condition.

any (*x*)   [Built-in Function]
any (*x*, *dim*)   [Built-in Function]

> For a vector argument, return true (logical 1) if any element of the vector is non-zero.
>
> For a matrix argument, return a row vector of logical ones and zeros with each element indicating whether any of the elements of the corresponding column of the matrix are non-zero. For example:
>
> ```
> any (eye (2, 4))
>    ⇒ [ 1, 1, 0, 0 ]
> ```
>
> If the optional argument *dim* is supplied, work along dimension *dim*. For example:
>
> ```
> any (eye (2, 4), 2)
>    ⇒ [ 1; 1 ]
> ```
>
> **See also:** [all], page 379.

all (*x*)   [Built-in Function]
all (*x*, *dim*)   [Built-in Function]

> For a vector argument, return true (logical 1) if all elements of the vector are non-zero.
>
> For a matrix argument, return a row vector of logical ones and zeros with each element indicating whether all of the elements of the corresponding column of the matrix are non-zero. For example:
>
> ```
> all ([2, 3; 1, 0]))
>    ⇒ [ 1, 0 ]
> ```
>
> If the optional argument *dim* is supplied, work along dimension *dim*.
>
> **See also:** [any], page 379.

Since the comparison operators (see Section 8.4 [Comparison Ops], page 140) return matrices of ones and zeros, it is easy to test a matrix for many things, not just whether the elements are nonzero. For example,

```
all (all (rand (5) < 0.9))
   ⇒ 0
```

tests a random 5 by 5 matrix to see if all of its elements are less than 0.9.

Note that in conditional contexts (like the test clause of **if** and **while** statements) Octave treats the test as if you had typed **all (all (condition))**.

z = xor (*x*, *y*)   [Mapping Function]
> Return the *exclusive or* of the entries of *x* and *y*. For boolean expressions *x* and *y*,
> xor (*x*, *y*) is true if and only if one of *x* or *y* is true. Otherwise, for *x* and *y* both
> true or both false, xor returns false.
>
> The truth table for the xor operation is

$$\begin{array}{ccc} x & y & z \\ 0 & 0 & 0 \\ 1 & 0 & 1 \\ 0 & 1 & 1 \\ 1 & 1 & 0 \end{array}$$

> **See also:** [and], page 142, [or], page 143, [not], page 143.

diff (*x*)   [Built-in Function]
diff (*x*, *k*)   [Built-in Function]
diff (*x*, *k*, *dim*)   [Built-in Function]
> If *x* is a vector of length n, diff (*x*) is the vector of first differences $x_2 - x_1, \ldots, x_n - x_{n-1}$.
>
> If *x* is a matrix, diff (*x*) is the matrix of column differences along the first non-singleton dimension.
>
> The second argument is optional. If supplied, diff (*x*, *k*), where *k* is a non-negative integer, returns the *k*-th differences. It is possible that *k* is larger than the first non-singleton dimension of the matrix. In this case, diff continues to take the differences along the next non-singleton dimension.
>
> The dimension along which to take the difference can be explicitly stated with the optional variable *dim*. In this case the *k*-th order differences are calculated along this dimension. In the case where *k* exceeds size (*x*, *dim*) an empty matrix is returned.
>
> **See also:** [sort], page 387, [merge], page 144.

isinf (*x*)   [Mapping Function]
> Return a logical array which is true where the elements of *x* are are infinite and false where they are not. For example:

```
isinf ([13, Inf, NA, NaN])
     ⇒ [ 0, 1, 0, 0 ]
```

> **See also:** [isfinite], page 380, [isnan], page 380, [isna], page 42.

isnan (*x*)   [Mapping Function]
> Return a logical array which is true where the elements of *x* are NaN values and false where they are not. NA values are also considered NaN values. For example:

```
isnan ([13, Inf, NA, NaN])
     ⇒ [ 0, 0, 1, 1 ]
```

> **See also:** [isna], page 42, [isinf], page 380, [isfinite], page 380.

isfinite (*x*)   [Mapping Function]
finite (*x*)   [Mapping Function]
> Return a logical array which is true where the elements of *x* are finite values and false where they are not. For example:

```
finite ([13, Inf, NA, NaN])
    ⇒ [ 1, 0, 0, 0 ]
```

See also: [isinf], page 380, [isnan], page 380, [isna], page 42.

[*err, y1, ...*] = common_size (*x1, ...*)   [Function File]

Determine if all input arguments are either scalar or of common size. If so, *err* is zero, and *yi* is a matrix of the common size with all entries equal to *xi* if this is a scalar or *xi* otherwise. If the inputs cannot be brought to a common size, *err* is 1, and *yi* is *xi*. For example:

```
[errorcode, a, b] = common_size ([1 2; 3 4], 5)
    ⇒ errorcode = 0
    ⇒ a = [ 1, 2; 3, 4 ]
    ⇒ b = [ 5, 5; 5, 5 ]
```

This is useful for implementing functions where arguments can either be scalars or of common size.

idx = find (*x*)   [Built-in Function]
idx = find (*x, n*)   [Built-in Function]
idx = find (*x, n, direction*)   [Built-in Function]
[*i, j*] = find (...)   [Built-in Function]
[*i, j, v*] = find (...)   [Built-in Function]

Return a vector of indices of nonzero elements of a matrix, as a row if *x* is a row vector or as a column otherwise. To obtain a single index for each matrix element, Octave pretends that the columns of a matrix form one long vector (like Fortran arrays are stored). For example:

```
find (eye (2))
    ⇒ [ 1; 4 ]
```

If two outputs are requested, `find` returns the row and column indices of nonzero elements of a matrix. For example:

```
[i, j] = find (2 * eye (2))
    ⇒ i = [ 1; 2 ]
    ⇒ j = [ 1; 2 ]
```

If three outputs are requested, `find` also returns a vector containing the nonzero values. For example:

```
[i, j, v] = find (3 * eye (2))
    ⇒ i = [ 1; 2 ]
    ⇒ j = [ 1; 2 ]
    ⇒ v = [ 3; 3 ]
```

If two inputs are given, *n* indicates the maximum number of elements to find from the beginning of the matrix or vector.

If three inputs are given, *direction* should be one of `"first"` or `"last"`, requesting only the first or last *n* indices, respectively. However, the indices are always returned in ascending order.

Note that this function is particularly useful for sparse matrices, as it extracts the non-zero elements as vectors, which can then be used to create the original matrix. For example:

```
sz = size (a);
[i, j, v] = find (a);
b = sparse (i, j, v, sz(1), sz(2));
```

See also: [nonzeros], page 494.

idx = lookup (*table*, *y*)   [Built-in Function]
idx = lookup (*table*, *y*, *opt*)   [Built-in Function]
Lookup values in a sorted table. Usually used as a prelude to interpolation.

If table is increasing and `idx = lookup (table, y)`, then `table(idx(i)) <= y(i) <`
`table(idx(i+1))` for all `y(i)` within the table. If `y(i) < table(1)` then `idx(i)` is
0. If `y(i) >= table(end)` or `isnan (y(i))` then `idx(i)` is n.

If the table is decreasing, then the tests are reversed. For non-strictly monotonic
tables, empty intervals are always skipped. The result is undefined if *table* is not
monotonic, or if *table* contains a NaN.

The complexity of the lookup is O(M*log(N)) where N is the size of *table* and
M is the size of *y*. In the special case when *y* is also sorted, the complexity is
O(min(M*log(N),M+N)).

table and *y* can also be cell arrays of strings (or *y* can be a single string). In this
case, string lookup is performed using lexicographical comparison.

If *opts* is specified, it must be a string with letters indicating additional options.

m   `table(idx(i)) == val(i)` if `val(i)` occurs in table; otherwise, `idx(i)`
             is zero.

b   `idx(i)` is a logical 1 or 0, indicating whether `val(i)` is contained in table
             or not.

l   For numeric lookups the leftmost subinterval shall be extended to infinity
             (i.e., all indices at least 1)

r   For numeric lookups the rightmost subinterval shall be extended to infin-
             ity (i.e., all indices at most n-1).

If you wish to check if a variable exists at all, instead of properties its elements may
have, consult Section 7.3 [Status of Variables], page 126.

16.2 Rearranging Matrices

fliplr (*x*)   [Function File]
Return a copy of *x* with the order of the columns reversed. In other words, *x* is
flipped left-to-right about a vertical axis. For example:

```
fliplr ([1, 2; 3, 4])
     ⇒   2   1
         4   3
```

Note that `fliplr` only works with 2-D arrays. To flip N-D arrays use `flipdim` instead.

See also: [flipud], page 383, [flipdim], page 383, [rot90], page 383, [rotdim], page 383.

flipud (*x*)   [Function File]

> Return a copy of *x* with the order of the rows reversed. In other words, *x* is flipped upside-down about a horizontal axis. For example:
>
>   flipud ([1, 2; 3, 4])
>   ⇒   3 4
>   1 2
>
> Note that flipud only works with 2-D arrays. To flip N-D arrays use flipdim instead.
>
> **See also:** [fliplr], page 382, [flipdim], page 383, [rot90], page 383, [rotdim], page 383.

flipdim (*x*)   [Function File]
flipdim (*x*, *dim*)   [Function File]

> Return a copy of *x* flipped about the dimension *dim*. *dim* defaults to the first non-singleton dimension. For example:
>
>   flipdim ([1, 2; 3, 4], 2)
>   ⇒   2 1
>   4 3
>
> **See also:** [fliplr], page 382, [flipud], page 383, [rot90], page 383, [rotdim], page 383.

rot90 (*A*)   [Function File]
rot90 (*A*, *k*)   [Function File]

> Return a copy of *A* with the elements rotated counterclockwise in 90-degree increments. The second argument is optional, and specifies how many 90-degree rotations are to be applied (the default value is 1). Negative values of *k* rotate the matrix in a clockwise direction. For example,
>
>   rot90 ([1, 2; 3, 4], -1)
>   ⇒   3 1
>   4 2
>
> rotates the given matrix clockwise by 90 degrees. The following are all equivalent statements:
>
>   rot90 ([1, 2; 3, 4], -1)
>   rot90 ([1, 2; 3, 4], 3)
>   rot90 ([1, 2; 3, 4], 7)
>
> Note that rot90 only works with 2-D arrays. To rotate N-D arrays use rotdim instead.
>
> **See also:** [rotdim], page 383, [flipud], page 383, [fliplr], page 382, [flipdim], page 383.

rotdim (*x*)   [Function File]
rotdim (*x*, *n*)   [Function File]
rotdim (*x*, *n*, *plane*)   [Function File]

> Return a copy of *x* with the elements rotated counterclockwise in 90-degree increments. The second argument *n* is optional, and specifies how many 90-degree rotations are to be applied (the default value is 1). The third argument is also optional and defines the plane of the rotation. If present, *plane* is a two element vector containing two different valid dimensions of the matrix. When *plane* is not given the first two non-singleton dimensions are used.
>
> Negative values of *n* rotate the matrix in a clockwise direction. For example,

```
rotdim ([1, 2; 3, 4], -1, [1, 2])
   ⇒  3  1
      4  2
```

rotates the given matrix clockwise by 90 degrees. The following are all equivalent statements:

```
rotdim ([1, 2; 3, 4], -1, [1, 2])
rotdim ([1, 2; 3, 4], 3, [1, 2])
rotdim ([1, 2; 3, 4], 7, [1, 2])
```

See also: [rot90], page 383, [flipud], page 383, [fliplr], page 382, [flipdim], page 383.

cat (*dim*, *array1*, *array2*, ..., *arrayN*)   [Built-in Function]
  Return the concatenation of N-D array objects, *array1*, *array2*, ..., *arrayN* along dimension *dim*.

```
A = ones (2, 2);
B = zeros (2, 2);
cat (2, A, B)
   ⇒ 1 1 0 0
     1 1 0 0
```

Alternatively, we can concatenate A and B along the second dimension in the following way:

```
[A, B]
```

dim can be larger than the dimensions of the N-D array objects and the result will thus have *dim* dimensions as the following example shows:

```
cat (4, ones (2, 2), zeros (2, 2))
   ⇒ ans(:,:,1,1) =

     1 1
     1 1

     ans(:,:,1,2) =

     0 0
     0 0
```

See also: [horzcat], page 384, [vertcat], page 385.

horzcat (*array1*, *array2*, ..., *arrayN*)   [Built-in Function]
  Return the horizontal concatenation of N-D array objects, *array1*, *array2*, ..., *arrayN* along dimension 2.

Arrays may also be concatenated horizontally using the syntax for creating new matrices. For example:

```
hcat = [ array1, array2, ... ]
```

See also: [cat], page 384, [vertcat], page 385.

vertcat (`array1`, `array2`, ..., `arrayN`)   [Built-in Function]
>   Return the vertical concatenation of N-D array objects, *array1*, *array2*, ..., *arrayN*
>   along dimension 1.
>
>   Arrays may also be concatenated vertically using the syntax for creating new matrices.
>   For example:
>
>   *vcat* = [`array1`; `array2`; ...]
>
>   **See also:** [cat], page 384, [horzcat], page 384.

permute (`A`, `perm`)   [Built-in Function]
>   Return the generalized transpose for an N-D array object *A*. The permutation vector
>   *perm* must contain the elements `1:ndims (A)` (in any order, but each element must
>   appear only once).
>
>   **See also:** [ipermute], page 385.

ipermute (`A`, `iperm`)   [Built-in Function]
>   The inverse of the `permute` function. The expression
>
>   `ipermute (permute (A, perm), perm)`
>
>   returns the original array *A*.
>
>   **See also:** [permute], page 385.

reshape (`A`, `m`, `n`, ...)   [Built-in Function]
reshape (`A`, `[m n ...]`)   [Built-in Function]
reshape (`A`, ..., `[]`, ...)   [Built-in Function]
reshape (`A`, `size`)   [Built-in Function]
>   Return a matrix with the specified dimensions (m, n, ...) whose elements are taken
>   from the matrix *A*. The elements of the matrix are accessed in column-major order
>   (like Fortran arrays are stored).
>
>   The following code demonstrates reshaping a 1x4 row vector into a 2x2 square matrix.
>
>   `reshape ([1, 2, 3, 4], 2, 2)`
>   ⇒   1   3
>   2   4
>
>   Note that the total number of elements in the original matrix (`prod (size (A))`)
>   must match the total number of elements in the new matrix (`prod ([m n ...])`).
>
>   A single dimension of the return matrix may be left unspecified and Octave will
>   determine its size automatically. An empty matrix ([]) is used to flag the unspecified
>   dimension.
>
>   **See also:** [resize], page 385, [vec], page 389, [postpad], page 390, [cat], page 384,
>   [squeeze], page 45.

resize (`x`, `m`)   [Built-in Function]
resize (`x`, `m`, `n`, ...)   [Built-in Function]
resize (`x`, `[m n ...]`)   [Built-in Function]
>   Resize *x* cutting off elements as necessary.
>
>   In the result, element with certain indices is equal to the corresponding element of *x*
>   if the indices are within the bounds of *x*; otherwise, the element is set to zero.
>
>   In other words, the statement

```
y = resize (x, dv)
```

is equivalent to the following code:

```
y = zeros (dv, class (x));
sz = min (dv, size (x));
for i = 1:length (sz)
  idx{i} = 1:sz(i);
endfor
y(idx{:}) = x(idx{:});
```

but is performed more efficiently.

If only m is supplied, and it is a scalar, the dimension of the result is m-by-m. If m, n, ... are all scalars, then the dimensions of the result are m-by-n-by-.... If given a vector as input, then the dimensions of the result are given by the elements of that vector.

An object can be resized to more dimensions than it has; in such case the missing dimensions are assumed to be 1. Resizing an object to fewer dimensions is not possible.

See also: [reshape], page 385, [postpad], page 390, [prepad], page 390, [cat], page 384.

y = circshift (x, n)   [Function File]

Circularly shift the values of the array x. n must be a vector of integers no longer than the number of dimensions in x. The values of n can be either positive or negative, which determines the direction in which the values or x are shifted. If an element of n is zero, then the corresponding dimension of x will not be shifted. For example:

```
x = [1, 2, 3; 4, 5, 6; 7, 8, 9];
circshift (x, 1)
⇒  7, 8, 9
   1, 2, 3
   4, 5, 6
circshift (x, -2)
⇒  7, 8, 9
   1, 2, 3
   4, 5, 6
circshift (x, [0,1])
⇒  3, 1, 2
   6, 4, 5
   9, 7, 8
```

See also: permute, ipermute, shiftdim.

shift (x, b)   [Function File]
shift (x, b, dim)   [Function File]

If x is a vector, perform a circular shift of length b of the elements of x.

If x is a matrix, do the same for each column of x. If the optional dim argument is given, operate along this dimension.

y = shiftdim (x, n)   [Function File]
[y, ns] = shiftdim (x)   [Function File]
> Shift the dimensions of x by n, where n must be an integer scalar. When n is positive, the dimensions of x are shifted to the left, with the leading dimensions circulated to the end. If n is negative, then the dimensions of x are shifted to the right, with n leading singleton dimensions added.
>
> Called with a single argument, shiftdim, removes the leading singleton dimensions, returning the number of dimensions removed in the second output argument ns.
>
> For example:
>
> ```
> x = ones (1, 2, 3);
> size (shiftdim (x, -1))
>     ⇒ [1, 1, 2, 3]
> size (shiftdim (x, 1))
>     ⇒ [2, 3]
> [b, ns] = shiftdim (x)
>     ⇒ b = [1, 1, 1; 1, 1, 1]
>     ⇒ ns = 1
> ```
>
> **See also:** reshape, permute, ipermute, circshift, squeeze.

[s, i] = sort (x)   [Built-in Function]
[s, i] = sort (x, dim)   [Built-in Function]
[s, i] = sort (x, $mode$)   [Built-in Function]
[s, i] = sort (x, dim, $mode$)   [Built-in Function]
> Return a copy of x with the elements arranged in increasing order. For matrices, sort orders the elements within columns
>
> For example:
>
> ```
> sort ([1, 2; 2, 3; 3, 1])
>     ⇒   1  1
>         2  2
>         3  3
> ```
>
> If the optional argument dim is given, then the matrix is sorted along the dimension defined by dim. The optional argument mode defines the order in which the values will be sorted. Valid values of mode are "ascend" or "descend".
>
> The sort function may also be used to produce a matrix containing the original row indices of the elements in the sorted matrix. For example:
>
> ```
> [s, i] = sort ([1, 2; 2, 3; 3, 1])
>     ⇒ s = 1  1
>           2  2
>           3  3
>     ⇒ i = 1  3
>           2  1
>           3  2
> ```
>
> For equal elements, the indices are such that equal elements are listed in the order in which they appeared in the original list.

Sorting of complex entries is done first by magnitude (abs (z)) and for any ties by phase angle (angle (z)). For example:

```
sort ([1+i; 1; 1-i])
    ⇒ 1 + 0i
      1 - 1i
      1 + 1i
```

NaN values are treated as being greater than any other value and are sorted to the end of the list.

The sort function may also be used to sort strings and cell arrays of strings, in which case ASCII dictionary order (uppercase 'A' precedes lowercase 'a') of the strings is used.

The algorithm used in sort is optimized for the sorting of partially ordered lists.

See also: [sortrows], page 388, [issorted], page 388.

[s, i] = sortrows (A)   [Function File]
[s, i] = sortrows (A, c)   [Function File]
    Sort the rows of the matrix A according to the order of the columns specified in c. If c is omitted, a lexicographical sort is used. By default ascending order is used however if elements of c are negative then the corresponding column is sorted in descending order.

    See also: [sort], page 387.

issorted (a)   [Built-in Function]
issorted (a, mode)   [Built-in Function]
issorted (a, "rows", mode)   [Built-in Function]
    Return true if the array is sorted according to mode, which may be either "ascending", "descending", or "either". By default, mode is "ascending". NaNs are treated in the same manner as sort.

    If the optional argument "rows" is supplied, check whether the array is sorted by rows as output by the function sortrows (with no options).

    This function does not support sparse matrices.

    See also: [sort], page 387, [sortrows], page 388.

nth_element (x, n)   [Built-in Function]
nth_element (x, n, dim)   [Built-in Function]
    Select the n-th smallest element of a vector, using the ordering defined by sort. In other words, the result is equivalent to sort(x)(n). n can also be a contiguous range, either ascending 1:u or descending u:-1:1, in which case a range of elements is returned. If x is an array, nth_element operates along the dimension defined by dim, or the first non-singleton dimension if dim is not given.

    nth_element encapsulates the C++ standard library algorithms nth_element and partial_sort. On average, the complexity of the operation is $O(M*\log(K))$, where M = size (x, dim) and K = length (n). This function is intended for cases where the ratio K/M is small; otherwise, it may be better to use sort.

    See also: [sort], page 387, [min], page 416, [max], page 416.

`tril (A)`	[Function File]
`tril (A, k)`	[Function File]
`tril (A, k, pack)`	[Function File]
`triu (A)`	[Function File]
`triu (A, k)`	[Function File]
`triu (A, k, pack)`	[Function File]

Return a new matrix formed by extracting the lower (`tril`) or upper (`triu`) triangular part of the matrix A, and setting all other elements to zero. The second argument is optional, and specifies how many diagonals above or below the main diagonal should also be set to zero.

The default value of k is zero, so that `triu` and `tril` normally include the main diagonal as part of the result.

If the value of k is nonzero integer, the selection of elements starts at an offset of k diagonals above or below the main diagonal; above for positive k and below for negative k.

The absolute value of k must not be greater than the number of sub-diagonals or super-diagonals.

For example:

```
tril (ones (3), -1)
    ⇒  0  0  0
       1  0  0
       1  1  0
```

and

```
tril (ones (3), 1)
    ⇒  1  1  0
       1  1  1
       1  1  1
```

If the option `"pack"` is given as third argument, the extracted elements are not inserted into a matrix, but rather stacked column-wise one above other.

See also: [diag], page 390.

`v = vec (x)`	[Built-in Function]
`v = vec (x, dim)`	[Built-in Function]

Return the vector obtained by stacking the columns of the matrix x one above the other. Without dim this is equivalent to `x(:)`. If dim is supplied, the dimensions of v are set to dim with all elements along the last dimension. This is equivalent to `shiftdim (x(:), 1-dim)`.

See also: [vech], page 389, [resize], page 385, [cat], page 384.

`vech (x)`	[Function File]

Return the vector obtained by eliminating all supradiagonal elements of the square matrix x and stacking the result one column above the other. This has uses in matrix calculus where the underlying matrix is symmetric and it would be pointless to keep values above the main diagonal.

See also: [vec], page 389.

prepad (*x, l*)   [Function File]
prepad (*x, l, c*)   [Function File]
prepad (*x, l, c, dim*)   [Function File]

> Prepend the scalar value *c* to the vector *x* until it is of length *l*. If *c* is not given, a value of 0 is used.
>
> If length (*x*) > *l*, elements from the beginning of *x* are removed until a vector of length *l* is obtained.
>
> If *x* is a matrix, elements are prepended or removed from each row.
>
> If the optional argument *dim* is given, operate along this dimension.
>
> **See also:** [postpad], page 390, [cat], page 384, [resize], page 385.

postpad (*x, l*)   [Function File]
postpad (*x, l, c*)   [Function File]
postpad (*x, l, c, dim*)   [Function File]

> Append the scalar value *c* to the vector *x* until it is of length *l*. If *c* is not given, a value of 0 is used.
>
> If length (*x*) > *l*, elements from the end of *x* are removed until a vector of length *l* is obtained.
>
> If *x* is a matrix, elements are appended or removed from each row.
>
> If the optional argument *dim* is given, operate along this dimension.
>
> **See also:** [prepad], page 390, [cat], page 384, [resize], page 385.

M = diag (*v*)   [Built-in Function]
M = diag (*v, k*)   [Built-in Function]
M = diag (*v, m, n*)   [Built-in Function]
v = diag (*M*)   [Built-in Function]
v = diag (*M, k*)   [Built-in Function]

> Return a diagonal matrix with vector *v* on diagonal *k*. The second argument is optional. If it is positive, the vector is placed on the *k*-th super-diagonal. If it is negative, it is placed on the -*k*-th sub-diagonal. The default value of *k* is 0, and the vector is placed on the main diagonal. For example:

```
diag ([1, 2, 3], 1)
  ⇒   0   1   0   0
      0   0   2   0
      0   0   0   3
      0   0   0   0
```

> The 3-input form returns a diagonal matrix with vector *v* on the main diagonal and the resulting matrix being of size *m* rows x *n* columns.
>
> Given a matrix argument, instead of a vector, **diag** extracts the *k*-th diagonal of the matrix.

blkdiag (*A, B, C, . . .*)   [Function File]

> Build a block diagonal matrix from *A, B, C, . . .* All the arguments must be numeric and are two-dimensional matrices or scalars. If any argument is of type sparse, the output will also be sparse.
>
> **See also:** [diag], page 390, [horzcat], page 384, [vertcat], page 385, [sparse], page 492.

16.3 Special Utility Matrices

eye (*n*)	[Built-in Function]
eye (*m*, *n*)	[Built-in Function]
eye ([*m n*])	[Built-in Function]
eye (..., *class*)	[Built-in Function]

Return an identity matrix. If invoked with a single scalar argument *n*, return a square NxN identity matrix. If supplied two scalar arguments (*m*, *n*), **eye** takes them to be the number of rows and columns. If given a vector with two elements, **eye** uses the values of the elements as the number of rows and columns, respectively. For example:

```
eye (3)
 ⇒  1  0  0
     0  1  0
     0  0  1
```

The following expressions all produce the same result:

```
eye (2)
≡
eye (2, 2)
≡
eye (size ([1, 2; 3, 4])
```

The optional argument *class*, allows **eye** to return an array of the specified type, like

```
val = zeros (n,m, "uint8")
```

Calling **eye** with no arguments is equivalent to calling it with an argument of 1. Any negative dimensions are treated as zero. These odd definitions are for compatibility with MATLAB.

See also: [speye], page 489, [ones], page 391, [zeros], page 392.

ones (*n*)	[Built-in Function]
ones (*m*, *n*)	[Built-in Function]
ones (*m*, *n*, *k*, ...)	[Built-in Function]
ones ([*m n* ...])	[Built-in Function]
ones (..., *class*)	[Built-in Function]

Return a matrix or N-dimensional array whose elements are all 1. If invoked with a single scalar integer argument *n*, return a square NxN matrix. If invoked with two or more scalar integer arguments, or a vector of integer values, return an array with the given dimensions.

If you need to create a matrix whose values are all the same, you should use an expression like

```
val_matrix = val * ones (m, n)
```

The optional argument *class* specifies the class of the return array and defaults to double. For example:

```
val = ones (m,n, "uint8")
```

See also: [zeros], page 392.

zeros (*n*)   [Built-in Function]
zeros (*m*, *n*)   [Built-in Function]
zeros (*m*, *n*, *k*, ...)   [Built-in Function]
zeros ([*m n* ...])   [Built-in Function]
zeros (..., *class*)   [Built-in Function]
> Return a matrix or N-dimensional array whose elements are all 0. If invoked with a
> single scalar integer argument, return a square NxN matrix. If invoked with two or
> more scalar integer arguments, or a vector of integer values, return an array with the
> given dimensions.
>
> The optional argument *class* specifies the class of the return array and defaults to
> double. For example:
>
>   val = zeros (m,n, "uint8")
>
> **See also:** [ones], page 391.

repmat (*A*, *m*)   [Function File]
repmat (*A*, *m*, *n*)   [Function File]
repmat (*A*, [*m n*])   [Function File]
repmat (*A*, [*m n p* ...])   [Function File]
> Form a block matrix of size *m* by *n*, with a copy of matrix *A* as each element. If
> *n* is not specified, form an *m* by *m* block matrix. For copying along more than two
> dimensions, specify the number of times to copy across each dimension *m*, *n*, *p*, ...,
> in a vector in the second argument.
>
> **See also:** [repelems], page 392.

repelems (*x*, *r*)   [Built-in Function]
> Construct a vector of repeated elements from *x*. *r* is a 2xN integer matrix specifying
> which elements to repeat and how often to repeat each element.
>
> Entries in the first row, $r(1,j)$, select an element to repeat. The corresponding entry
> in the second row, $r(2,j)$, specifies the repeat count. If *x* is a matrix then the columns
> of *x* are imagined to be stacked on top of each other for purposes of the selection
> index. A row vector is always returned.
>
> Conceptually the result is calculated as follows:
>
> ```
> y = [];
> for i = 1:columns (r)
>   y = [y, x(r(1,i)*ones(1, r(2,i)))];
> endfor
> ```
>
> **See also:** [repmat], page 392, [cat], page 384.

The functions `linspace` and `logspace` make it very easy to create vectors with evenly
or logarithmically spaced elements. See Section 4.2 [Ranges], page 52.

linspace (*base*, *limit*)   [Built-in Function]
linspace (*base*, *limit*, *n*)   [Built-in Function]
> Return a row vector with *n* linearly spaced elements between *base* and *limit*. If
> the number of elements is greater than one, then the endpoints *base* and *limit* are

always included in the range. If *base* is greater than *limit*, the elements are stored in decreasing order. If the number of points is not specified, a value of 100 is used.

The `linspace` function always returns a row vector if both *base* and *limit* are scalars. If one, or both, of them are column vectors, `linspace` returns a matrix.

For compatibility with MATLAB, return the second argument (*limit*) if fewer than two values are requested.

See also: [logspace], page 393.

logspace (*a*, *b*)   [Function File]
logspace (*a*, *b*, *n*)   [Function File]
logspace (*a*, *pi*, *n*)   [Function File]
Return a row vector with *n* elements logarithmically spaced from 10^a to 10^b. If *n* is unspecified it defaults to 50.

If *b* is equal to π, the points are between 10^a and π, *not* 10^a and 10^π, in order to be compatible with the corresponding MATLAB function.

Also for compatibility with MATLAB, return the second argument *b* if fewer than two values are requested.

See also: [linspace], page 392.

rand (*n*)   [Built-in Function]
rand (*m*, *n*, ...)   [Built-in Function]
rand ([*m n* ...])   [Built-in Function]
v = rand ("*state*")   [Built-in Function]
rand ("*state*", *v*)   [Built-in Function]
rand ("*state*", "*reset*")   [Built-in Function]
v = rand ("*seed*")   [Built-in Function]
rand ("*seed*", *v*)   [Built-in Function]
rand ("*seed*", "*reset*")   [Built-in Function]
rand (..., "*single*")   [Built-in Function]
rand (..., "*double*")   [Built-in Function]
Return a matrix with random elements uniformly distributed on the interval (0, 1). The arguments are handled the same as the arguments for `eye`.

You can query the state of the random number generator using the form

        v = rand ("state")

This returns a column vector *v* of length 625. Later, you can restore the random number generator to the state *v* using the form

        rand ("state", v)

You may also initialize the state vector from an arbitrary vector of length ≤ 625 for *v*. This new state will be a hash based on the value of *v*, not *v* itself.

By default, the generator is initialized from `/dev/urandom` if it is available, otherwise from CPU time, wall clock time, and the current fraction of a second. Note that this differs from MATLAB, which always initializes the state to the same state at startup. To obtain behavior comparable to MATLAB, initialize with a deterministic state vector in Octave's startup files (see Section 2.1.2 [Startup Files], page 19).

To compute the pseudo-random sequence, **rand** uses the Mersenne Twister with a period of $2^{19937} - 1$ (See M. Matsumoto and T. Nishimura, *Mersenne Twister: A 623-dimensionally equidistributed uniform pseudorandom number generator*, ACM Trans. on Modeling and Computer Simulation Vol. 8, No. 1, pp. 3-30, January 1998, http://www.math.sci.hiroshima-u.ac.jp/~m-mat/MT/emt.html). Do **not** use for cryptography without securely hashing several returned values together, otherwise the generator state can be learned after reading 624 consecutive values.

Older versions of Octave used a different random number generator. The new generator is used by default as it is significantly faster than the old generator, and produces random numbers with a significantly longer cycle time. However, in some circumstances it might be desirable to obtain the same random sequences as used by the old generators. To do this the keyword **"seed"** is used to specify that the old generators should be use, as in

```
rand ("seed", val)
```

which sets the seed of the generator to *val*. The seed of the generator can be queried with

```
s = rand ("seed")
```

However, it should be noted that querying the seed will not cause **rand** to use the old generators, only setting the seed will. To cause **rand** to once again use the new generators, the keyword **"state"** should be used to reset the state of the **rand**.

The state or seed of the generator can be reset to a new random value using the **"reset"** keyword.

The class of the value returned can be controlled by a trailing **"double"** or **"single"** argument. These are the only valid classes.

See also: [randn], page 395, [rande], page 395, [randg], page 396, [randp], page 396.

---

randi (*imax*)   [Function File]
randi (*imax*, *n*)   [Function File]
randi (*imax*, *m*, *n*, ...)   [Function File]
randi ([*imin imax*], ...)   [Function File]
randi (..., "*class*")   [Function File]

Return random integers in the range 1:*imax*.

Additional arguments determine the shape of the return matrix. When no arguments are specified a single random integer is returned. If one argument *n* is specified then a square matrix (*n* x *n*) is returned. Two or more arguments will return a multi-dimensional matrix (*m* x *n* x ...).

The integer range may optionally be described by a two element matrix with a lower and upper bound in which case the returned integers will be on the interval [*imin*, *imax*].

The optional argument *class* will return a matrix of the requested type. The default is "double".

The following example returns 150 integers in the range 1-10.

```
ri = randi (10, 150, 1)
```

Implementation Note: `randi` relies internally on `rand` which uses class `"double"` to represent numbers. This limits the maximum integer (*imax*) and range (*imax* - *imin*) to the value returned by the `bitmax` function. For IEEE floating point numbers this value is $2^{53} - 1$.

See also: [rand], page 393.

`randn (n)`	[Built-in Function]
`randn (m, n, ...)`	[Built-in Function]
`randn ([m n ...])`	[Built-in Function]
`v = randn ("state")`	[Built-in Function]
`randn ("state", v)`	[Built-in Function]
`randn ("state", "reset")`	[Built-in Function]
`v = randn ("seed")`	[Built-in Function]
`randn ("seed", v)`	[Built-in Function]
`randn ("seed", "reset")`	[Built-in Function]
`randn (..., "single")`	[Built-in Function]
`randn (..., "double")`	[Built-in Function]

Return a matrix with normally distributed random elements having zero mean and variance one. The arguments are handled the same as the arguments for `rand`.

By default, `randn` uses the Marsaglia and Tsang "Ziggurat technique" to transform from a uniform to a normal distribution.

The class of the value returned can be controlled by a trailing `"double"` or `"single"` argument. These are the only valid classes.

Reference: G. Marsaglia and W.W. Tsang, *Ziggurat Method for Generating Random Variables*, J. Statistical Software, vol 5, 2000, `http://www.jstatsoft.org/v05/i08/`)

See also: [rand], page 393, [rande], page 395, [randg], page 396, [randp], page 396.

`rande (n)`	[Built-in Function]
`rande (m, n, ...)`	[Built-in Function]
`rande ([m n ...])`	[Built-in Function]
`v = rande ("state")`	[Built-in Function]
`rande ("state", v)`	[Built-in Function]
`rande ("state", "reset")`	[Built-in Function]
`v = rande ("seed")`	[Built-in Function]
`rande ("seed", v)`	[Built-in Function]
`rande ("seed", "reset")`	[Built-in Function]
`rande (..., "single")`	[Built-in Function]
`rande (..., "double")`	[Built-in Function]

Return a matrix with exponentially distributed random elements. The arguments are handled the same as the arguments for `rand`.

By default, `randn` uses the Marsaglia and Tsang "Ziggurat technique" to transform from a uniform to an exponential distribution.

The class of the value returned can be controlled by a trailing `"double"` or `"single"` argument. These are the only valid classes.

Reference: G.   Marsaglia   and   W.W.   Tsang,   *Ziggurat   Method   for
Generating   Random   Variables,   J.   Statistical   Software,   vol 5,   2000,
http://www.jstatsoft.org/v05/i08/)

See also: [rand], page 393, [randn], page 395, [randg], page 396, [randp], page 396.

randp (*l*, *n*)   [Built-in Function]
randp (*l*, *m*, *n*, ...)   [Built-in Function]
randp (*l*, [*m* *n* ...])   [Built-in Function]
v = randp ("*state*")   [Built-in Function]
randp ("*state*", *v*)   [Built-in Function]
randp ("*state*", "*reset*")   [Built-in Function]
v = randp ("*seed*")   [Built-in Function]
randp ("*seed*", *v*)   [Built-in Function]
randp ("*seed*", "*reset*")   [Built-in Function]
randp (..., "*single*")   [Built-in Function]
randp (..., "*double*")   [Built-in Function]

Return a matrix with Poisson distributed random elements with mean value parameter given by the first argument, *l*. The arguments are handled the same as the arguments for rand, except for the argument *l*.

Five different algorithms are used depending on the range of *l* and whether or not *l* is a scalar or a matrix.

For scalar $l \leq 12$, use direct method.

> W.H. Press, et al., *Numerical Recipes in C*, Cambridge University Press, 1992.

For scalar $l > 12$, use rejection method.[1]

> W.H. Press, et al., *Numerical Recipes in C*, Cambridge University Press, 1992.

For matrix $l \leq 10$, use inversion method.[2]

> E. Stadlober, et al., WinRand source code, available via FTP.

For matrix $l > 10$, use patchwork rejection method.

> E. Stadlober, et al., WinRand source code, available via FTP, or H. Zechner, *Efficient sampling from continuous and discrete unimodal distributions*, Doctoral Dissertation, 156pp., Technical University Graz, Austria, 1994.

For $l > 1e8$, use normal approximation.

> L. Montanet, et al., *Review of Particle Properties*, Physical Review D 50 p1284, 1994.

The class of the value returned can be controlled by a trailing "**double**" or "**single**" argument. These are the only valid classes.

See also: [rand], page 393, [randn], page 395, [rande], page 395, [randg], page 396.

randg (*n*)   [Built-in Function]
randg (*m*, *n*, ...)   [Built-in Function]
randg ([*m* *n* ...])   [Built-in Function]

v = randg ("*state*")	[Built-in Function]
randg ("*state*", v)	[Built-in Function]
randg ("*state*", "*reset*")	[Built-in Function]
v = randg ("*seed*")	[Built-in Function]
randg ("*seed*", v)	[Built-in Function]
randg ("*seed*", "*reset*")	[Built-in Function]
randg (..., "*single*")	[Built-in Function]
randg (..., "*double*")	[Built-in Function]

Return a matrix with gamma (a,1) distributed random elements. The arguments are handled the same as the arguments for rand, except for the argument a.

This can be used to generate many distributions:

gamma (a, b) for a > -1, b > 0

```
            r = b * randg (a)
```

beta (a, b) for a > -1, b > -1

```
            r1 = randg (a, 1)
            r = r1 / (r1 + randg (b, 1))
```

Erlang (a, n)

```
            r = a * randg (n)
```

chisq (df) for df > 0

```
            r = 2 * randg (df / 2)
```

t (df) for 0 < df < inf (use randn if df is infinite)

```
            r = randn () / sqrt (2 * randg (df / 2) / df)
```

F (n1, n2) for 0 < n1, 0 < n2

```
            ## r1 equals 1 if n1 is infinite
            r1 = 2 * randg (n1 / 2) / n1
            ## r2 equals 1 if n2 is infinite
            r2 = 2 * randg (n2 / 2) / n2
            r = r1 / r2
```

negative binomial (n, p) for n > 0, 0 < p <= 1

```
            r = randp ((1 - p) / p * randg (n))
```

non-central chisq (df, L), for df >= 0 and L > 0
    (use chisq if L = 0)

```
            r = randp (L / 2)
            r(r > 0) = 2 * randg (r(r > 0))
            r(df > 0) += 2 * randg (df(df > 0)/2)
```

Dirichlet (a1, ... ak)

```
            r = (randg (a1), ..., randg (ak))
            r = r / sum (r)
```

The class of the value returned can be controlled by a trailing "double" or "single" argument. These are the only valid classes.

See also: [rand], page 393, [randn], page 395, [rande], page 395, [randp], page 396.

The generators operate in the new or old style together, it is not possible to mix the two. Initializing any generator with "state" or "seed" causes the others to switch to the same style for future calls.

The state of each generator is independent and calls to different generators can be interleaved without affecting the final result. For example,

```
rand ("state", [11, 22, 33]);
randn ("state", [44, 55, 66]);
u = rand (100, 1);
n = randn (100, 1);
```

and

```
rand ("state", [11, 22, 33]);
randn ("state", [44, 55, 66]);
u = zeros (100, 1);
n = zeros (100, 1);
for i = 1:100
  u(i) = rand ();
  n(i) = randn ();
end
```

produce equivalent results. When the generators are initialized in the old style with "seed" only rand and randn are independent, because the old rande, randg and randp generators make calls to rand and randn.

The generators are initialized with random states at start-up, so that the sequences of random numbers are not the same each time you run Octave.[1] If you really do need to reproduce a sequence of numbers exactly, you can set the state or seed to a specific value.

If invoked without arguments, rand and randn return a single element of a random sequence.

The original rand and randn functions use Fortran code from RANLIB, a library of Fortran routines for random number generation, compiled by Barry W. Brown and James Lovato of the Department of Biomathematics at The University of Texas, M.D. Anderson Cancer Center, Houston, TX 77030.

randperm (n)   [Built-in Function]
randperm (n, m)   [Built-in Function]
Return a row vector containing a random permutation of 1:n. If m is supplied, return m unique entries, sampled without replacement from 1:n. The complexity is $O(n)$ in memory and $O(m)$ in time, unless $m < n/5$, in which case $O(m)$ memory is used as well. The randomization is performed using rand(). All permutations are equally likely.

See also: [perms], page 568.

---

[1] The old versions of rand and randn obtain their initial seeds from the system clock.

16.4 Famous Matrices

The following functions return famous matrix forms.

gallery (*name*)   [Function File]
gallery (*name*, *args*)   [Function File]
  Create interesting matrices for testing.

c = gallery ("*cauchy*", *x*)   [Function File]
c = gallery ("*cauchy*", *x*, *y*)   [Function File]
  Create a Cauchy matrix.

c = gallery ("*chebspec*", *n*)   [Function File]
c = gallery ("*chebspec*", *n*, *k*)   [Function File]
  Create a Chebyshev spectral differentiation matrix.

c = gallery ("*chebvand*", *p*)   [Function File]
c = gallery ("*chebvand*", *m*, *p*)   [Function File]
  Create a Vandermonde-like matrix for the Chebyshev polynomials.

a = gallery ("*chow*", *n*)   [Function File]
a = gallery ("*chow*", *n*, *alpha*)   [Function File]
a = gallery ("*chow*", *n*, *alpha*, *delta*)   [Function File]
  Create a Chow matrix – a singular Toeplitz lower Hessenberg matrix.

c = gallery ("*circul*", *v*)   [Function File]
  Create a circulant matrix.

a = gallery ("*clement*", *n*)   [Function File]
a = gallery ("*clement*", *n*, *k*)   [Function File]
  Create a tridiagonal matrix with zero diagonal entries.

c = gallery ("*compar*", *a*)   [Function File]
c = gallery ("*compar*", *a*, *k*)   [Function File]
  Create a comparison matrix.

a = gallery ("*condex*", *n*)   [Function File]
a = gallery ("*condex*", *n*, *k*)   [Function File]
a = gallery ("*condex*", *n*, *k*, *theta*)   [Function File]
  Create a 'counterexample' matrix to a condition estimator.

a = gallery ("*cycol*", [*m n*])   [Function File]
a = gallery ("*cycol*", *n*)   [Function File]
a = gallery (..., *k*)   [Function File]
  Create a matrix whose columns repeat cyclically.

[c, d, e] = gallery ("*dorr*", *n*)   [Function File]
[c, d, e] = gallery ("*dorr*", *n*, *theta*)   [Function File]
a = gallery ("*dorr*", ...)   [Function File]
  Create a diagonally dominant, ill conditioned, tridiagonal matrix.

`a = gallery ("dramadah", n)`   [Function File]
`a = gallery ("dramadah", n, k)`   [Function File]
    Create a (0, 1) matrix whose inverse has large integer entries.

`a = gallery ("fiedler", c)`   [Function File]
    Create a symmetric Fiedler matrix.

`a = gallery ("forsythe", n)`   [Function File]
`a = gallery ("forsythe", n, alpha)`   [Function File]
`a = gallery ("forsythe", n, alpha, lambda)`   [Function File]
    Create a Forsythe matrix (a perturbed Jordan block).

`f = gallery ("frank", n)`   [Function File]
`f = gallery ("frank", n, k)`   [Function File]
    Create a Frank matrix (ill conditioned eigenvalues).

`c = gallery ("gcdmat", n)`   [Function File]
    Create a greatest common divisor matrix.

    c is an n-by-n matrix whose values correspond to the greatest common divisor of its coordinate values, i.e., $c(i,j)$ correspond `gcd (i, j)`.

`a = gallery ("gearmat", n)`   [Function File]
`a = gallery ("gearmat", n, i)`   [Function File]
`a = gallery ("gearmat", n, i, j)`   [Function File]
    Create a Gear matrix.

`g = gallery ("grcar", n)`   [Function File]
`g = gallery ("grcar", n, k)`   [Function File]
    Create a Toeplitz matrix with sensitive eigenvalues.

`a = gallery ("hanowa", n)`   [Function File]
`a = gallery ("hanowa", n, d)`   [Function File]
    Create a matrix whose eigenvalues lie on a vertical line in the complex plane.

`v = gallery ("house", x)`   [Function File]
`[v, beta] = gallery ("house", x)`   [Function File]
    Create a householder matrix.

`a = gallery ("integerdata", imax, [M N ...], j)`   [Function File]
`a = gallery ("integerdata", imax, M, N, ..., j)`   [Function File]
`a = gallery ("integerdata", [imin, imax], [M N ...], j)`   [Function File]
`a = gallery ("integerdata", [imin, imax], M, N, ..., j)`   [Function File]
`a = gallery ("integerdata", ..., "class")`   [Function File]
    Create a matrix with random integers in the range [1, $imax$]. If $imin$ is given then the integers are in the range [$imin$, $imax$].

    The second input is a matrix of dimensions describing the size of the output. The dimensions can also be input as comma-separated arguments.

    The input j is an integer index in the range [0, 2^32-1]. The values of the output matrix are always exactly the same (reproducibility) for a given size input and j index.

The final optional argument determines the class of the resulting matrix. Possible values for *class*: "uint8", "uint16", "uint32", "int8", "int16", int32", "single", "double". The default is "double".

a = gallery ("*invhess*", *x*)   [Function File]
a = gallery ("*invhess*", *x*, *y*)   [Function File]
    Create the inverse of an upper Hessenberg matrix.

a = gallery ("*invol*", *n*)   [Function File]
    Create an involutory matrix.

a = gallery ("*ipjfact*", *n*)   [Function File]
a = gallery ("*ipjfact*", *n*, *k*)   [Function File]
    Create an Hankel matrix with factorial elements.

a = gallery ("*jordbloc*", *n*)   [Function File]
a = gallery ("*jordbloc*", *n*, *lambda*)   [Function File]
    Create a Jordan block.

u = gallery ("*kahan*", *n*)   [Function File]
u = gallery ("*kahan*", *n*, *theta*)   [Function File]
u = gallery ("*kahan*", *n*, *theta*, *pert*)   [Function File]
    Create a Kahan matrix (upper trapezoidal).

a = gallery ("*kms*", *n*)   [Function File]
a = gallery ("*kms*", *n*, *rho*)   [Function File]
    Create a Kac-Murdock-Szego Toeplitz matrix.

b = gallery ("*krylov*", *a*)   [Function File]
b = gallery ("*krylov*", *a*, *x*)   [Function File]
b = gallery ("*krylov*", *a*, *x*, *j*)   [Function File]
    Create a Krylov matrix.

a = gallery ("*lauchli*", *n*)   [Function File]
a = gallery ("*lauchli*", *n*, *mu*)   [Function File]
    Create a Lauchli matrix (rectangular).

a = gallery ("*lehmer*", *n*)   [Function File]
    Create a Lehmer matrix (symmetric positive definite).

t = gallery ("*lesp*", *n*)   [Function File]
    Create a tridiagonal matrix with real, sensitive eigenvalues.

a = gallery ("*lotkin*", *n*)   [Function File]
    Create a Lotkin matrix.

a = gallery ("*minij*", *n*)   [Function File]
    Create a symmetric positive definite matrix MIN(i,j).

a = gallery ("*moler*", *n*)   [Function File]
a = gallery ("*moler*", *n*, *alpha*)   [Function File]
    Create a Moler matrix (symmetric positive definite).

[a, t] = gallery ("*neumann*", *n*)   [Function File]
> Create a singular matrix from the discrete Neumann problem (sparse).

a = gallery ("*normaldata*", [*M N* ...], *j*)   [Function File]
a = gallery ("*normaldata*", *M, N, ..., j*)   [Function File]
a = gallery ("*normaldata*", ..., "class")   [Function File]
> Create a matrix with random samples from the standard normal distribution (mean = 0, std = 1).
>
> The first input is a matrix of dimensions describing the size of the output. The dimensions can also be input as comma-separated arguments.
>
> The input *j* is an integer index in the range [0, 2^32-1]. The values of the output matrix are always exactly the same (reproducibility) for a given size input and *j* index.
>
> The final optional argument determines the class of the resulting matrix. Possible values for *class*: "single", "double". The default is "double".

q = gallery ("*orthog*", *n*)   [Function File]
q = gallery ("*orthog*", *n, k*)   [Function File]
> Create orthogonal and nearly orthogonal matrices.

a = gallery ("*parter*", *n*)   [Function File]
> Create a Parter matrix (a Toeplitz matrix with singular values near pi).

p = gallery ("*pei*", *n*)   [Function File]
p = gallery ("*pei*", *n, alpha*)   [Function File]
> Create a Pei matrix.

a = gallery ("*Poisson*", *n*)   [Function File]
> Create a block tridiagonal matrix from Poisson's equation (sparse).

a = gallery ("*prolate*", *n*)   [Function File]
a = gallery ("*prolate*", *n, w*)   [Function File]
> Create a prolate matrix (symmetric, ill-conditioned Toeplitz matrix).

h = gallery ("*randhess*", *x*)   [Function File]
> Create a random, orthogonal upper Hessenberg matrix.

a = gallery ("*rando*", *n*)   [Function File]
a = gallery ("*rando*", *n, k*)   [Function File]
> Create a random matrix with elements -1, 0 or 1.

a = gallery ("*randsvd*", *n*)   [Function File]
a = gallery ("*randsvd*", *n, kappa*)   [Function File]
a = gallery ("*randsvd*", *n, kappa, mode*)   [Function File]
a = gallery ("*randsvd*", *n, kappa, mode, kl*)   [Function File]
a = gallery ("*randsvd*", *n, kappa, mode, kl, ku*)   [Function File]
> Create a random matrix with pre-assigned singular values.

a = gallery ("*redheff*", *n*)   [Function File]
> Create a zero and ones matrix of Redheffer associated with the Riemann hypothesis.

a = gallery ("*riemann*", *n*)   [Function File]
    Create a matrix associated with the Riemann hypothesis.

a = gallery ("*ris*", *n*)   [Function File]
    Create a symmetric Hankel matrix.

a = gallery ("*smoke*", *n*)   [Function File]
a = gallery ("*smoke*", *n*, *k*)   [Function File]
    Create a complex matrix, with a 'smoke ring' pseudospectrum.

t = gallery ("*toeppd*", *n*)   [Function File]
t = gallery ("*toeppd*", *n*, *m*)   [Function File]
t = gallery ("*toeppd*", *n*, *m*, *w*)   [Function File]
t = gallery ("*toeppd*", *n*, *m*, *w*, theta)   [Function File]
    Create a symmetric positive definite Toeplitz matrix.

p = gallery ("*toeppen*", *n*)   [Function File]
p = gallery ("*toeppen*", *n*, a)   [Function File]
p = gallery ("*toeppen*", *n*, a, b)   [Function File]
p = gallery ("*toeppen*", *n*, a, b, c)   [Function File]
p = gallery ("*toeppen*", *n*, a, b, c, d)   [Function File]
p = gallery ("*toeppen*", *n*, a, b, c, d, e)   [Function File]
    Create a pentadiagonal Toeplitz matrix (sparse).

a = gallery ("*tridiag*", x, y, z)   [Function File]
a = gallery ("*tridiag*", *n*)   [Function File]
a = gallery ("*tridiag*", *n*, c, d, e)   [Function File]
    Create a tridiagonal matrix (sparse).

t = gallery ("*triw*", *n*)   [Function File]
t = gallery ("*triw*", *n*, alpha)   [Function File]
t = gallery ("*triw*", *n*, alpha, k)   [Function File]
    Create an upper triangular matrix discussed by Kahan, Golub and Wilkinson.

a = gallery ("*uniformdata*", [*M N* ...], *j*)   [Function File]
a = gallery ("*uniformdata*", *M*, *N*, ..., *j*)   [Function File]
a = gallery ("*uniformdata*", ..., "class")   [Function File]
    Create a matrix with random samples from the standard uniform distribution (range
    [0,1]).

    The first input is a matrix of dimensions describing the size of the output. The
    dimensions can also be input as comma-separated arguments.

    The input j is an integer index in the range [0, 2^32-1]. The values of the output
    matrix are always exactly the same (reproducibility) for a given size input and j index.

    The final optional argument determines the class of the resulting matrix. Possible
    values for *class*: "single", "double". The default is "double".

a = gallery ("*wathen*", *nx*, *ny*)   [Function File]
a = gallery ("*wathen*", *nx*, *ny*, *k*)   [Function File]
    Create the Wathen matrix.

`[a, b] = gallery ("`*wilk*`", n)`   [Function File]
   Create various specific matrices devised/discussed by Wilkinson.

`hadamard (n)`   [Function File]
   Construct a Hadamard matrix (Hn) of size n-by-n. The size n must be of the form
   $2^k * p$ in which p is one of 1, 12, 20 or 28. The returned matrix is normalized, meaning
   `Hn(:,1) == 1` and `Hn(1,:) == 1`.

   Some of the properties of Hadamard matrices are:

   - `kron (Hm, Hn)` is a Hadamard matrix of size m-by-n.

   - `Hn * Hn' = n * eye (n)`.

   - The rows of Hn are orthogonal.

   - `det (A) <= abs (det (Hn))` for all A with `abs (A(i, j)) <= 1`.

   - Multiplying any row or column by -1 and the matrix will remain a Hadamard
     matrix.

   See also: [compan], page 602, [hankel], page 404, [toeplitz], page 405.

`hankel (c)`   [Function File]
`hankel (c, r)`   [Function File]
   Return the Hankel matrix constructed from the first column c, and (optionally) the
   last row r. If the last element of c is not the same as the first element of r, the last
   element of c is used. If the second argument is omitted, it is assumed to be a vector
   of zeros with the same size as c.

   A Hankel matrix formed from an m-vector c, and an n-vector r, has the elements

$$H(i,j) = \begin{cases} c_{i+j-1}, & i+j-1 \le m; \\ r_{i+j-m}, & \text{otherwise.} \end{cases}$$

   See also: [hadamard], page 404, [toeplitz], page 405.

`hilb (n)`   [Function File]
   Return the Hilbert matrix of order n. The i, j element of a Hilbert matrix is defined
   as

$$H(i,j) = \frac{1}{(i+j-1)}$$

   Hilbert matrices are close to being singular which make them difficult to invert with
   numerical routines. Comparing the condition number of a random matrix 5x5 matrix
   with that of a Hilbert matrix of order 5 reveals just how difficult the problem is.

```
       cond (rand (5))
          ⇒ 14.392
       cond (hilb (5))
          ⇒ 4.7661e+05
```

   See also: [invhilb], page 405.

`invhilb` (n)   [Function File]

Return the inverse of the Hilbert matrix of order n. This can be computed exactly using

$$A_{ij} = -1^{i+j}(i+j-1) \binom{n+i-1}{n-j} \binom{n+j-1}{n-i} \binom{i+j-2}{i-2}^2$$

$$= \frac{p(i)p(j)}{(i+j-1)}$$

where

$$p(k) = -1^k \binom{k+n-1}{k-1} \binom{n}{k}$$

The validity of this formula can easily be checked by expanding the binomial coefficients in both formulas as factorials. It can be derived more directly via the theory of Cauchy matrices. See J. W. Demmel, *Applied Numerical Linear Algebra*, p. 92.

Compare this with the numerical calculation of `inverse (hilb (n))`, which suffers from the ill-conditioning of the Hilbert matrix, and the finite precision of your computer's floating point arithmetic.

See also: [hilb], page 404.

`magic` (n)   [Function File]

Create an n-by-n magic square. A magic square is an arrangement of the integers `1:n^2` such that the row sums, column sums, and diagonal sums are all equal to the same value.

Note: n must be greater than 2 for the magic square to exist.

`pascal` (n)   [Function File]
`pascal` (n, t)   [Function File]

Return the Pascal matrix of order n if t = 0. t defaults to 0. Return the pseudo-lower triangular Cholesky factor of the Pascal matrix if t = 1 (The sign of some columns may be negative). This matrix is its own inverse, that is `pascal (n, 1) ^ 2 == eye (n)`. If t = -1, return the true Cholesky factor with strictly positive values on the diagonal. If t = 2, return a transposed and permuted version of `pascal (n, 1)`, which is the cube root of the identity matrix. That is, `pascal (n, 2) ^ 3 == eye (n)`.

See also: [chol], page 442.

`rosser` ()   [Function File]

Return the Rosser matrix. This is a difficult test case used to evaluate eigenvalue algorithms.

See also: [wilkinson], page 406, [eig], page 437.

`toeplitz` (c)   [Function File]
`toeplitz` (c, r)   [Function File]

Return the Toeplitz matrix constructed from the first column c, and (optionally) the first row r. If the first element of r is not the same as the first element of c, the first element of c is used. If the second argument is omitted, the first row is taken to be the same as the first column.

A square Toeplitz matrix has the form:

$$\begin{bmatrix} c_0 & r_1 & r_2 & \cdots & r_n \\ c_1 & c_0 & r_1 & \cdots & r_{n-1} \\ c_2 & c_1 & c_0 & \cdots & r_{n-2} \\ \vdots & \vdots & \vdots & \ddots & \vdots \\ c_n & c_{n-1} & c_{n-2} & \cdots & c_0 \end{bmatrix}$$

See also: [hankel], page 404.

vander (*c*)   [Function File]
vander (*c*, *n*)   [Function File]
  Return the Vandermonde matrix whose next to last column is *c*. If *n* is specified, it determines the number of columns; otherwise, *n* is taken to be equal to the length of *c*.

  A Vandermonde matrix has the form:

$$\begin{bmatrix} c_1^{n-1} & \cdots & c_1^2 & c_1 & 1 \\ c_2^{n-1} & \cdots & c_2^2 & c_2 & 1 \\ \vdots & \ddots & \vdots & \vdots & \vdots \\ c_n^{n-1} & \cdots & c_n^2 & c_n & 1 \end{bmatrix}$$

  See also: [polyfit], page 607.

wilkinson (*n*)   [Function File]
  Return the Wilkinson matrix of order *n*. Wilkinson matrices are symmetric and tridiagonal with pairs of nearly, but not exactly, equal eigenvalues. They are useful in testing the behavior and performance of eigenvalue solvers.

  See also: [rosser], page 405, [eig], page 437.

17 Arithmetic

Unless otherwise noted, all of the functions described in this chapter will work for real and complex scalar, vector, or matrix arguments. Functions described as *mapping functions* apply the given operation individually to each element when given a matrix argument. For example:

```
sin ([1, 2; 3, 4])
    ⇒   0.84147    0.90930
        0.14112   -0.75680
```

17.1 Exponents and Logarithms

exp (x)   [Mapping Function]

> Compute e^x for each element of x. To compute the matrix exponential, see Chapter 18 [Linear Algebra], page 435.
>
> **See also:** [log], page 407.

expm1 (x)   [Mapping Function]

> Compute $e^x - 1$ accurately in the neighborhood of zero.
>
> **See also:** [exp], page 407.

log (x)   [Mapping Function]

> Compute the natural logarithm, $\ln(x)$, for each element of x. To compute the matrix logarithm, see Chapter 18 [Linear Algebra], page 435.
>
> **See also:** [exp], page 407, [log1p], page 407, [log2], page 407, [log10], page 407, [logspace], page 393.

reallog (x)   [Function File]

> Return the real-valued natural logarithm of each element of x. Report an error if any element results in a complex return value.
>
> **See also:** [log], page 407, [realpow], page 408, [realsqrt], page 408.

log1p (x)   [Mapping Function]

> Compute $\ln(1 + x)$ accurately in the neighborhood of zero.
>
> **See also:** [log], page 407, [exp], page 407, [expm1], page 407.

log10 (x)   [Mapping Function]

> Compute the base-10 logarithm of each element of x.
>
> **See also:** [log], page 407, [log2], page 407, [logspace], page 393, [exp], page 407.

log2 (x)   [Mapping Function]
[f, e] = log2 (x)   [Mapping Function]

> Compute the base-2 logarithm of each element of x.
>
> If called with two output arguments, split x into binary mantissa and exponent so that $\frac{1}{2} \le |f| < 1$ and e is an integer. If $x = 0$, $f = e = 0$.
>
> **See also:** [pow2], page 408, [log], page 407, [log10], page 407, [exp], page 407.

pow2 (x)   [Mapping Function]
pow2 (f, e)   [Mapping Function]
>   With one argument, computes 2^x for each element of x.
>
>   With two arguments, returns $f \cdot 2^e$.
>
>   **See also:** [log2], page 407, [nextpow2], page 408.

nextpow2 (x)   [Function File]
>   If x is a scalar, return the first integer n such that $2^n \geq |x|$.
>
>   If x is a vector, return `nextpow2 (length (x))`.
>
>   **See also:** [pow2], page 408, [log2], page 407.

realpow (x, y)   [Function File]
>   Compute the real-valued, element-by-element power operator. This is equivalent to
>   `x .^ y`, except that `realpow` reports an error if any return value is complex.
>
>   **See also:** [reallog], page 407, [realsqrt], page 408.

sqrt (x)   [Mapping Function]
>   Compute the square root of each element of x. If x is negative, a complex result
>   is returned. To compute the matrix square root, see Chapter 18 [Linear Algebra],
>   page 435.
>
>   **See also:** [realsqrt], page 408, [nthroot], page 408.

realsqrt (x)   [Function File]
>   Return the real-valued square root of each element of x. Report an error if any
>   element results in a complex return value.
>
>   **See also:** [sqrt], page 408, [realpow], page 408, [reallog], page 407.

cbrt (x)   [Mapping Function]
>   Compute the real cube root of each element of x. Unlike `x^(1/3)`, the result will be
>   negative if x is negative.
>
>   **See also:** [nthroot], page 408.

nthroot (x, n)   [Function File]
>   Compute the n-th root of x, returning real results for real components of x. For
>   example:
>
>    ```
>    nthroot (-1, 3)
>    ⇒ -1
>    (-1) ^ (1 / 3)
>    ⇒ 0.50000 - 0.86603i
>    ```
>
>   x must have all real entries. n must be a scalar. If n is an even integer and X has
>   negative entries, an error is produced.
>
>   **See also:** [realsqrt], page 408, [sqrt], page 408, [cbrt], page 408.

17.2 Complex Arithmetic

In the descriptions of the following functions, z is the complex number $x + iy$, where i is defined as $\sqrt{-1}$.

abs (*z*) [Mapping Function]

> Compute the magnitude of z, defined as $|z| = \sqrt{x^2 + y^2}$.
>
> For example:
>
>   abs (3 + 4i)
>   ⇒ 5

arg (*z*) [Mapping Function]
angle (*z*) [Mapping Function]

> Compute the argument of z, defined as, $\theta = atan2(y, x)$, in radians.
>
> For example:
>
>   arg (3 + 4i)
>   ⇒ 0.92730

conj (*z*) [Mapping Function]

> Return the complex conjugate of z, defined as $\bar{z} = x - iy$.
>
> **See also:** [real], page 409, [imag], page 409.

cplxpair (*z*) [Function File]
cplxpair (*z*, *tol*) [Function File]
cplxpair (*z*, *tol*, *dim*) [Function File]

> Sort the numbers z into complex conjugate pairs ordered by increasing real part. Place the negative imaginary complex number first within each pair. Place all the real numbers (those with `abs (imag (z) / z) < tol`)) after the complex pairs.
>
> If *tol* is unspecified the default value is 100*`eps`.
>
> By default the complex pairs are sorted along the first non-singleton dimension of z. If *dim* is specified, then the complex pairs are sorted along this dimension.
>
> Signal an error if some complex numbers could not be paired. Signal an error if all complex numbers are not exact conjugates (to within *tol*). Note that there is no defined order for pairs with identical real parts but differing imaginary parts.
>
>   cplxpair (exp(2i*pi*[0:4]'/5)) == exp(2i*pi*[3; 2; 4; 1; 0]/5)

imag (*z*) [Mapping Function]

> Return the imaginary part of z as a real number.
>
> **See also:** [real], page 409, [conj], page 409.

real (*z*) [Mapping Function]

> Return the real part of z.
>
> **See also:** [imag], page 409, [conj], page 409.

17.3 Trigonometry

Octave provides the following trigonometric functions where angles are specified in radians. To convert from degrees to radians multiply by $\pi/180$ (e.g., `sin (30 * pi/180)` returns the sine of 30 degrees). As an alternative, Octave provides a number of trigonometric functions which work directly on an argument specified in degrees. These functions are named after the base trigonometric function with a 'd' suffix. For example, `sin` expects an angle in radians while `sind` expects an angle in degrees.

Octave uses the C library trigonometric functions. It is expected that these functions are defined by the ISO/IEC 9899 Standard. This Standard is available at: `http://www.open-std.org/jtc1/sc22/wg14/www/docs/n1124.pdf`. Section F.9.1 deals with the trigonometric functions. The behavior of most of the functions is relatively straightforward. However, there are some exceptions to the standard behavior. Many of the exceptions involve the behavior for -0. The most complex case is atan2. Octave exactly implements the behavior given in the Standard. Including $atan2(\pm 0, -0)$ returns $\pm\pi$.

It should be noted that MATLAB uses different definitions which apparently do not distinguish -0.

`sin (x)`   [Mapping Function]
    Compute the sine for each element of x in radians.

    See also: [asin], page 410, [sind], page 412, [sinh], page 411.

`cos (x)`   [Mapping Function]
    Compute the cosine for each element of x in radians.

    See also: [acos], page 411, [cosd], page 412, [cosh], page 411.

`tan (z)`   [Mapping Function]
    Compute the tangent for each element of x in radians.

    See also: [atan], page 411, [tand], page 412, [tanh], page 411.

`sec (x)`   [Mapping Function]
    Compute the secant for each element of x in radians.

    See also: [asec], page 411, [secd], page 413, [sech], page 411.

`csc (x)`   [Mapping Function]
    Compute the cosecant for each element of x in radians.

    See also: [acsc], page 411, [cscd], page 413, [csch], page 411.

`cot (x)`   [Mapping Function]
    Compute the cotangent for each element of x in radians.

    See also: [acot], page 411, [cotd], page 413, [coth], page 411.

`asin (x)`   [Mapping Function]
    Compute the inverse sine in radians for each element of x.

    See also: [sin], page 410, [asind], page 413.

acos (x) [Mapping Function]
> Compute the inverse cosine in radians for each element of x.
>
> **See also:** [cos], page 410, [acosd], page 413.

atan (x) [Mapping Function]
> Compute the inverse tangent in radians for each element of x.
>
> **See also:** [tan], page 410, [atand], page 413.

asec (x) [Mapping Function]
> Compute the inverse secant in radians for each element of x.
>
> **See also:** [sec], page 410, [asecd], page 413.

acsc (x) [Mapping Function]
> Compute the inverse cosecant in radians for each element of x.
>
> **See also:** [csc], page 410, [acscd], page 413.

acot (x) [Mapping Function]
> Compute the inverse cotangent in radians for each element of x.
>
> **See also:** [cot], page 410, [acotd], page 413.

sinh (x) [Mapping Function]
> Compute the hyperbolic sine for each element of x.
>
> **See also:** [asinh], page 411, [cosh], page 411, [tanh], page 411.

cosh (x) [Mapping Function]
> Compute the hyperbolic cosine for each element of x.
>
> **See also:** [acosh], page 412, [sinh], page 411, [tanh], page 411.

tanh (x) [Mapping Function]
> Compute hyperbolic tangent for each element of x.
>
> **See also:** [atanh], page 412, [sinh], page 411, [cosh], page 411.

sech (x) [Mapping Function]
> Compute the hyperbolic secant of each element of x.
>
> **See also:** [asech], page 412.

csch (x) [Mapping Function]
> Compute the hyperbolic cosecant of each element of x.
>
> **See also:** [acsch], page 412.

coth (x) [Mapping Function]
> Compute the hyperbolic cotangent of each element of x.
>
> **See also:** [acoth], page 412.

asinh (x) [Mapping Function]
> Compute the inverse hyperbolic sine for each element of x.
>
> **See also:** [sinh], page 411.

acosh (*x*)   [Mapping Function]
>   Compute the inverse hyperbolic cosine for each element of *x*.
>
>   **See also:** [cosh], page 411.

atanh (*x*)   [Mapping Function]
>   Compute the inverse hyperbolic tangent for each element of *x*.
>
>   **See also:** [tanh], page 411.

asech (*x*)   [Mapping Function]
>   Compute the inverse hyperbolic secant of each element of *x*.
>
>   **See also:** [sech], page 411.

acsch (*x*)   [Mapping Function]
>   Compute the inverse hyperbolic cosecant of each element of *x*.
>
>   **See also:** [csch], page 411.

acoth (*x*)   [Mapping Function]
>   Compute the inverse hyperbolic cotangent of each element of *x*.
>
>   **See also:** [coth], page 411.

atan2 (*y*, *x*)   [Mapping Function]
>   Compute atan (*y* / *x*) for corresponding elements of *y* and *x*. Signal an error if *y* and *x* do not match in size and orientation.
>
>   **See also:** [tan], page 410, [tand], page 412, [tanh], page 411, [atanh], page 412.

Octave provides the following trigonometric functions where angles are specified in degrees. These functions produce true zeros at the appropriate intervals rather than the small round-off error that occurs when using radians. For example:

```
cosd (90)
    ⇒ 0
cos (pi/2)
    ⇒ 6.1230e-17
```

sind (*x*)   [Function File]
>   Compute the sine for each element of *x* in degrees. Returns zero for elements where *x*/180 is an integer.
>
>   **See also:** [asind], page 413, [sin], page 410.

cosd (*x*)   [Function File]
>   Compute the cosine for each element of *x* in degrees. Returns zero for elements where (*x*-90)/180 is an integer.
>
>   **See also:** [acosd], page 413, [cos], page 410.

tand (*x*)   [Function File]
>   Compute the tangent for each element of *x* in degrees. Returns zero for elements where *x*/180 is an integer and **Inf** for elements where (*x*-90)/180 is an integer.
>
>   **See also:** [atand], page 413, [tan], page 410.

secd (*x*)   [Function File]
>   Compute the secant for each element of *x* in degrees.
>
>   **See also:** [asecd], page 413, [sec], page 410.

cscd (*x*)   [Function File]
>   Compute the cosecant for each element of *x* in degrees.
>
>   **See also:** [acscd], page 413, [csc], page 410.

cotd (*x*)   [Function File]
>   Compute the cotangent for each element of *x* in degrees.
>
>   **See also:** [acotd], page 413, [cot], page 410.

asind (*x*)   [Function File]
>   Compute the inverse sine in degrees for each element of *x*.
>
>   **See also:** [sind], page 412, [asin], page 410.

acosd (*x*)   [Function File]
>   Compute the inverse cosine in degrees for each element of *x*.
>
>   **See also:** [cosd], page 412, [acos], page 411.

atand (*x*)   [Function File]
>   Compute the inverse tangent in degrees for each element of *x*.
>
>   **See also:** [tand], page 412, [atan], page 411.

atan2d (*y*, *x*)   [Function File]
>   Compute atan2 (*y* / *x*) in degrees for corresponding elements from *y* and *x*.
>
>   **See also:** [tand], page 412, [atan2], page 412.

asecd (*x*)   [Function File]
>   Compute the inverse secant in degrees for each element of *x*.
>
>   **See also:** [secd], page 413, [asec], page 411.

acscd (*x*)   [Function File]
>   Compute the inverse cosecant in degrees for each element of *x*.
>
>   **See also:** [cscd], page 413, [acsc], page 411.

acotd (*x*)   [Function File]
>   Compute the inverse cotangent in degrees for each element of *x*.
>
>   **See also:** [cotd], page 413, [acot], page 411.

17.4 Sums and Products

sum (*x*)   [Built-in Function]
sum (*x*, *dim*)   [Built-in Function]
sum (..., "*native*")   [Built-in Function]
sum (..., "*double*")   [Built-in Function]
sum (..., "*extra*")   [Built-in Function]

   Sum of elements along dimension *dim*. If *dim* is omitted, it defaults to the first
   non-singleton dimension.

   If the optional argument "native" is given, then the sum is performed in the same
   type as the original argument, rather than in the default double type. For example:

```
sum ([true, true])
   ⇒ 2
sum ([true, true], "native")
   ⇒ true
```

   On the contrary, if "double" is given, the sum is performed in double precision even
   for single precision inputs.

   For double precision inputs, "extra" indicates that a more accurate algorithm than
   straightforward summation is to be used. For single precision inputs, "extra" is the
   same as "double". Otherwise, "extra" has no effect.

   See also: [cumsum], page 414, [sumsq], page 415, [prod], page 414.

prod (*x*)   [Built-in Function]
prod (*x*, *dim*)   [Built-in Function]

   Product of elements along dimension *dim*. If *dim* is omitted, it defaults to the first
   non-singleton dimension.

   See also: [cumprod], page 414, [sum], page 414.

cumsum (*x*)   [Built-in Function]
cumsum (*x*, *dim*)   [Built-in Function]
cumsum (..., "*native*")   [Built-in Function]
cumsum (..., "*double*")   [Built-in Function]
cumsum (..., "*extra*")   [Built-in Function]

   Cumulative sum of elements along dimension *dim*. If *dim* is omitted, it defaults to
   the first non-singleton dimension.

   See sum for an explanation of the optional parameters "native", "double", and
   "extra".

   See also: [sum], page 414, [cumprod], page 414.

cumprod (*x*)   [Built-in Function]
cumprod (*x*, *dim*)   [Built-in Function]

   Cumulative product of elements along dimension *dim*. If *dim* is omitted, it defaults
   to the first non-singleton dimension.

   See also: [prod], page 414, [cumsum], page 414.

sumsq (*x*)   [Built-in Function]
sumsq (*x*, *dim*)   [Built-in Function]
>   Sum of squares of elements along dimension *dim*. If *dim* is omitted, it defaults to the
>   first non-singleton dimension.
>
>   This function is conceptually equivalent to computing
>
>   sum (x .* conj (x), dim)
>
>   but it uses less memory and avoids calling `conj` if *x* is real.
>
>   **See also:** [sum], page 414, [prod], page 414.

17.5 Utility Functions

ceil (*x*)   [Mapping Function]
>   Return the smallest integer not less than *x*. This is equivalent to rounding towards
>   positive infinity. If *x* is complex, return ceil (real (*x*)) + ceil (imag (*x*)) * I.
>
>   ceil ([-2.7, 2.7])
>   ⇒ -2   3
>
>   **See also:** [floor], page 415, [round], page 415, [fix], page 415.

fix (*x*)   [Mapping Function]
>   Truncate fractional portion of *x* and return the integer portion. This is equivalent to
>   rounding towards zero. If *x* is complex, return fix (real (*x*)) + fix (imag (*x*)) *
>   I.
>
>   fix ([-2.7, 2.7])
>   ⇒ -2   2
>
>   **See also:** [ceil], page 415, [floor], page 415, [round], page 415.

floor (*x*)   [Mapping Function]
>   Return the largest integer not greater than *x*. This is equivalent to rounding towards
>   negative infinity. If *x* is complex, return floor (real (*x*)) + floor (imag (*x*)) *
>   I.
>
>   floor ([-2.7, 2.7])
>   ⇒ -3   2
>
>   **See also:** [ceil], page 415, [round], page 415, [fix], page 415.

round (*x*)   [Mapping Function]
>   Return the integer nearest to *x*. If *x* is complex, return round (real (*x*)) + round
>   (imag (*x*)) * I. If there are two nearest integers, return the one further away from
>   zero.
>
>   round ([-2.7, 2.7])
>   ⇒ -3   3
>
>   **See also:** [ceil], page 415, [floor], page 415, [fix], page 415, [roundb], page 415.

roundb (*x*)   [Mapping Function]
>   Return the integer nearest to *x*. If there are two nearest integers, return the even
>   one (banker's rounding). If *x* is complex, return roundb (real (*x*)) + roundb (imag
>   (*x*)) * I.

See also: [round], page 415.

max (*x*)   [Built-in Function]
max (*x*, *y*)   [Built-in Function]
max (*x*, [], *dim*)   [Built-in Function]
max (*x*, *y*, *dim*)   [Built-in Function]
[*w*, *iw*] = max (*x*)   [Built-in Function]
> For a vector argument, return the maximum value. For a matrix argument, return
> the maximum value from each column, as a row vector, or over the dimension *dim* if
> defined, in which case *y* should be set to the empty matrix (it's ignored otherwise).
> For two matrices (or a matrix and scalar), return the pair-wise maximum. Thus,
>
>   max (max (x))
>
> returns the largest element of the matrix *x*, and
>
>   max (2:5, pi)
>   ⇒   3.1416   3.1416   4.0000   5.0000
>
> compares each element of the range 2:5 with pi, and returns a row vector of the
> maximum values.
>
> For complex arguments, the magnitude of the elements are used for comparison.
>
> If called with one input and two output arguments, max also returns the first index
> of the maximum value(s). Thus,
>
>   [x, ix] = max ([1, 3, 5, 2, 5])
>   ⇒   x = 5
>   ix = 3

See also: [min], page 416, [cummax], page 417, [cummin], page 417.

min (*x*)   [Built-in Function]
min (*x*, *y*)   [Built-in Function]
min (*x*, [], *dim*)   [Built-in Function]
min (*x*, *y*, *dim*)   [Built-in Function]
[*w*, *iw*] = min (*x*)   [Built-in Function]
> For a vector argument, return the minimum value. For a matrix argument, return
> the minimum value from each column, as a row vector, or over the dimension *dim* if
> defined, in which case *y* should be set to the empty matrix (it's ignored otherwise).
> For two matrices (or a matrix and scalar), return the pair-wise minimum. Thus,
>
>   min (min (x))
>
> returns the smallest element of *x*, and
>
>   min (2:5, pi)
>   ⇒   2.0000   3.0000   3.1416   3.1416
>
> compares each element of the range 2:5 with pi, and returns a row vector of the
> minimum values.
>
> For complex arguments, the magnitude of the elements are used for comparison.
>
> If called with one input and two output arguments, min also returns the first index
> of the minimum value(s). Thus,

```
[x, ix] = min ([1, 3, 0, 2, 0])
    ⇒   x = 0
        ix = 3
```

See also: [max], page 416, [cummin], page 417, [cummax], page 417.

cummax (*x*)	[Built-in Function]
cummax (*x*, *dim*)	[Built-in Function]
[*w*, *iw*] = cummax (...)	[Built-in Function]

Return the cumulative maximum values along dimension *dim*.

If *dim* is unspecified it defaults to column-wise operation. For example:

```
cummax ([1 3 2 6 4 5])
    ⇒  1  3  3  6  6  6
```

If called with two output arguments the index of the maximum value is also returned.

```
[w, iw] = cummax ([1 3 2 6 4 5])
⇒
w  = 1  3  3  6  6  6
iw = 1  2  2  4  4  4
```

See also: [cummin], page 417, [max], page 416, [min], page 416.

cummin (*x*)	[Built-in Function]
cummin (*x*, *dim*)	[Built-in Function]
[*w*, *iw*] = cummin (*x*)	[Built-in Function]

Return the cumulative minimum values along dimension *dim*.

If *dim* is unspecified it defaults to column-wise operation. For example:

```
cummin ([5 4 6 2 3 1])
    ⇒  5  4  4  2  2  1
```

If called with two output arguments the index of the minimum value is also returned.

```
[w, iw] = cummin ([5 4 6 2 3 1])
⇒
w  = 5  4  4  2  2  1
iw = 1  2  2  4  4  6
```

See also: [cummax], page 417, [min], page 416, [max], page 416.

hypot (*x*, *y*)	[Built-in Function]
hypot (*x*, *y*, *z*, ...)	[Built-in Function]

Compute the element-by-element square root of the sum of the squares of *x* and *y*. This is equivalent to `sqrt (x.^2 + y.^2)`, but calculated in a manner that avoids overflows for large values of *x* or *y*. `hypot` can also be called with more than 2 arguments; in this case, the arguments are accumulated from left to right:

```
hypot (hypot (x, y), z)
hypot (hypot (hypot (x, y), z), w), etc.
```

dx = gradient (*m*)	[Function File]
[*dx*, *dy*, *dz*, ...] = gradient (*m*)	[Function File]
[...] = gradient (*m*, *s*)	[Function File]

[...] = gradient (*m*, *x*, *y*, *z*, ...)   [Function File]
[...] = gradient (*f*, *x0*)   [Function File]
[...] = gradient (*f*, *x0*, *s*)   [Function File]
[...] = gradient (*f*, *x0*, *x*, *y*, ...)   [Function File]
> Calculate the gradient of sampled data or a function. If *m* is a vector, calculate the one-dimensional gradient of *m*. If *m* is a matrix the gradient is calculated for each dimension.
>
> [*dx*, *dy*] = gradient (*m*) calculates the one dimensional gradient for *x* and *y* direction if *m* is a matrix. Additional return arguments can be use for multi-dimensional matrices.
>
> A constant spacing between two points can be provided by the *s* parameter. If *s* is a scalar, it is assumed to be the spacing for all dimensions. Otherwise, separate values of the spacing can be supplied by the *x*, ... arguments. Scalar values specify an equidistant spacing. Vector values for the *x*, ... arguments specify the coordinate for that dimension. The length must match their respective dimension of *m*.
>
> At boundary points a linear extrapolation is applied. Interior points are calculated with the first approximation of the numerical gradient
>
>> y'(i) = 1/(x(i+1)-x(i-1)) * (y(i-1)-y(i+1)).
>
> If the first argument *f* is a function handle, the gradient of the function at the points in *x0* is approximated using central difference. For example, gradient (@cos, 0) approximates the gradient of the cosine function in the point $x0 = 0$. As with sampled data, the spacing values between the points from which the gradient is estimated can be set via the *s* or *dx*, *dy*, ... arguments. By default a spacing of 1 is used.
>
> **See also:** [diff], page 380, [del2], page 419.

dot (*x*, *y*, *dim*)   [Built-in Function]
> Compute the dot product of two vectors. If *x* and *y* are matrices, calculate the dot products along the first non-singleton dimension. If the optional argument *dim* is given, calculate the dot products along this dimension.
>
> This is equivalent to sum (conj (*X*) .* *Y*, *dim*), but avoids forming a temporary array and is faster. When *X* and *Y* are column vectors, the result is equivalent to *X*' * *Y*.
>
> **See also:** [cross], page 418, [divergence], page 419.

cross (*x*, *y*)   [Function File]
cross (*x*, *y*, *dim*)   [Function File]
> Compute the vector cross product of two 3-dimensional vectors *x* and *y*.
>
>> cross ([1,1,0], [0,1,1])
>>   ⇒ [1; -1; 1]
>
> If *x* and *y* are matrices, the cross product is applied along the first dimension with 3 elements. The optional argument *dim* forces the cross product to be calculated along the specified dimension.
>
> **See also:** [dot], page 418, [curl], page 419, [divergence], page 419.

```
div = divergence (x, y, z, fx, fy, fz)                      [Function File]
div = divergence (fx, fy, fz)                               [Function File]
div = divergence (x, y, fx, fy)                             [Function File]
div = divergence (fx, fy)                                   [Function File]
```
    Calculate divergence of a vector field given by the arrays *fx*, *fy*, and *fz* or *fx*, *fy* respectively.

$$divF(x, y, z) = \partial_x F + \partial_y F + \partial_z F$$

    The coordinates of the vector field can be given by the arguments *x*, *y*, *z* or *x*, *y* respectively.

    See also: [curl], page 419, [gradient], page 417, [del2], page 419, [dot], page 418.

```
[cx, cy, cz, v] = curl (x, y, z, fx, fy, fz)               [Function File]
[cz, v] = curl (x, y, fx, fy)                               [Function File]
[...] = curl (fx, fy, fz)                                   [Function File]
[...] = curl (fx, fy)                                       [Function File]
v = curl (...)                                              [Function File]
```
    Calculate curl of vector field given by the arrays *fx*, *fy*, and *fz* or *fx*, *fy* respectively.

$$curlF(x, y, z) = \left(\frac{\partial d}{\partial y} F_z - \frac{\partial d}{\partial z} F_y, \frac{\partial d}{\partial z} F_x - \frac{\partial d}{\partial x} F_z, \frac{\partial d}{\partial x} F_y - \frac{\partial d}{\partial y} F_x \right)$$

    The coordinates of the vector field can be given by the arguments *x*, *y*, *z* or *x*, *y* respectively. *v* calculates the scalar component of the angular velocity vector in direction of the z-axis for two-dimensional input. For three-dimensional input the scalar rotation is calculated at each grid point in direction of the vector field at that point.

    See also: [divergence], page 419, [gradient], page 417, [del2], page 419, [cross], page 418.

```
d = del2 (M)                                                [Function File]
d = del2 (M, h)                                             [Function File]
d = del2 (M, dx, dy, ...)                                   [Function File]
```
    Calculate the discrete Laplace operator (∇^2). For a 2-dimensional matrix M this is defined as

$$d = \frac{1}{4} \left(\frac{d^2}{dx^2} M(x, y) + \frac{d^2}{dy^2} M(x, y) \right)$$

    For N-dimensional arrays the sum in parentheses is expanded to include second derivatives over the additional higher dimensions.

    The spacing between evaluation points may be defined by *h*, which is a scalar defining the equidistant spacing in all dimensions. Alternatively, the spacing in each dimension may be defined separately by *dx*, *dy*, etc. A scalar spacing argument defines equidistant spacing, whereas a vector argument can be used to specify variable spacing. The length of the spacing vectors must match the respective dimension of *M*. The default spacing value is 1.

    At least 3 data points are needed for each dimension. Boundary points are calculated from the linear extrapolation of interior points.

    See also: [gradient], page 417, [diff], page 380.

factorial (*n*)   [Function File]

> Return the factorial of *n* where *n* is a positive integer. If *n* is a scalar, this is equivalent to prod (1:*n*). For vector or matrix arguments, return the factorial of each element in the array. For non-integers see the generalized factorial function gamma.

> **See also:** [prod], page 414, [gamma], page 426.

p = factor (*q*)   [Function File]
[*p*, *n*] = factor (*q*)   [Function File]

> Return the prime factorization of *q*. That is, prod (*p*) == *q* and every element of *p* is a prime number. If *q* == 1, return 1.

> With two output arguments, return the unique primes *p* and their multiplicities. That is, prod (*p* .^ *n*) == *q*.

> Implementation Note: The input *q* must not be greater than bitmax (9.0072e+15) in order to factor correctly.

> **See also:** [gcd], page 420, [lcm], page 420, [isprime], page 63.

g = gcd (*a1*, *a2*, ...)   [Built-in Function]
[*g*, *v1*, ...] = gcd (*a1*, *a2*, ...)   [Built-in Function]

> Compute the greatest common divisor of *a1*, *a2*, ... If more than one argument is given all arguments must be the same size or scalar. In this case the greatest common divisor is calculated for each element individually. All elements must be ordinary or Gaussian (complex) integers. Note that for Gaussian integers, the gcd is not unique up to units (multiplication by 1, -1, *i* or -*i*), so an arbitrary greatest common divisor amongst four possible is returned.

> Example code:

```
gcd ([15, 9], [20, 18])
    ⇒  5  9
```

> Optional return arguments *v1*, etc., contain integer vectors such that,

$$g = v_1 a_1 + v_2 a_2 + \cdots$$

> **See also:** [lcm], page 420, [factor], page 420.

lcm (*x*, *y*)   [Mapping Function]
lcm (*x*, *y*, ...)   [Mapping Function]

> Compute the least common multiple of *x* and *y*, or of the list of all arguments. All elements must be the same size or scalar.

> **See also:** [factor], page 420, [gcd], page 420.

chop (*x*, *ndigits*, *base*)   [Function File]

> Truncate elements of *x* to a length of *ndigits* such that the resulting numbers are exactly divisible by *base*. If *base* is not specified it defaults to 10.

```
chop (-pi, 5, 10)
    ⇒ -3.14200000000000
chop (-pi, 5, 5)
    ⇒ -3.14150000000000
```

rem (*x*, *y*)   [Mapping Function]
fmod (*x*, *y*)   [Mapping Function]

> Return the remainder of the division *x* / *y*, computed using the expression
>
>   x - y .* fix (x ./ y)
>
> An error message is printed if the dimensions of the arguments do not agree, or if either of the arguments is complex.
>
> **See also:** [mod], page 421.

mod (*x*, *y*)   [Mapping Function]

> Compute the modulo of *x* and *y*. Conceptually this is given by
>
>   x - y .* floor (x ./ y)
>
> and is written such that the correct modulus is returned for integer types. This function handles negative values correctly. That is, `mod (-1, 3)` is 2, not -1, as `rem (-1, 3)` returns. `mod (x, 0)` returns *x*.
>
> An error results if the dimensions of the arguments do not agree, or if either of the arguments is complex.
>
> **See also:** [rem], page 421.

primes (*n*)   [Function File]

> Return all primes up to *n*.
>
> The algorithm used is the Sieve of Eratosthenes.
>
> Note that if you need a specific number of primes you can use the fact that the distance from one prime to the next is, on average, proportional to the logarithm of the prime. Integrating, one finds that there are about k primes less than $k \log(5k)$.
>
> **See also:** [list_primes], page 421, [isprime], page 63.

list_primes ()   [Function File]
list_primes (*n*)   [Function File]

> List the first *n* primes. If *n* is unspecified, the first 25 primes are listed.
>
> The algorithm used is from page 218 of the TEXbook.
>
> **See also:** [primes], page 421, [isprime], page 63.

sign (*x*)   [Mapping Function]

> Compute the *signum* function, which is defined as
>
> $$\mathrm{sign}(x) = \begin{cases} 1, & x > 0; \\ 0, & x = 0; \\ -1, & x < 0. \end{cases}$$
>
> For complex arguments, `sign` returns x ./ abs (*x*).
>
> Note that `sign (-0.0)` is 0. Although IEEE 754 floating point allows zero to be signed, 0.0 and -0.0 compare equal. If you must test whether zero is signed, use the `signbit` function.
>
> **See also:** [signbit], page 422.

`signbit (x)`   [Mapping Function]

Return logical true if the value of *x* has its sign bit set. Otherwise return logical false. This behavior is consistent with the other logical functions. SeeSection 4.6 [Logical Values], page 59. The behavior differs from the C language function which returns non-zero if the sign bit is set.

This is not the same as x < 0.0, because IEEE 754 floating point allows zero to be signed. The comparison -0.0 < 0.0 is false, but `signbit` (-0.0) will return a nonzero value.

See also: [sign], page 421.

17.6 Special Functions

`[a, ierr] = airy (k, z, opt)`   [Built-in Function]

Compute Airy functions of the first and second kind, and their derivatives.

```
K    Function   Scale factor (if "opt" is supplied)
---  --------   ------------------------------------
0    Ai (Z)     exp ((2/3) * Z * sqrt (Z))
1    dAi(Z)/dZ  exp ((2/3) * Z * sqrt (Z))
2    Bi (Z)     exp (-abs (real ((2/3) * Z * sqrt (Z))))
3    dBi(Z)/dZ  exp (-abs (real ((2/3) * Z * sqrt (Z))))
```

The function call `airy` (*z*) is equivalent to `airy` (0, *z*).

The result is the same size as *z*.

If requested, *ierr* contains the following status information and is the same size as the result.

0. Normal return.

1. Input error, return `NaN`.

2. Overflow, return `Inf`.

3. Loss of significance by argument reduction results in less than half of machine accuracy.

4. Complete loss of significance by argument reduction, return `NaN`.

5. Error—no computation, algorithm termination condition not met, return `NaN`.

`[j, ierr] = besselj (alpha, x, opt)`   [Built-in Function]
`[y, ierr] = bessely (alpha, x, opt)`   [Built-in Function]
`[i, ierr] = besseli (alpha, x, opt)`   [Built-in Function]
`[k, ierr] = besselk (alpha, x, opt)`   [Built-in Function]
`[h, ierr] = besselh (alpha, k, x, opt)`   [Built-in Function]

Compute Bessel or Hankel functions of various kinds:

besselj   Bessel functions of the first kind. If the argument *opt* is supplied, the result is multiplied by `exp` (-abs (imag (*x*))).

bessely   Bessel functions of the second kind. If the argument *opt* is supplied, the result is multiplied by `exp` (-abs (imag (*x*))).

besseli

> Modified Bessel functions of the first kind. If the argument *opt* is supplied, the result is multiplied by `exp (-abs (real (x)))`.

besselk

> Modified Bessel functions of the second kind. If the argument *opt* is supplied, the result is multiplied by `exp (x)`.

besselh Compute Hankel functions of the first ($k = 1$) or second ($k = 2$) kind. If the argument *opt* is supplied, the result is multiplied by `exp (-I*x)` for $k = 1$ or `exp (I*x)` for $k = 2$.

If *alpha* is a scalar, the result is the same size as *x*. If *x* is a scalar, the result is the same size as *alpha*. If *alpha* is a row vector and *x* is a column vector, the result is a matrix with `length (x)` rows and `length (alpha)` columns. Otherwise, *alpha* and *x* must conform and the result will be the same size.

The value of *alpha* must be real. The value of *x* may be complex.

If requested, *ierr* contains the following status information and is the same size as the result.

0. Normal return.

1. Input error, return `NaN`.

2. Overflow, return `Inf`.

3. Loss of significance by argument reduction results in less than half of machine accuracy.

4. Complete loss of significance by argument reduction, return `NaN`.

5. Error—no computation, algorithm termination condition not met, return `NaN`.

beta (a, b)   [Mapping Function]
> For real inputs, return the Beta function,

$$B(a,b) = \frac{\Gamma(a)\Gamma(b)}{\Gamma(a+b)}.$$

See also: [betaln], page 424, [betainc], page 423.

betainc (x, a, b)   [Mapping Function]
> Return the regularized incomplete Beta function,

$$I(x,a,b) = \frac{1}{B(a,b)} \int_0^x t^{(a-z)}(1-t)^{(b-1)}dt.$$

If *x* has more than one component, both *a* and *b* must be scalars. If *x* is a scalar, *a* and *b* must be of compatible dimensions.

See also: [betaincinv], page 423, [beta], page 423, [betaln], page 424.

betaincinv (y, a, b)   [Mapping Function]
> Compute the inverse of the incomplete Beta function, i.e., *x* such that

```
y == betainc (x, a, b)
```

See also: [betainc], page 423, [beta], page 423, [betaln], page 424.

betaln (a, b)   [Mapping Function]
> Return the natural logarithm of the Beta function,

$$\text{betaln}(a, b) = \ln(B(a, b)) \equiv \ln(\frac{\Gamma(a)\Gamma(b)}{\Gamma(a + b)}).$$

> calculated in a way to reduce the occurrence of underflow.

> **See also:** [beta], page 423, [betainc], page 423, [gammaln], page 428.

bincoeff (n, k)   [Mapping Function]
> Return the binomial coefficient of n and k, defined as

$$\binom{n}{k} = \frac{n(n - 1)(n - 2)\cdots(n - k + 1)}{k!}$$

> For example:

```
bincoeff (5, 2)
    ⇒ 10
```

> In most cases, the nchoosek function is faster for small scalar integer arguments. It also warns about loss of precision for big arguments.

> **See also:** [nchoosek], page 568.

commutation_matrix (m, n)   [Function File]
> Return the commutation matrix $K_{m,n}$ which is the unique $mn \times mn$ matrix such that $K_{m,n} \cdot \text{vec}(A) = \text{vec}(A^T)$ for all $m \times n$ matrices A.

> If only one argument m is given, $K_{m,m}$ is returned.

> See Magnus and Neudecker (1988), *Matrix Differential Calculus with Applications in Statistics and Econometrics*.

duplication_matrix (n)   [Function File]
> Return the duplication matrix D_n which is the unique $n^2 \times n(n + 1)/2$ matrix such that $D_n * \text{vech}(A) = \text{vec}(A)$ for all symmetric $n \times n$ matrices A.

> See Magnus and Neudecker (1988), Matrix differential calculus with applications in statistics and econometrics.

dawson (z)   [Mapping Function]
> Compute the Dawson (scaled imaginary error) function,

$$\frac{\sqrt{\pi}}{2} e^{-z^2} \text{erfi}(z) \equiv -i\frac{\sqrt{\pi}}{2} e^{-z^2} \text{erf}(iz)$$

> **See also:** [erfc], page 425, [erf], page 425, [erfcx], page 426, [erfi], page 426, [erfinv], page 426, [erfcinv], page 426.

`[sn, cn, dn, err] = ellipj (u, m)`   [Built-in Function]
`[sn, cn, dn, err] = ellipj (u, m, tol)`   [Built-in Function]
> Compute the Jacobi elliptic functions sn, cn, and dn of complex argument u and real parameter m.
>
> If m is a scalar, the results are the same size as u. If u is a scalar, the results are the same size as m. If u is a column vector and m is a row vector, the results are matrices with `length (u)` rows and `length (m)` columns. Otherwise, u and m must conform in size and the results will be the same size as the inputs.
>
> The value of u may be complex. The value of m must be $0 \leq m \leq 1$.
>
> The optional input *tol* is currently ignored (MATLAB uses this to allow faster, less accurate approximation).
>
> If requested, *err* contains the following status information and is the same size as the result.
>
> 0. Normal return.
>
> 1. Error—no computation, algorithm termination condition not met, return `NaN`.
>
> Reference: Milton Abramowitz and Irene A Stegun, *Handbook of Mathematical Functions*, Chapter 16 (Sections 16.4, 16.13, and 16.15), Dover, 1965.
>
> **See also:** [ellipke], page 425.

`k = ellipke (m)`   [Function File]
`k = ellipke (m, tol)`   [Function File]
`[k, e] = ellipke (...)`   [Function File]
> Compute complete elliptic integrals of the first K(m) and second E(m) kind.
>
> m must be a scalar or real array with -Inf $\leq m \leq 1$.
>
> The optional input *tol* is currently ignored (MATLAB uses this to allow a faster, less accurate approximation).
>
> Called with only one output, elliptic integrals of the first kind are returned.
>
> Reference: Milton Abramowitz and Irene A. Stegun, *Handbook of Mathematical Functions*, Chapter 17, Dover, 1965.
>
> **See also:** [ellipj], page 425.

`erf (z)`   [Mapping Function]
> Compute the error function,
>
> $$\operatorname{erf}(z) = \frac{2}{\sqrt{\pi}} \int_0^z e^{-t^2} dt$$
>
> **See also:** [erfc], page 425, [erfcx], page 426, [erfi], page 426, [dawson], page 424, [erfinv], page 426, [erfcinv], page 426.

`erfc (z)`   [Mapping Function]
> Compute the complementary error function, $1 - \operatorname{erf}(z)$.
>
> **See also:** [erfcinv], page 426, [erfcx], page 426, [erfi], page 426, [dawson], page 424, [erf], page 425, [erfinv], page 426.

erfcx (*z*)   [Mapping Function]
    Compute the scaled complementary error function,

$$e^{z^2}\mathrm{erfc}(z) \equiv e^{z^2}(1 - \mathrm{erf}(z))$$

    See also: [erfc], page 425, [erf], page 425, [erfi], page 426, [dawson], page 424, [erfinv], page 426, [erfcinv], page 426.

erfi (*z*)   [Mapping Function]
    Compute the imaginary error function,

$$-i\mathrm{erf}(iz)$$

    See also: [erfc], page 425, [erf], page 425, [erfcx], page 426, [dawson], page 424, [erfinv], page 426, [erfcinv], page 426.

erfinv (*x*)   [Mapping Function]
    Compute the inverse error function, i.e., *y* such that
        erf (y) == x

    See also: [erf], page 425, [erfc], page 425, [erfcx], page 426, [erfi], page 426, [dawson], page 424, [erfcinv], page 426.

erfcinv (*x*)   [Mapping Function]
    Compute the inverse complementary error function, i.e., *y* such that
        erfc (y) == x

    See also: [erfc], page 425, [erf], page 425, [erfcx], page 426, [erfi], page 426, [dawson], page 424, [erfinv], page 426.

expint (*x*)   [Function File]
    Compute the exponential integral:

$$\mathrm{E}_1(x) = \int_x^\infty \frac{e^{-t}}{t} dt$$

    Note: For compatibility, this functions uses the MATLAB definition of the exponential integral. Most other sources refer to this particular value as $E_1(x)$, and the exponential integral is

$$\mathrm{Ei}(x) = -\int_{-x}^\infty \frac{e^{-t}}{t} dt.$$

    The two definitions are related, for positive real values of x, by $E_1(-x) = -\mathrm{Ei}(x) - i\pi$.

gamma (*z*)   [Mapping Function]
    Compute the Gamma function,

$$\Gamma(z) = \int_0^\infty t^{z-1}e^{-t}dt.$$

    See also: [gammainc], page 427, [lgamma], page 428.

gammainc (*x*, *a*)   [Mapping Function]
gammainc (*x*, *a*, "*lower*")   [Mapping Function]
gammainc (*x*, *a*, "*upper*")   [Mapping Function]
Compute the normalized incomplete gamma function,

$$\gamma(x, a) = \frac{1}{\Gamma(a)} \int_0^x t^{a-1} e^{-t} dt$$

with the limiting value of 1 as x approaches infinity. The standard notation is $P(a, x)$, e.g., Abramowitz and Stegun (6.5.1).

If a is scalar, then gammainc (*x*, *a*) is returned for each element of x and vice versa.

If neither x nor a is scalar, the sizes of x and a must agree, and gammainc is applied element-by-element.

By default the incomplete gamma function integrated from 0 to x is computed. If "upper" is given then the complementary function integrated from x to infinity is calculated. It should be noted that

gammainc (*x*, *a*) $\equiv$ 1 - gammainc (*x*, *a*, "*upper*")

See also: [gamma], page 426, [lgamma], page 428.

l = legendre (*n*, *x*)   [Function File]
l = legendre (*n*, *x*, *normalization*)   [Function File]
Compute the Legendre function of degree n and order $m = 0 \ldots$ N. The optional argument, *normalization*, may be one of "unnorm", "sch", or "norm". The default is "unnorm". The value of n must be a non-negative scalar integer.

If the optional argument *normalization* is missing or is "unnorm", compute the Legendre function of degree n and order m and return all values for $m = 0 \ldots n$. The return value has one dimension more than x.

The Legendre Function of degree n and order m:

$$P_n^m(x) = (-1)^m (1 - x^2)^{m/2} \frac{d^m}{dx^m} P_n(x)$$

with Legendre polynomial of degree n:

$$P(x) = \frac{1}{2^n n!} \left(\frac{d^n}{dx^n} (x^2 - 1)^n \right)$$

legendre (3, [-1.0, -0.9, -0.8]) returns the matrix:

```
x  |   -1.0   |   -0.9   |   -0.8
-----------------------------------
m=0 | -1.00000 | -0.47250 | -0.08000
m=1 |  0.00000 | -1.99420 | -1.98000
m=2 |  0.00000 | -2.56500 | -4.32000
m=3 |  0.00000 | -1.24229 | -3.24000
```

If the optional argument **normalization** is "sch", compute the Schmidt semi-normalized associated Legendre function. The Schmidt semi-normalized associated Legendre function is related to the unnormalized Legendre functions by the following:

For Legendre functions of degree n and order 0:

$$SP_n^0(x) = P_n^0(x)$$

For Legendre functions of degree n and order m:

$$SP_n^m(x) = P_n^m(x)(-1)^m \left(\frac{2(n-m)!}{(n+m)!}\right)^{0.5}$$

If the optional argument *normalization* is "norm", compute the fully normalized associated Legendre function. The fully normalized associated Legendre function is related to the unnormalized Legendre functions by the following:

For Legendre functions of degree *n* and order *m*

$$NP_n^m(x) = P_n^m(x)(-1)^m \left(\frac{(n+0.5)(n-m)!}{(n+m)!}\right)^{0.5}$$

lgamma (*x*)   [Mapping Function]
gammaln (*x*)   [Mapping Function]
    Return the natural logarithm of the gamma function of *x*.

    See also: [gamma], page 426, [gammainc], page 427.

17.7 Rational Approximations

s = rat (*x*, *tol*)   [Function File]
[n, d] = rat (*x*, *tol*)   [Function File]
    Find a rational approximation to *x* within the tolerance defined by *tol* using a continued fraction expansion. For example:

```
rat (pi) = 3 + 1/(7 + 1/16) = 355/113
rat (e) = 3 + 1/(-4 + 1/(2 + 1/(5 + 1/(-2 + 1/(-7)))))
        = 1457/536
```

    Called with two arguments returns the numerator and denominator separately as two matrices.

    See also: [rats], page 428.

rats (*x*, *len*)   [Built-in Function]
    Convert *x* into a rational approximation represented as a string. You can convert the string back into a matrix as follows:

```
r = rats (hilb (4));
x = str2num (r)
```

    The optional second argument defines the maximum length of the string representing the elements of *x*. By default *len* is 9.

    See also: [format], page 224, [rat], page 428.

17.8 Coordinate Transformations

[*theta*, *r*] = cart2pol (*x*, *y*)   [Function File]
[*theta*, *r*, *z*] = cart2pol (*x*, *y*, *z*)   [Function File]
[*theta*, *r*] = cart2pol (*C*)   [Function File]
[*theta*, *r*, *z*] = cart2pol (*C*)   [Function File]
P = cart2pol (...)   [Function File]
    Transform Cartesian to polar or cylindrical coordinates.

    theta describes the angle relative to the positive x-axis. *r* is the distance to the z-axis $(0, 0, z)$. x, y (, and z) must be the same shape, or scalar. If called with a single matrix argument then each row of *C* represents the Cartesian coordinate $(x, y \ (, z))$.

    If only a single return argument is requested then return a matrix *P* where each row represents one polar/(cylindrical) coordinate $(theta, phi \ (, z))$.

    See also: [pol2cart], page 429, [cart2sph], page 429, [sph2cart], page 429.

[*x*, *y*] = pol2cart (*theta*, *r*)   [Function File]
[*x*, *y*, *z*] = pol2cart (*theta*, *r*, *z*)   [Function File]
[*x*, *y*] = pol2cart (*P*)   [Function File]
[*x*, *y*, *z*] = pol2cart (*P*)   [Function File]
C = pol2cart (...)   [Function File]
    Transform polar or cylindrical to Cartesian coordinates.

    theta, *r*, (and *z*) must be the same shape, or scalar. *theta* describes the angle relative to the positive x-axis. *r* is the distance to the z-axis $(0, 0, z)$. If called with a single matrix argument then each row of *P* represents the polar/(cylindrical) coordinate $(theta, r \ (, z))$.

    If only a single return argument is requested then return a matrix *C* where each row represents one Cartesian coordinate $(x, y \ (, z))$.

    See also: [cart2pol], page 429, [sph2cart], page 429, [cart2sph], page 429.

[*theta*, *phi*, *r*] = cart2sph (*x*, *y*, *z*)   [Function File]
[*theta*, *phi*, *r*] = cart2sph (*C*)   [Function File]
S = cart2sph (...)   [Function File]
    Transform Cartesian to spherical coordinates.

    theta describes the angle relative to the positive x-axis. *phi* is the angle relative to the xy-plane. *r* is the distance to the origin $(0, 0, 0)$. x, y, and z must be the same shape, or scalar. If called with a single matrix argument then each row of *C* represents the Cartesian coordinate (x, y, z).

    If only a single return argument is requested then return a matrix *S* where each row represents one spherical coordinate $(theta, phi, r)$.

    See also: [sph2cart], page 429, [cart2pol], page 429, [pol2cart], page 429.

[*x*, *y*, *z*] = sph2cart (*theta*, *phi*, *r*)   [Function File]
[*x*, *y*, *z*] = sph2cart (*S*)   [Function File]
C = sph2cart (...)   [Function File]
    Transform spherical to Cartesian coordinates.

theta describes the angle relative to the positive x-axis. *phi* is the angle relative to the xy-plane. *r* is the distance to the origin $(0, 0, 0)$. *theta*, *phi*, and *r* must be the same shape, or scalar. If called with a single matrix argument then each row of S represents the spherical coordinate (*theta*, *phi*, *r*).

If only a single return argument is requested then return a matrix C where each row represents one Cartesian coordinate (*x*, *y*, *z*).

See also: [cart2sph], page 429, [pol2cart], page 429, [cart2pol], page 429.

17.9 Mathematical Constants

e	[Built-in Function]
e (*n*)	[Built-in Function]
e (*n*, *m*)	[Built-in Function]
e (*n*, *m*, *k*, ...)	[Built-in Function]
e (..., *class*)	[Built-in Function]

Return a scalar, matrix, or N-dimensional array whose elements are all equal to the base of natural logarithms. The constant e satisfies the equation $\log(e) = 1$.

When called with no arguments, return a scalar with the value e. When called with a single argument, return a square matrix with the dimension specified. When called with more than one scalar argument the first two arguments are taken as the number of rows and columns and any further arguments specify additional matrix dimensions. The optional argument *class* specifies the return type and may be either `"double"` or `"single"`.

See also: [log], page 407, [exp], page 407, [pi], page 430, [I], page 430.

pi	[Built-in Function]
pi (*n*)	[Built-in Function]
pi (*n*, *m*)	[Built-in Function]
pi (*n*, *m*, *k*, ...)	[Built-in Function]
pi (..., *class*)	[Built-in Function]

Return a scalar, matrix, or N-dimensional array whose elements are all equal to the ratio of the circumference of a circle to its diameter(π). Internally, `pi` is computed as '`4.0 * atan (1.0)`'.

When called with no arguments, return a scalar with the value of π. When called with a single argument, return a square matrix with the dimension specified. When called with more than one scalar argument the first two arguments are taken as the number of rows and columns and any further arguments specify additional matrix dimensions. The optional argument *class* specifies the return type and may be either `"double"` or `"single"`.

See also: [e], page 430, [I], page 430.

I	[Built-in Function]
I (*n*)	[Built-in Function]
I (*n*, *m*)	[Built-in Function]
I (*n*, *m*, *k*, ...)	[Built-in Function]

I (..., *class*)   [Built-in Function]
>   Return a scalar, matrix, or N-dimensional array whose elements are all equal to the pure imaginary unit, defined as $\sqrt{-1}$.
>
>   I, and its equivalents i, j, and J, are functions so any of the names may be reused for other purposes (such as i for a counter variable).
>
>   When called with no arguments, return a scalar with the value i. When called with a single argument, return a square matrix with the dimension specified. When called with more than one scalar argument the first two arguments are taken as the number of rows and columns and any further arguments specify additional matrix dimensions. The optional argument *class* specifies the return type and may be either "double" or "single".
>
>   **See also:** [e], page 430, [pi], page 430, [log], page 407, [exp], page 407.

Inf   [Built-in Function]
Inf (*n*)   [Built-in Function]
Inf (*n, m*)   [Built-in Function]
Inf (*n, m, k, ...*)   [Built-in Function]
Inf (..., *class*)   [Built-in Function]
>   Return a scalar, matrix or N-dimensional array whose elements are all equal to the IEEE representation for positive infinity.
>
>   Infinity is produced when results are too large to be represented using the the IEEE floating point format for numbers. Two common examples which produce infinity are division by zero and overflow.
>
> ```
>         [ 1/0 e^800 ]
>       ⇒ Inf    Inf
> ```
>
>   When called with no arguments, return a scalar with the value 'Inf'. When called with a single argument, return a square matrix with the dimension specified. When called with more than one scalar argument the first two arguments are taken as the number of rows and columns and any further arguments specify additional matrix dimensions. The optional argument *class* specifies the return type and may be either "double" or "single".
>
>   **See also:** [isinf], page 380, [NaN], page 431.

NaN   [Built-in Function]
NaN (*n*)   [Built-in Function]
NaN (*n, m*)   [Built-in Function]
NaN (*n, m, k, ...*)   [Built-in Function]
NaN (..., *class*)   [Built-in Function]
>   Return a scalar, matrix, or N-dimensional array whose elements are all equal to the IEEE symbol NaN (Not a Number). NaN is the result of operations which do not produce a well defined numerical result. Common operations which produce a NaN are arithmetic with infinity ($\infty - \infty$), zero divided by zero (0/0), and any operation involving another NaN value (5 + NaN).
>
>   Note that NaN always compares not equal to NaN (NaN != NaN). This behavior is specified by the IEEE standard for floating point arithmetic. To find NaN values, use the isnan function.

When called with no arguments, return a scalar with the value 'NaN'. When called with a single argument, return a square matrix with the dimension specified. When called with more than one scalar argument the first two arguments are taken as the number of rows and columns and any further arguments specify additional matrix dimensions. The optional argument *class* specifies the return type and may be either `"double"` or `"single"`.

See also: [isnan], page 380, [Inf], page 431.

eps	[Built-in Function]
eps (*x*)	[Built-in Function]
eps (*n*, *m*)	[Built-in Function]
eps (*n*, *m*, *k*, ...)	[Built-in Function]
eps (..., *class*)	[Built-in Function]

Return a scalar, matrix or N-dimensional array whose elements are all eps, the machine precision. More precisely, `eps` is the relative spacing between any two adjacent numbers in the machine's floating point system. This number is obviously system dependent. On machines that support IEEE floating point arithmetic, `eps` is approximately 2.2204×10^{-16} for double precision and 1.1921×10^{-7} for single precision.

When called with no arguments, return a scalar with the value `eps (1.0)`. Given a single argument *x*, return the distance between *x* and the next largest value. When called with more than one argument the first two arguments are taken as the number of rows and columns and any further arguments specify additional matrix dimensions. The optional argument *class* specifies the return type and may be either `"double"` or `"single"`.

See also: [realmax], page 432, [realmin], page 432, [intmax], page 55, [bitmax], page 57.

realmax	[Built-in Function]
realmax (*n*)	[Built-in Function]
realmax (*n*, *m*)	[Built-in Function]
realmax (*n*, *m*, *k*, ...)	[Built-in Function]
realmax (..., *class*)	[Built-in Function]

Return a scalar, matrix or N-dimensional array whose elements are all equal to the largest floating point number that is representable. The actual value is system dependent. On machines that support IEEE floating point arithmetic, `realmax` is approximately 1.7977×10^{308} for double precision and 3.4028×10^{38} for single precision.

When called with no arguments, return a scalar with the value `realmax ("double")`. When called with a single argument, return a square matrix with the dimension specified. When called with more than one scalar argument the first two arguments are taken as the number of rows and columns and any further arguments specify additional matrix dimensions. The optional argument *class* specifies the return type and may be either `"double"` or `"single"`.

See also: [realmin], page 432, [intmax], page 55, [bitmax], page 57, [eps], page 432.

realmin	[Built-in Function]
realmin (*n*)	[Built-in Function]

`realmin (n, m)`   [Built-in Function]
`realmin (n, m, k, ...)`   [Built-in Function]
`realmin (..., class)`   [Built-in Function]

Return a scalar, matrix or N-dimensional array whose elements are all equal to the smallest normalized floating point number that is representable. The actual value is system dependent. On machines that support IEEE floating point arithmetic, `realmin` is approximately 2.2251×10^{-308} for double precision and 1.1755×10^{-38} for single precision.

When called with no arguments, return a scalar with the value `realmin ("double")`. When called with a single argument, return a square matrix with the dimension specified. When called with more than one scalar argument the first two arguments are taken as the number of rows and columns and any further arguments specify additional matrix dimensions. The optional argument *class* specifies the return type and may be either `"double"` or `"single"`.

See also: [realmax], page 432, [intmin], page 55, [eps], page 432.